Lecture Notes in Physics

Edited by H. Araki, Kyoto, J. Ehlers, München, K. Hepp, Zürich
R. Kippenhahn, München, H. A. Weidenmüller, Heidelberg
J. Wess, Karlsruhe and J. Zittartz, Köln
Managing Editor: W. Beiglböck

288

J. Arbocz M. Potier-Ferry
J. Singer V. Tvergaard

Buckling and Post-Buckling

Four Lectures in Experimental, Numerical
and Theoretical Solid Mechanics
Based on Talks Given at the CISM-Meeting
Held in Udine, Italy, September 29–October 3, 1985

Springer-Verlag Berlin Heidelberg GmbH

Authors

Johann Arbocz
Department of Aerospace Engineering, Delft University of Technology
Delft, The Netherlands

Michel Potier-Ferry
Laboratoire de Physique et Mécanique des Matériaux, Université de Metz
Ile du Saulcy, F-57045 Metz, France

Josef Singer
Department of Aeronautical Engineering, Technion, Israel Institute of Technology
Haifa 32000, Israel

Viggo Tvergaard
Department of Solid Mechanics, The Technical University of Denmark
Lyngby, Denmark

ISBN 978-3-662-13619-5 ISBN 978-3-540-47875-1 (eBook)
DOI 10.1007/978-3-540-47875-1

Originally published by Springer-Verlag Berlin Heidelberg New York in 1987
Softcover reprint of the hardcover 1st edition 1987

PREFACE

This book contains four lectures delivered at the International Centre for Mechanical Sciences, UDINE, Italy, in 1985.

Slender and thin-walled structures are frequently used in aeronautics and also in mechanical, nuclear and civil engineering. The computation of their practical carrying capacity requires a nonlinear analysis. For many shells in the elastic or elastoplastic range, geometric nonlinearities are needed to calculate the reduction of the critical load due, for instance, to imperfections. These lectures aim to give a comprehensive account for the theoretical, numerical and experimental methods which are useful to analyse the buckling and post-buckling behavior of structures.

The book is intended for postgraduate students, researchers and engineers who are interested in this field.

The first lecture by M. POTIER-FERRY is devoted to the elastic post-buckling theory and some more or less simple applications. The ideas of recent mathematical approaches like bifurcation, catastrophe and singularity theories are included in this lecture, but without advanced mathematics. J. ARBOCZ is an aeronautical engineer. In his paper, he discusses the influence of imperfections and boundary conditions on the maximum load of stiffened shells of revolution. By coupling expertise and simple numerical methods, he sets examples how to design real structures. The state of the art in plastic buckling is presented in the lecture by V. TVERGAARD. It essentially contains bifurcation criteria for plates and shells, asymptotic and finite-element analyses of plastic post-buckling and discussions of the influence of the constitutive laws. In spite of tremendous progress, numerical and mechanical models are not sufficiently reliable, especially to predict a very sensitive phenomenon like buckling. The last lecture by J. SINGER is an excellent guide to the experimental methods used in testing compressed structures.

We express our sincere thanks to the authorities and staff of the CISM for their efficiency in the organization of the course.

W.T. KOITER M. POTIER-FERRY

TABLE OF CONTENTS

M. Potier-Ferry

J. Arbocz

POST-BUCKLING BEHAVIOUR OF STRUCTURES

NUMERICAL TECHNIQUES FOR MORE COMPLICATED STRUCTURES

V. Tvergaard

EFFECT OF PLASTICITY ON POST-BUCKLING BEHAVIOUR

J. Singer

EXPERIMENTAL TECHNIQUES AND COMPARISON WITH THEORETICAL RESULTS

FOUNDATIONS OF ELASTIC POSTBUCKLING THEORY

by Michel POTIER-FERRY
Laboratoire de Physique et Mécanique des Matériaux
Université de METZ, Ile du Saulcy
57045 - METZ, France

ABSTRACT

In this paper we present the elastic postbuckling theory, but we introduce two modifications with respect to its classical statement. First, we account for the influence of symmetries. Second, the singularities will be classified according to their robustness, what was introduced by Catastrophe Theory. Various recent progresses are discussed ; mathematical foundations of the energy criterion, buckling of structures with a large aspect ratio, contribution of catastrophe theory ...

1. INTRODUCTION

The long story of postbuckling theory began with a study of the flexible compressed beam by L. Euler /20/ in 1744. The research in the field of elastic stability became important only a century later, when the extensive use of steel permitted the introduction of slender structures, especially in civil engineering. The practical interest of a nonlinear theory was recognized in the 1930's and the general postbuckling theory of elastic structures was enunciated by Koiter in his thesis /31/, 1945. He established that imperfections may give rise to significant reductions of the critical load, which explained the discrepancy between linear theories and experiments on shells. The development of aeronautical and aerospace programs motivated a considerable amount of research in this field during the 1960's. From that time, the research activity is remained important and new applications appeared such as off-shore structures /48/ /62/, nuclear engineering /3/ ...

Developments subsequent to Koiter's thesis had not fundamentally modified the theory itself. The British school (Sewell /57/, Thompson-Hunt /60/, Huseyin /25/...) has considered mainly discrete systems. The Harvard's school (Budiansky - Hutchinson /11/, Budiansky /9/ ...) relies rather on the principle of virtual work than on the potential energy, rather on a priori set expansions than on the elimination of "passive variables".

In another respect, bifurcation and stability theory has been the subject of many mathematical studies. Since the pioneer works of Poincaré (1885), Lyapunov (1892) and Schmidt (1908), mathematical analysis has played a great part in bifurcation theory of continuous systems, the basic tool being the Inverse Function Theorem. Although postbuckling and bifurcation theories have progressed in a mutual ignorance, the two points of view often are similar, mathematical analysis providing quite rigorous justifications to the asymptotic computations of postbuckling theory. Accounts of these researches can be found in Vainberg - Trenogin /67/, Keller - Antman /28/, Sattinger /49/, Iooss - Joseph /27/. More recent tools of analysis, such as Sobolev spaces or semi-groups, have contributed to a better understanding of the dynamical foundations of the energy criterion (Ball /2/, Potier-Ferry /44/). Catastrophe theory (Thom /58/) and singularity theory (Golubitsky and Schaeffer /22/) put forward a more qualitative approach than that of the analysts. These theories are issued from topology and differential geometry. Perhaps the most interesting idea of catastrophe theory is to classify the singularities according to their "robustness", what we shall simply include in postbuckling theory. One can also modify the statement of postbuckling theory in a simple way, by accounting for the symmetries of the considered structures. Recently, this gave rise to a number of mathematical works (Sattinger /51/, Golubitsky and Schaeffer /52/ /53/).

Roughly, we shall follow Koiter's original statement. As we have just explained, some recent ideas shall be introduced in the classical theory, but we shall avoid the advanced mathematical techniques that are often used to present those ideas. In addition, the contribution of mathematics to postbuckling will be discussed in a separate chapter. Practical applications are beyond the scope of this short introduction. We shall only discuss simple models in order to present or to illustrate the basic notions. Especially, we shall give at least one example of each type of singular point that appears in our classifications. The paper is closed with a new postbuckling analysis for structures such that postbuckling patterns have a cellular shape with a large number of cells. Initially, the latter methods had been developed for problems of fluid mechanics/68/

The thermodynamical foundations of the energy criterion are well understood, the main contributions being those of Duhem /17/ and Ericksen /19/. This will not be presented here and we refer to recent accounts/33/ /7/. We neither discuss non-conservative systems /4/ /16/ /55/ that may lead to dynamic instabilities /27/. A complete bibliography should include thousands of references. For this purpose, we refer to textbooks /8/ /25/ /48/ /60/ /63/, review papers /9/ /26/ /29/ /33/ /64/and symposium proceedings /10/ /38/ /47/ /55/.

2. SIMPLE MODELS

The most typical feature of instability theory is that its fundamental characteristics can be found in very simple models. Moreover, any complicated structural system is equivalent, in some sense, to one of these simple models, at least in the neighbourhood of a critical state. We present here two simple models and the slightly more intricate example of a beam on a Winckler foundation, that will be useful for illustrating the general theory.

2.1 Amplitude equation

To explain the instability phenomena in fluid mechanics, one considers usually the so-called Landau equation /18/, which can be written as

$$\frac{da}{dt} = (\lambda - \lambda_c)\, a - C_3\, a^3 \qquad (2\text{-}1)$$

or as

$$\frac{da}{dt} = (\lambda - \lambda_c)\, a - C_2\, a^2 \qquad (2\text{-}2)$$

The function $a(t)$ is the amplitude of the disturbance, which implies that the spatial shape of this disturbance does not vary significantly. The control parameter is denoted by λ : in fluid mechanics, it may be the Reynolds number or the Rayleigh number ... In this course, it will be a loading parameter. The critical value of λ is designated by λ_c.

Look at the *equilibrium paths* in the a - λ plane for Eqs (2-1) or (2-2). One finds a *fundamental solution* $a_I(\lambda) = 0$ and a secondary branch of solutions $a_{II}(\lambda)$, which is given by :

$$a_{II}^2 = (\lambda - \lambda_c)/C_3 \quad \text{or} \quad a_{II} = (\lambda - \lambda_c)/C_2 \qquad (2\text{-}3)$$

These two branches intersect at the point $(a,\lambda) = (0,\lambda_c)$, that is called *bifurcation point*. The secondary branch (2-3) is called *bifurcating branch*. The stability of an equilibrium solution is governed by the linearized equation. One obtains, for the fundamental solution,

$$\frac{da}{dt} = (\lambda - \lambda_c)a \tag{2-4}$$

and for the solution (2-3),

$$a = a_{II} + b$$

$$\frac{db}{dt} = -\,2(\lambda - \lambda_c)b \text{ or } \frac{db}{dt} = -\,(\lambda - \lambda_c)\,b \tag{2-5}$$

By comparison of (2-4) and (2-5), one concludes that the secondary solutions are stable only for λ greater than λ_c, i.e. when the fundamental solution is unstable. Remark also that the bifurcation point coincides with the point where the fundamental solution loses its stability. The appellation "*principle of exchange of stability*" is used to designate the two latter properties, but the terminology is rather inappropriate except with Eq. (2-2).

The equilibrium paths are pictured in Figure 1, where the dashed curves represent unstable solutions. One distinguishes three cases according to the equation (2-1) or (2-2) and to the sign of the constant C_3. The bifurcation points of Figures 1-a, 1-b, 1-c are called respectively *stable-symmetric, unstable-symmetric and asymmetric.*

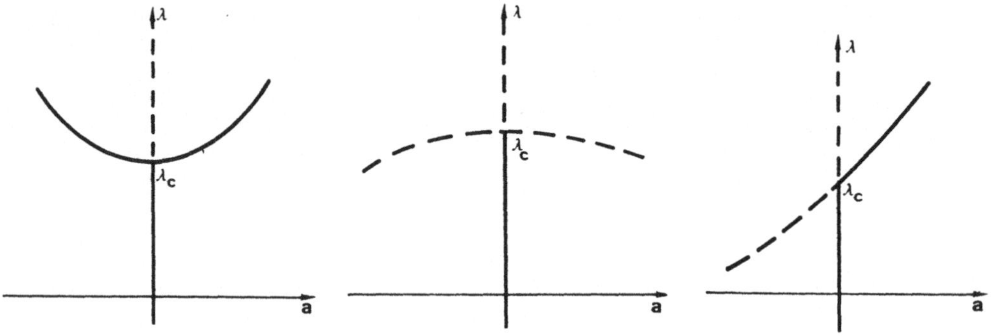

Fig. 1a, Stable - Symmetric bifurcation

Fig. 1b, Unstable - Symmetric bifurcation

Fig. 1c, Asymmetric bifurcation

Fig. 1, Amplitude - parameter curves for various bifurcation points

2.2 An elementary buckling model

Several recent expositions of postbuckling theory begin with this simple model (for instance /9/). A rigid rod of length L is subjected at the upper end to a force whose size λ is prescribed and whose direction remains vertical. The lower end is elastically restrained by a spring that supplies a restoring moment

$$f(a) = K_1 a + K_2 a^2 + K_3 a^3 + \ldots \tag{2-6}$$

where a is the angle between the rod and its unloaded direction. If this direction is vertical, we shall say that the structure is perfect (Figure 2-a). Of course such a situation cannot be carried out exactly. In the real structure, the unloaded state presents a small initial deviation a_0 from the vertical (Figure 2-b). Let I be the moment of inertia. The equation of motion is :

$$I \frac{d^2 a}{dt^2} + f(a) - \lambda L \sin(a + a_0) = 0 \tag{2-7}$$

First, consider the perfect structure, i.e. $a_0 = 0$. In that case, there exists a fundamental solution a = 0, which loses its stability at the following *critical load* :

$$\lambda_c = K_1/L \tag{2-8}$$

Fig. 2a, Perfect Model

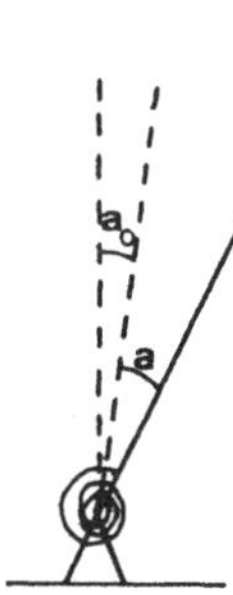

Fig. 2b, The model with an initial imperfection a_0

Fig 2

Remark that this load is reached when the linearized equilibrium equation has at least one solution a, which is different from zero. That is also a general feature, which leads to the so-called *method of adjacent equilibrium* /8/.

Introduce a new time $t(L/I)^{1/2}$, that will be again denoted by t and the following constants :

$$C_2 = K_2/L \ ; \ C_3 = K_3/L + \lambda_c/6$$

Then, after expansion with respect to a and $\lambda - \lambda_c$, Eq. (2-7) becomes :

$$\frac{d^2a}{dt^2} - (\lambda - \lambda_c)a + C_2 \ a^2 + C_3 \ a^3 + \text{h.o.t.} = 0 \qquad (2\text{-}9)$$

where the higher order terms (h.o.t.) are such that

$$\text{h.o.t.} = O(a^4) + O(a^3(\lambda - \lambda_c)).$$

At the equilibrium, Eq. (2-9) differs from the amplitude equation (2-1) or (2-2) only by those h.o.t. But this does not alter the shape of the bifurcating branch, that is given by :

$$\lambda - \lambda_c = C_2 a + C_3 a^2 + \text{h.o.t.} \qquad (2\text{-}10)$$

One finds again an asymmetric bifurcation (Figure 1-c) if the response of the spring is asymmetric ($K_2 \neq 0$). For a symmetric spring ($K_2 = 0$), the bifurcation is symmetric and may be stable or unstable according to the sign of C_3. The phrase *initial postbuckling behaviour* refers to the shape of the equilibrium paths. In the case of Figure 1-a (resp. 1-b), one talks about a *stable* (*resp. unstable*) *postbuckling*. When the system is symmetric ($K_2 = 0$) and the postbuckling is stable, the loss of stability is followed by a *symmetry breaking*.

Eq. (2-9 differs also from the amplitude equations of fluid mechanics (2-1) or (2-2) by its conservative dynamical behaviour. But the conclusions of the stability discussion are identical. The *potential energy* of the present system is :

$$P(a,\lambda) = L \{- (\lambda - \lambda_c) \ a^2/2 + C_2 a^3/3 + C_3 a^4/4 + \text{h.o.t.} \qquad (2\text{-}11)$$

Close to the bifurcation point, the shape of the potential energy varies as pictured in Figures 3, 4, 5. Of course, stable equilibria correspond to local minima of the potential energy with respect to a. At the critical load, the quadratic term of the potential energy vanishes and the series begin with terms in a^3 or a^4. When the load deviates slightly from λ_c, the degenerate extremum splits into several extrema (one, two or three in the present case). This behaviour is the basis of postbuckling theory, as well as of the catastrophe theory.

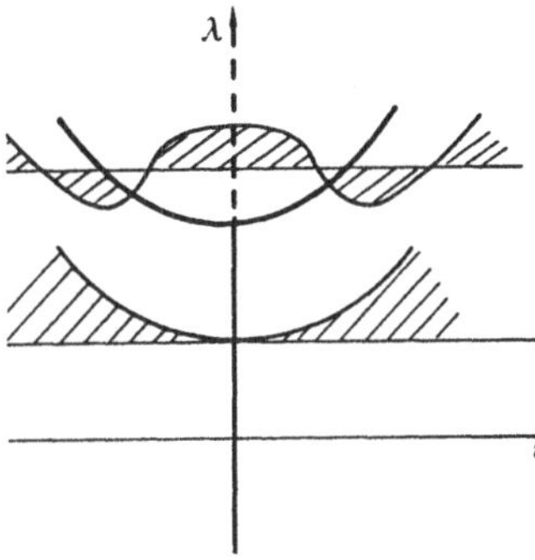

Fig. 3, Potential energy before and after the critical load : the case of the stable symmetric bifurcation

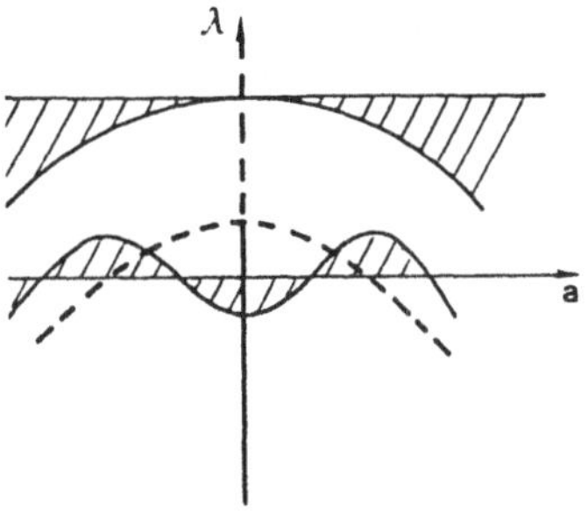

Fig. 4, Potential energy before and after the critical load : for the unstable-symmetric bifurcation

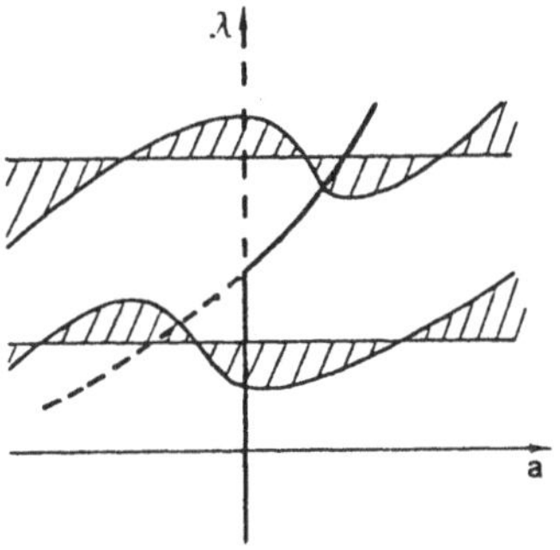

Fig. 5, Potential energy before and after the critical load : for the asymmetric bifurcation

Now consider the equilibria of Eq. (2-7) with imperfections. In addition to the amplitude a and to the load increment $\lambda - \lambda_c$, a third small parameter appears, which is the initial deviation a_0. At the lower orders there are terms in a^3 (or a^2), $(\lambda - \lambda_c)$ a and a_0. Each of the h.o.t. in a^4 (or a^3), $(\lambda - \lambda_c)a^2$, $a_0 a$, $a_0(\lambda - \lambda_c)$ is small with respect to at least one of the lower order terms. It has been established that those h.o.t. may be neglected [12], [42] § 4. This leads to the approximate equation :

$$C_3\, a^3 - (\lambda - \lambda_c)\, a - \lambda_c\, a_0 = 0 \qquad (2\text{-}12)$$

or

$$C_2\, a^2 - (\lambda - \lambda_c)\, a - \lambda_c\, a_0 = 0 \qquad (2\text{-}13)$$

It is straightforward to draw the equilibrium paths of (2-12) or (2-13), see Figure 7. The initial imperfection a_0 destroys the bifurcation point and two distinct branches appear. In the most of the cases (6-a, 6-b, 6-c), the equilibrium curves have extrema. These singular points are called *limit points*. In the cases of Figure 6-b and 6-c, there is a maximal load point λ_m on the fundamental path, i.e. the path which is close to a = 0 at vanishing load. When one loads the structure beyond λ_m, a dynamical process begins. What happens during and after this process is outside the scope of initial postbuckling theory. Real structures go to an alternative equilibrium state at the end of this dynamical process, but often in the plastic range. This phenomenon is called *snap through or snap buckling*.

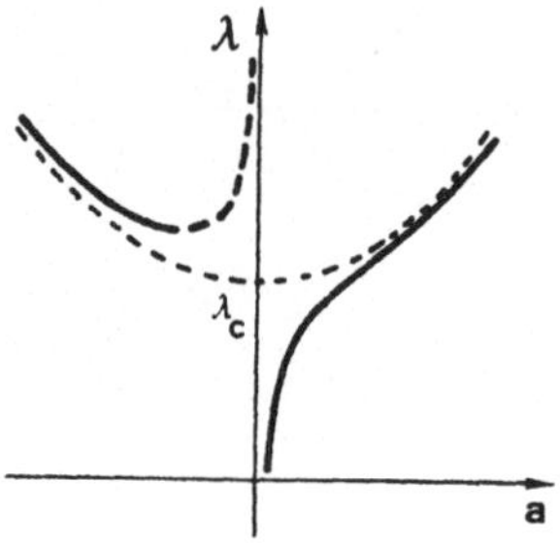

Fig. 6a, Stable - symmetric

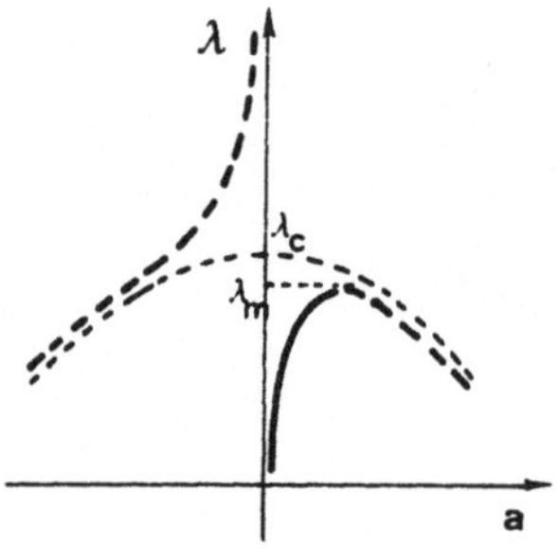

Fig. 6b, Unstable - symmetric

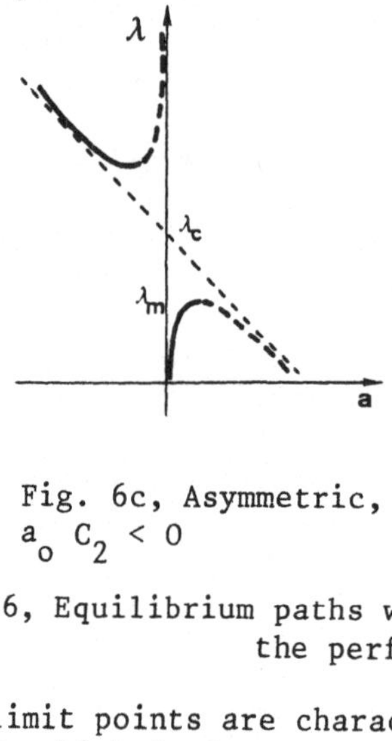

Fig. 6c, Asymmetric, $a_o\ C_2 < 0$

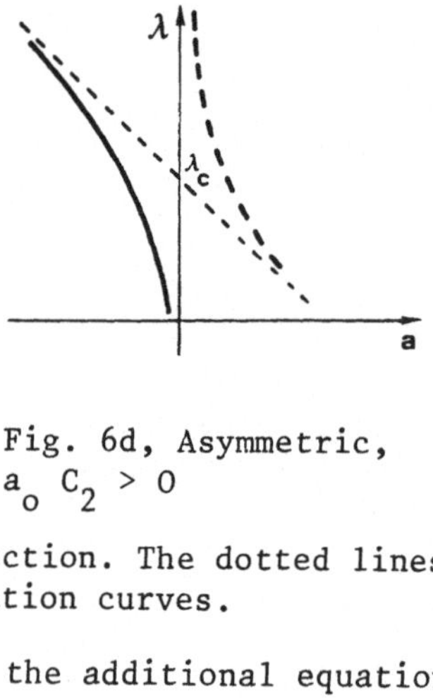

Fig. 6d, Asymmetric, $a_o\ C_2 > 0$

Fig. 6, Equilibrium paths with imperfection. The dotted lines are the perfect bifurcation curves.

The limit points are characterized by the additional equation $d\lambda/da = 0$, which leads to

$$3C_3\, a^2 - (\lambda_m - \lambda_c) = 0 \qquad (2\text{-}15)$$

or

$$2C_2 a - (\lambda_m - \lambda_c) = 0 \qquad (2\text{-}16)$$

One finds a maximal load point which is lower than λ_c in the cases 6-b or 6-c. The corresponding *reductions of the critical load* are given respectively by the formulae :

$$\lambda_c - \lambda_m = 3(\lambda_c \ a_0/2)^{2/3}(|C_3|)^{1/3} \qquad (2\text{-}17)$$

$$\lambda_c - \lambda_m = 2(-\lambda_c \ a_0 \ C_2)^{1/2} \qquad (2\text{-}18)$$

Structures in which this reduction may occur are said to be *imperfection sensitive*. On the contrary, with a stable-symmetric bifurcation point, the structure is not imperfection sensitive. Even though the theory requires a small $\lambda - \lambda_c$, the reduction may be substantial, especially with cylindrical shells. Nevertheless, the asymptotic formulae (2-17) (2-18) generally give a good approximate for the reduction of the critical load. Remark that, in the case of an unstable symmetric bifurcation, the $a_0 - \lambda_m$ curve corresponds to the *cusp* of elementary catastrophe theory /58/ (see Figure 7). There is another implication of this elementary study. Indeed the bifurcation points are destroyed by imperfections. These singularities are not *robust* or *structurally stable*, with the language of differential geometry, while a limit point is robust. Hence, the singular points may have different characters and it is necessary to classify them according to their robustness.

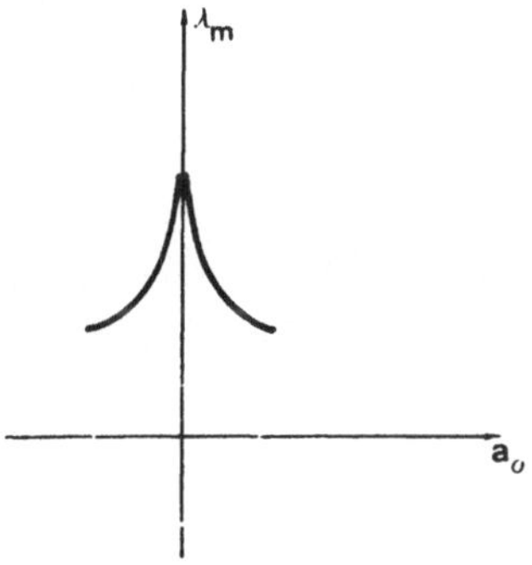

Fig. 7, Maximum load point as a function of the initial deviation.

2.3 Beam on foundation

Consider a flexible elastic beam of bending stiffness EI. It is subjected to a compressive axial force F. The lateral displacement $u(x)$ is restrained by an elastic foundation that provides a force $f(x)$ per unit length. As for the simple model, an initial imperfection $u_o(x)$ is taken into account (Figure 8). For the sake of simplicity, it is assumed that the beam and the foundation both are unstressed for $u = 0$. The balance of shear force and moment leads to :

$$M'' + F(\bar{u} + \bar{u}_o)'' + f(\bar{x}) = 0 \qquad (2\text{-}19)$$

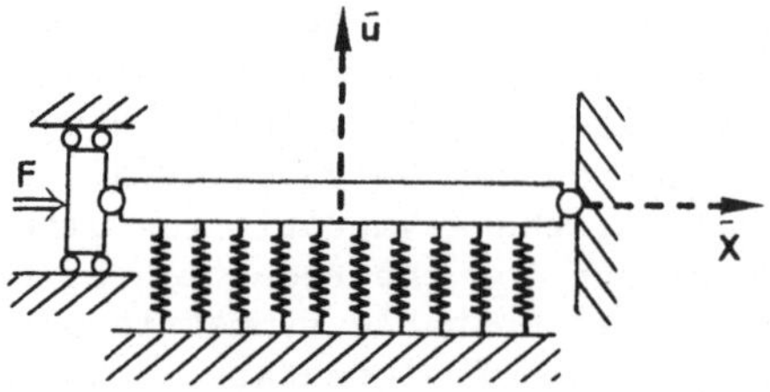

Fig. 8a, Beam on foundation

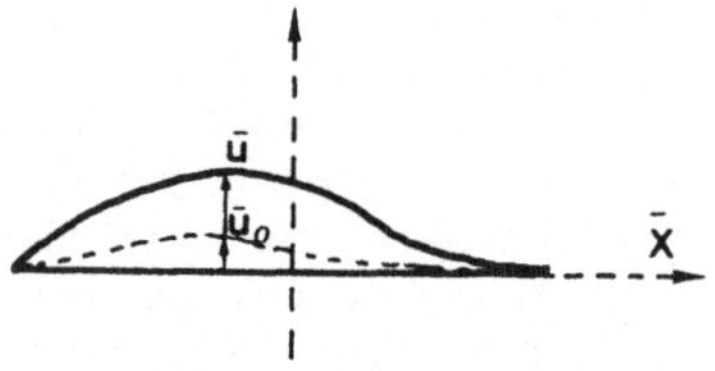

Fig. 8b, Definition of the initial imperfection.

One postulates constitutive laws in the form ($K_1 > 0$)

$$M = EI\,\bar{u}'', \quad f = K_1\,\bar{u} + K_3\,\bar{u}^3 \qquad (2\text{-}20)$$

Introduce a characteristic length L_0 and nondimensional quantities

$$L_0 = (EI/K_1)^{1/4}$$

$$x = \bar{x}/L_0 \quad , \; L = \bar{L}/L_0$$

$$u = \bar{u}(|K_3|/K_1)^{1/2}, \quad u_0 = \bar{u}_0(|K_3|/K_1)^{1/2}$$

$$\lambda = F/L_0^2\,K_1$$

Hence, from (2-19) and (2-20), the deflection u(x) is solution of the differential equation

$$u^4 + \lambda u'' + u + (\operatorname{sgn} K_3)\, u^3 = -\lambda u_0'' \qquad (2\text{-}21)$$

Here the beam is assumed to be simply supported :

$$u(0) = u''(0) = 0 \qquad (2\text{-}22)$$

$$u(L) = u''(L) = 0 \qquad (2\text{-}23)$$

In this section, we only seek the critical load whereas postbuckling and imperfection sensitivity will be discussed later on. As suggested by the study of the simple model, the critical load λ_c is the smallest value of the load for which the linearized equations have non zero solutions :

$$u^{(4)} + \lambda u'' + u = 0 \tag{2-24}$$

For λ smaller than 2, Eqs (2-22) to (2-24) have not such solutions, we omit the details of the computations. For λ greater than 2, any solution of (2-22), (2-24) is in the form :

$$\begin{aligned} &u = a_1 \sin(q_1 x) + a_2 \sin(q_2 x) \\ &q_1 q_2 = 1, \quad q_1^2 + q_2^2 = \lambda \end{aligned} \tag{2-25}$$

Inserting (2-25) into the last equation (2-23), one gets :

$$\begin{aligned} &a_1 \sin(q_1 L) + a_2 \sin(q_2 L) = 0 \\ &a_1 q_1^2 \sin(q_1 L) + a_2 q_2^2 \sin(q_2 L) = 0 \end{aligned} \tag{2-26}$$

The determinant of (2-26) has to be zero and this happens only if one of the sines is zero. Therefore, a_1 or a_2 is zero and, by dropping the remain ding arbitrary constant, one finds :

$$u(x) = \sin(qx) \tag{2-27}$$

$$\lambda = q^2 + 1/q^2 \tag{2-28}$$

$$q = n\,\pi/L, \ n \text{ integer} \tag{2-29}$$

The curve (2-28) in the (q, λ) plane is called the *neutral stability curve* (Figure 9). The minimum of this curve is 2 and it is reached for $q = 1$. But, by Eq. (2-29), the wavenumber q may only take discrete values. The minimum 2 is a good approximation of the critical load if L/π is nearly an integer or if the aspect ratio is large (L large, i.e. the true length $\bar{L}$ large with respect to the characteristic length L_0). The problem of discrete minimization has the following solution :

$$\begin{aligned} &\text{if} \quad L/\pi < 2^{1/2}, \quad n = 1, \ \lambda_c = \pi^2/L^2 + L^2/\pi^2 \\ &\text{if} \quad 2^{1/2} < L/\pi < 6^{1/2}, \quad n = 2, \ \lambda_c = 4\pi^2/L^2 + L^2/4\pi^2 \end{aligned} \tag{2-30}$$

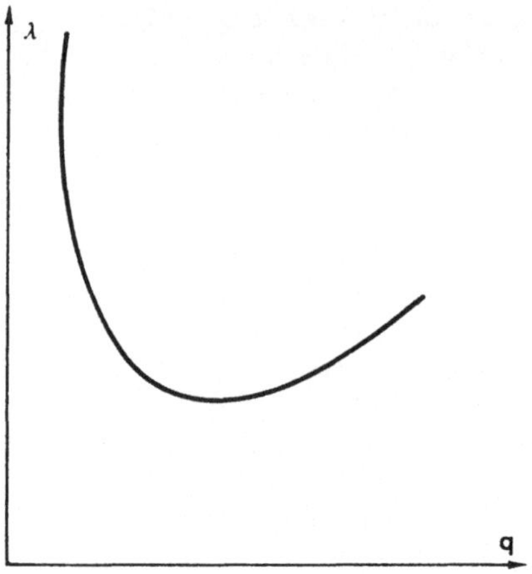

Fig. 9, Neutral stability curve for the beam on foundation.

The solution (2-27) is called the *buckling mode*. In the cases given in (2-30), it is unique after normalisation. In the special case $L = \pi\, 2^{1/2}$, there are two independant buckling modes corresponding to $n = 1, 2$ (of course, one will talk about mode 1, mode 2 ...). Remark that, for a large aspect ratio L, there are a great number of nearly coincident modes and the relevant postbuckling theory will be discussed in Section 6.

Remember that the linear analysis provides not only the buckling load, but also the buckling mode. The postbuckling theory will establish that the buckling mode gives the *shape of the exact solutions*. More exactly, for sufficiently small u_0, the solutions of (2-21) to (2-23) are taken in the form

$$u(x) = a\ (\text{buckling mode}) + \text{h.o.t.} \qquad (2\text{-}31)$$

the amplitude a being related to the load by an equation similar to (2-12). Remark also that (2-21) is the Euler equation of the following potential energy :

$$P(u,\lambda) = \int_0^L \{u''^2/2 - \lambda u'^2/2 + u^2/2 + \operatorname{sgn} K_3\, u^4/4 - \lambda u_0'\, u'\}\, dx \qquad (2\text{-}32)$$

3. CLASSICAL POSTBUCKLING THEORY

The study of a simple model (Section 2.2) has established that the imperfections may reduce the critical load of some structures, that are called imperfection sensitive. The first purpose of postbuckling theory is to determine if a structure is imperfection sensitive or not. If it is, this theory permits one to compute the reduction of the critical load, as in formulae (2-17) (2-18). Remark that such a computation involves both the nonlinearities and the imperfections. The general postbuckling theory is due to Koiter /31/. In these lectures, I roughly follow his exposure and his notations, especially that of ref. /32/. With respect to Koiter's papers, the main original points are the account of symmetries (Section 3.5) and the classification of singularities according to their robustness (one will say : their codimension). In this chapter, the proofs will be quite rigorous for discrete systems, even if some items will be omitted. But the theory is also established for continuous systems, additional informations being given in Section 5.1.

Let S(u) be a functional of the vector field u, that will be a generalized displacement. We assume that the functionals have as many derivatives as it is necessary. The Taylor expansions about the state u = 0 will be denoted as :

$$S(u) = S_o + S_1(u) + S_2(u) + \dots \tag{3-1}$$

The functional $S_m(u)$ is homogeneous of degree m. It can be itself expanded as :

$$S_m(v+w) = S_m(v) + S_{m-1,1}(v,w) + S_{m-2,2}(v,w) + \dots + S_m(w) \tag{3-2}$$

where $S_{p,q}(.,.)$ is homogeneous of degree p (resp. q) with respect to its first (resp. second) argument. Remark the comparison with the more classical formula :

$$\left(\frac{d}{dt}\right)^p S_m(v+tw)_{t=o} = p!\, S_{m-p,p}(v,w)$$

Homogeneous functions satisfy the Euler identity :

$$\begin{aligned} S_{m-p,p}(u,u) &= C_m^p\, S_m(u) \\ S_{11}(u,u) &= 2\, S_2(u) \end{aligned} \tag{3-3}$$

$$C_m^p = m!/p!\,(m-p)!$$

The functional $S_{11}(.,.)$ is bilinear and symmetric. We also shall use trilinear functionals $S_{111}(.,.,.)$ etc ... When we shall expand about a state $u_o \neq 0$, we shall adopt the following notations :

$$S(u_o + v) = S(u_o) + S_1(u_o;v) + \dots$$

$$S_m(u_o;v+w) = S_m(u_o;v) + S_{m-1,1}(u_o;v,w) + \dots$$

3.1 Energy criterion. Neutral stability

Because we deal with nonlinear elastic bodies subjected to dead loadings, the dynamical equations can be derived from a *potential energy* $P(u)$ from a *kinetic energy* $T(\partial_t u)$ and from the *space of admissible displacements*, denoted by U. For our purpose, it is not restrictive to assume that the kinetic energy does not depend on u, is quadratic (i.e. $T = T_2 = T_{11}/2$) positive definite and that U is a vectorial space.

Consider the example of the supported beam on a foundation, as presented in Section 2.3. The dynamical equation may be written in the form (case $K_3 > 0$)

$$\partial_t^2 u + \partial_x^4 u + \lambda \partial_x^2 u + u + u^3 = 0 \tag{3-4}$$

The potential energy has been given in (2-32). The kinetic energy and the space U are :

$$2\,T = \int_0^L (\partial_t u)^2 \, dx$$

$$U = \{u(x) \,|\, u(o) = u(L) = 0\}$$

From Lagrange variational principle, the equations of motion may be written as :

$$T_{11}\,(\partial_t^2 u, \delta u) + P_1(u;\delta u) = 0 \;\; \forall \; \delta u \in U \tag{3-5}$$

From now on, the symbol "for any δu in the space U" will be omitted in variational equations. If u_o is an equilibrium state, then one has :

$$P_1\,(u_o;\delta u) = 0 \tag{3-6}$$

For discrete systems (i.e. U finite dimensional) as in (3-5), the *stability criterion* is given by Lejeune-Dirichlet theorem, that has been roughly stated by Lagrange at the end of the 18th century and proved in 1846. According to this theorem, an equilibrium state u_o is stable if it achieves a *proper minimum of the potential energy*. According to the expansion (3-1) and to the equilibrium equation (3-6), this condition is realized if :

$$P_2(u_o;\delta u) > 0 \quad \forall \; \delta u \in U \quad , \; \delta u \neq 0 \tag{3-7}$$

On the contrary, one has not a proper minimum if

$$P_2\,(u_o;\delta u) < 0 \quad \text{for one} \quad \delta u \in U \tag{3-8}$$

The name "energy criterion" will be used to designate the stability test (3-7) and the instability test (3-8). Of course, the energy criterion is also used for continuous, even if a complete proof has not yet been given. Additional informations on this problem will be presented in Section 5.3.

One can easily sketch a proof of the energy criterion within the framework of linearized stability, i.e. by admitting that an equilibrium state is stable - or unstable - with respect to the nonlinear equation (3-5), if it is so with respect to the linearized equation :

$$T_{11}(\partial_t^2 w,\delta u) + P_{11}(u_o;w,\delta u) = 0 \qquad (3\text{-}9)$$

As usual in vibration theory, a natural pulsation ω and the corresponding vibration mode W(x) are such that Eq. (3-9) has solutions in the form :

$$w(x,t) = W(x) \exp i\omega t$$

Hence, they are characterized by :

$$P_{11}(u_o;W,\delta u) - \omega^2 T_{11}(W,u) = 0 \qquad (3\text{-}10)$$

Let ω^2 be the smallest of the ω^2's. By standard calculus of variations, it is characterized by the so-called quotient of Rayleigh ;

$$\omega_1^2 = \underset{W \in U}{\text{Min}} \frac{P_2(u_o;W)}{T(W)} = \underset{W \in U}{\text{Min}} \frac{P_{11}(u_o;W,W)}{(W,W)} \qquad (3\text{-}11)$$

If ω_1^2 is negative, the equilibrium state is (linearly) unstable. Admit that it is stable if ω_1^2 is positive and that there is a minimizer of (3-11). Hence, it follows from (3-11) that (3-7) and (3-8) are sufficient conditions for stability and instability, respectively. A complete proof of this statement has been written in /6/ within three-dimensional elasto dynamics under the assumptions of a bounded domain and of strong ellipticity.

Now suppose that the potential energy depends also on a *loading parameter* λ. Within postbuckling and bifurcation theory, the custom is to handle a single loading parameter, but there is at least one notable exception /25/. Additional parameters may be introduced later on. The notations are modified as follows :

$$P(\lambda,u),\quad P_1(\lambda,u;\delta u)\ \ldots$$

First and second derivatives with respect to λ will be denoted by :

$$P'(\lambda,u),\quad P_1'(\lambda,u;\delta u)\ \ldots \text{ and } P_1''(\lambda,u;\delta u)\ \ldots$$

A *critical load* λ_c is defined as the smallest value of λ such that an equilibrium path $u_o(\lambda)$ loses its stability. Because of the energy criterion (3-7), (3-8) and of (3-10), a critical load is characterized by the existence of at least one $u_1 \neq 0$ such that :

$$P_{11}(\lambda_c, u_o; u_1, \delta u) = 0 \qquad (3\text{-}12)$$

Eq. (3-12) means that the linearized equilibrium equation has a non-trivial solution, which is the basis of method of adjacent equilibrium. According to the terminology introduced in Section 2.3, u_1 is called the *buckling mode*. In what follows, we assume that there exists a finite number of buckling modes, denoted by $u_1, u_2, \ldots, u_n$. With the previously mentionned hypothesises (bounded domain, strong ellipticity), this can be established by using the theory of compact operators. To simplify the notations, we shall drop the dependence of any functional with respect to the state (λ, u) if it is taken at the critical state. For instance, Eq. (3-12) will be written as :

$$P_{11}(u_1, \delta u) = 0 \qquad (3\text{-}13)$$

If the system is discrete. the bilinear form $P_{11}(.,.)$ can be represented by a matrix, denoted again by P_{11}. Hence, (3-13) is equivalent to :

$$\det(P_{11}) = 0 \qquad (3\text{-}14)$$

Consider a non-critical equilibrium state (λ_o, u_o), which means that this determinant is not equal to zero. By the *Inverse Function Theorem*, a unique equilibrium path $(\lambda, u(\lambda))$ passes through (λ_o, u_o). Hence, a critical state is a state where, locally, the equilibrium may be not unique. If one knows an equilibrium path $u_o(\lambda)$ before and after λ_c, a secondary branch may cross it at λ_c. In this sense, *loss of stability coincides with possible bifurcations* ; either of these properties may be used to define the critical states. This analysis can be extended to continuous systems (see Section 5.1) but the coincidence does not hold within plasticity theory.

Remark that the energy criterion (3-7), (3-8) involves only the potential energy, even if stability is a dynamical concept. This important property is typical of the conservative systems that are studied here. That is why, from now one, we no longer need kinetic energy and dynamics.

3.2 Lyapunov and Schmidt reduction

We now begin with the study of initial postbuckling theory. At this stage, we do not account for the imperfections, but we include the non linear terms that set a more difficult problem. In the present Section and the next one, we suppose, first that *there exists a fundamental equilibrium path* $u_o(\lambda)$ for any load λ in an open interval, second that there is a critical load λ_c inside this interval. Furthermore, this path is assumed to depend smoothly on the load. This assumption is standard within the fields of elastic stability and of bifurcation theory. Nevertheless, it is rather *restrictive* because such a fundamental path does not persist when one introduces imperfections, as shown by the simple model of Section 2.2. In most applications, the existence of a fundamental path follows from symmetry properties. In this respect, the present assumption is not sufficiently restrictive. That is why other cases are studied in Section 3.5 and 3.6.

Introduce new variables $\bar{u}$ and $\bar{P}$

$$\bar{u} = u - u_o(\lambda), \quad \bar{P}(\lambda,\bar{u}) = P(\lambda, u_o(\lambda) + \bar{u}) \tag{3-15}$$

in such a way that $\bar{u} = 0$ is the fundamental equilibrium path. In other to simplify the notations, we drop the bars in (3-15), but we shall mind that a derivative with respect to λ (denoted by ') has to include the variation of the fundamental equilibrium with respect to the load. For instance, we write ;

$$P'(\lambda,u) \text{ instead of } P'(\lambda,u_o(\lambda)+u) + P_1(\lambda,u_o(\lambda)+u;u'_o(\lambda)) \tag{3-16}$$

The assumption of an equilibrium path $u_o(\lambda) = 0$ is written as (see (3-6)).

$$P_1(\lambda,0;\delta u) = 0 \tag{3-17}$$

If the fundamental path loses its stability at λ_c, the solutions of the linearized equation

$$P_{11}(\lambda_c, 0;u,\delta u) = 0 \tag{3-18}$$

are linear combinations of n linearly independent buckling modes u_1, u_2, ..., u_n :

$$u = a_i u_i \tag{3-19}$$

where we have followed the usual summation convention to imply summation for a repeated subscript. Let

$$\varepsilon = \lambda - \lambda_c \tag{3-20}$$

be the *load increment*. Because of (3-17), the Taylor expansion (3-3) may be written as :

$$\begin{aligned} P(\lambda_c + \varepsilon,u) &= P_2(u) + \varepsilon P_2'(u) + P_3(u) + P_4(u) + \text{h.o.t.} \\ \text{h.o.t.} &= O(u^5 + \varepsilon^2 u^2 + \varepsilon u^3) \end{aligned} \tag{3-21}$$

The aim of initial postbuckling theory is to seek all the solutions of (3-6) close to the critical state (λ_c, 0). First, one splits the space of kinematically displacements U as follows :

$$u = a_i\ u_i + v, \qquad v \in \mathcal{V} \tag{3-22}$$

where the space $\mathcal{V}$ is a supplementary of the n-dimensional subspace generated by the buckling modes. For instance, $\mathcal{V}$ may be the orthogonal space to the buckling modes with respect to the bilinear form $T_{11}(.,.)$, as done in /32/ :

$$\mathcal{V} = \{v \in U \mid T_{11}\ (v,u_i) = 0 \quad \text{for} \quad i = 1,2, \ldots,n\}$$

but this choice is not exclusive. Postbuckling theory is only interested in studying critical states that correspond to loss of stability. At such critical states, the second variation P_2 (u) is non-negative, but, of course, not positive definite. It is zero only if u is in the form (3-19). Hence, we may assume that:

$$P_2(v) > 0 \quad \text{for } v \quad \text{in } \mathcal{V},\ v \neq 0 \tag{3-23}$$

This assumption is not fundamental, except for questions of stability.

The variational equation (3-6) is also splitted by choosing as for δu, first any of the buckling modes, second any element of $\mathcal{V}$. One begins with the second choice :

$$P_1(\lambda, a_i\ u_i + v\ ;\ \delta v) = 0 \qquad \forall\ \delta v \in \mathcal{V} \tag{3-24}$$

In what follows, the symbol "for any δv in $\mathcal{V}$" will be omitted. Because of (3-23), the variational problem (3-24) corresponds to the minimization problem

$$\underset{v \in \mathcal{V}}{\text{Min}}\ P(\lambda,\ a_i\ u_i + v) \tag{3-25}$$

With account of (3-18) (3-21), Eq. (3-24) can be expanded as :

$$P_{11}(v,\delta v) + \varepsilon P'_{11}(a_i\ u_i + v,\ \delta v) + P_{21}(a_i\ u_i + v,\delta v) + \text{h.o.t.} = 0 \tag{3-26}$$

For discrete problems, (3-23) implies that the restriction to $\mathcal{V}$ of the matrix P_{11} is invertible. Hence, by virtue of the Inverse Function Theorem, Eq. (3-26) has a unique solution $\hat{v}(\varepsilon,a_i)$. By considering Eq.(3-26), one sees that this solution is in the form :

$$\hat{v}(\varepsilon,a_i) = \varepsilon a_i(\hat{v}_i + \text{h.o.t.}) + a_i\ a_j(\hat{v}_{ij} + \text{h.o.t.}) \tag{3-27}$$

the first terms $\hat{v}_i$, $\hat{v}_{ij}$ being solutions of the following variational problems :

$$P_{11}(\hat{v}_i,\delta v) + P'_{11}(u_i,\delta v) = 0 \quad (3\text{-}28)$$

$$P_{11}(\hat{v}_{ij},\delta v) + P_{111}(u_i,u_j,\delta v)/2 = 0 \quad (3\text{-}29)$$

where the identity (3-3) has been used for the computation of the second term in (3-29).

Because ε and a_i are small, the orthogonal part v of the displacement is small with respect to the in-mode part (see (3-22) (3-27)). Hence any postbuckling path is tangent to the space generated by the modes. If one neglects the higher order terms, a_i is the *amplitude of the displacement in the direction of the mode* u_i. In the bifurcation literature, this step is called *Lyapunov and Schmidt reduction*. For the english school [60] it is called elimination of the "passive coordinates" v, the active coordinates being the amplitudes a_i. Details on this procedure in the continuous case will be given in Section 5.1.

It remains to insert the minimum (3-27) into the potential energy. This defines a reduced potential energy :

$$F(\varepsilon,a_i) = P(\lambda_c + \varepsilon, a_j u_j + v(\varepsilon,a_i)) \quad (3\text{-}30)$$

Because of (3-17) (3-18) (3-21), the function F has the following expansion :

$$\begin{aligned} F(\varepsilon, a_i) &= \varepsilon F'_2(a_i) + F_3(a_i) + F_4(a_i) + \text{h.o.t.} \\ \text{h.o.t.} &= O(\varepsilon^2 a^2 + \varepsilon a^3 + a^5) \end{aligned} \quad (3\text{-}31)$$

Of course, there are no quadratic terms at the critical load ($\varepsilon = 0$). The functions F'_2, F_3, F_4 of the amplitudes can be computed from the potential energy, from the buckling modes u_i and from the solutions $\hat{v}_{ij}$ of the variational equation (3-29) :

$$F'_2 = P'_2(a_i u_i) \quad (3\text{-}32)$$

$$F_3 = P_3(a_i u_i) \quad (3\text{-}33)$$

$$F_4 = P_4(a_i u_i) - P_2(a_i a_j \hat{v}_{ij}) \quad (3\text{-}34)$$

The only difficulty lies in the computation of F_4. Easy algebra from (3-18) (3-21) (3-22) (3-27) and from the rule (3-2) leads to :

$$F_4 = P_4(a_i u_i) + P_{21}(a_i u_i, a_j a_k \hat{v}_{jk}) + P_2(a_i a_j \hat{v}_{ij})$$

and from the rules (3-3) :

$$F_4 = P_4(a_i\, u_i) + P_{111}\,(a_i u_i, a_j u_j, a_k a_l \hat{v}_{kl})/2$$

$$+\, P_{11}(a_i a_j \hat{v}_{ij}, a_k a_l \hat{v}_{kl})/2 \qquad (3\text{-}35)$$

Now choose $\delta v = \hat{v}_{kl}$ in the variational equation (3-29). This gives :

$$P_{111}(u_i, u_j, \hat{v}_{kl}) = -\,2\, P_{11}(\hat{v}_{ij}, \hat{v}_{kl}) \qquad (3\text{-}36)$$

and (3-35) (3-36) leads to F_4 as in (3-34).

After the elimination,(3-6) is reduced to the n following equations:

$$\partial F/\partial a_i = 0, \qquad i = 1,n \qquad (3\text{-}37)$$

Because of the assumption (3-23), $F(\varepsilon, a_i)$ is the minimum of the potential energy with respect to v. Hence, the stability of an equilibrium state can be characterized by the second variation of F. For a detailed proof, see /41/ p. 600-601.

3.3 First case : bifurcation from a fundamental path

In this Section, we finish the discussion of possible bifurcation points from a fundamental path $u_o = 0$. The most simple singularity is obtained when the following three conditions hold :

There is only one independent buckling mode u_1 (3-38)

$$P_2'(u_1) \neq 0 \qquad (3\text{-}39)$$

$$P_3(u_1) \neq 0 \qquad (3\text{-}40)$$

These three conditions persist after any perturbation that does not alter the existence of a smooth fundamental path (which excludes general imperfections, as in Section 2.2). Let us introduce the three following coefficients :

$$\alpha_{11} = 2\, P_2'\,(u_1)$$

$$(3\text{-}41)$$

$$\alpha_{20} = 3\, P_3\,(u_1) \quad ; \qquad \alpha_{30} = 4(P_4(u_1) - P_2(\hat{v}_{11}))$$

The reduced potential energy is a function of the load increment ε and of only one amplitude, denoted by a instead of a_1 :

$$F(\varepsilon,a) = \alpha_{20}\, a^3/3 + \alpha_{30}\, a^4/4 + \varepsilon a^2\, \alpha_{11}/2 + \text{h.o.t.}$$

Because of the form of the higher order terms in (3-31), the amplitude can be factorized in the bifurcation equation :

$$\partial F/\partial a = a(\alpha_{20}\, a + \alpha_{30}\, a^2 + \varepsilon\alpha_{11} + \text{h.o.t.}) = 0 \qquad (3\text{-}42)$$

The coefficients α_{20} and α_{11} are different from zero, by the assumption (3-39) (3-40). (In the applications, the fundamental equilibrium is stable for a load lower than λ_c and unstable for a load greater than λ_c, which implies that α_{11} is negative). Then, after division by a, Eq. (3-42) can be solved in a or ε, by virtue of (3-39) (3-40) and of the Inverse Function Theorem :

$$\lambda - \lambda_c = \varepsilon = -\, a\, \alpha_{20}/\alpha_{11} + O(a^2) \qquad (3\text{-}43)$$

Eq. (3-43) is the equation of a secondary path, which crosses the fundamental one at the critical state $(\lambda_c,0)$. Hence, the singular point is the *asymmetric bifurcation point,* as pictured in Figure 1-c.

The asymmetric bifurcation point is robust in the present conditions, i.e. when the existence of a fundamental path is ensured. Indeed, if one introduces a perturbation of the potential energy, the critical state $(\lambda_c,0)$ is slightly modified, but the buckling mode remains simple and the conditions (3-39) (3-40) still hold. Other singular points will be obtained by relaxing one of the three conditions (3-38) to (3-40). But the corresponding singular points will be not robust with respect to any perturbation that maintains the existence of a fundamental path. Remark that the asymmetric bifurcation point itself is not robust with respect to more general perturbations, as indicated by the simple example (Figure 6).

Now we turn to the study of singularities that are not robust. First, we rule out (3-40), but assume instead that :

$$a/\ \alpha_{20} = 0\ , \qquad b/\ \alpha_{30} \neq 0 \qquad (3\text{-}44)$$

Bifurcating equilibrium paths are obtained from Eq. (3-42) in the same manner as previously. There is again only one bifurcating path, that is given by :

$$\varepsilon = \lambda - \lambda_c = -\, a^2\, \alpha_{30}/\alpha_{11} + O(a^3) \qquad (3\text{-}45)$$

Therefore, the singular point is a *symmetric bifurcation point,* as pictured in Figure 1-a, 1-b. Since α_{11} generally is negative, the sign of α_{30} permits one to conclude that a structure is imperfection sensitive ($\alpha_{30}<0$) or not ($\alpha_{30} > 0$). Remark that α_{30} is given in (3-41), where $\hat{v}_{11}$ is the solution of a variational problem (see (3-29)).

$$P_{11}(\hat{v}_{11},\delta v) + P_{111}(u_1,u_1,\delta v)/2 = 0 \qquad (3\text{-}46)$$

Now assume that (3-39) no longer holds, i.e. that :

$$\alpha_{11} = 0 \qquad (3\text{-}47)$$

although this case is rather uncommon in elastic stability. Then, the reduced potential energy is in the form :

$$\begin{aligned} F(\varepsilon,a) &= \alpha_{20}\, a^3/3 + \alpha_{12}\, a^2\, \varepsilon^2/2 + \text{h.o.t.} \\ \text{h.o.t.} &= O(a^4 + a^3\varepsilon + a^2\, \varepsilon^3) \end{aligned} \qquad (3\text{-}48)$$

As in (3-41) (3-42), we have denoted by α_{rs} the coefficients in the expansions of F, the subscripts r and s being the degrees of a and ε in the bifurcation equation. The new coefficient α_{12}, that appears in (3-48) is given by :

$$\alpha_{12} = P'_{11}(u_1,\hat{v}_1) + P''_2(u_1) \qquad (3\text{-}49)$$

where " designates two derivatives with respect to the load and $\hat{v}_1$ is the solution of the variational problem (3-28). The proof of (3-49) is easy and similar to that of (3-34) ; it requires to choose $\delta v = \hat{v}_1$ in (3-28). We further assume that :

$$\alpha_{20} \neq 0, \quad \alpha_{12} \neq 0 \qquad (3\text{-}50)$$

The bifurcation equation is :

$$a(\alpha_{20}\, a + \alpha_{12}\, \varepsilon^2 + \text{h.o.t.}) = 0 \qquad (3\text{-}51)$$

The bifurcating branch can be solved with respect to a :

$$a = -\, \varepsilon^2\, \alpha_{12}/\alpha_{20} + O(\varepsilon^3) \qquad (3\text{-}52)$$

We obtain a tangent bifurcation point (Figure 10-b).

To understand the nature of non-robust singularity, it is necessary to look at its behaviour after a perturbation. In the present case, this means reintroduction of a small, but non-zero, coefficient α_{11}. As shown by figure 10, a tangent bifurcation point is the coincidence of two bifurcation points, one with destabilization, the other with restabilization of the fundamental path (Fig. 10-a). After the coincidence, this path loses no longer its stability and the two branches have no intersections (Fig.10-c). An example of tangent bifurcation will be given in Section 4.2

We have found two non-robust singularities that are defined, in addition to Eq. (3-14), by a single supplementary equation (3-44a) or (3-47). According to the terminology of catastrophe theory or of singularity theory (version Golubitsky and Schaeffer /23/), one says that they are of *codimension one*. Any other singularity would be of codimension two, which means that they are defined by at least two supplementary equations.

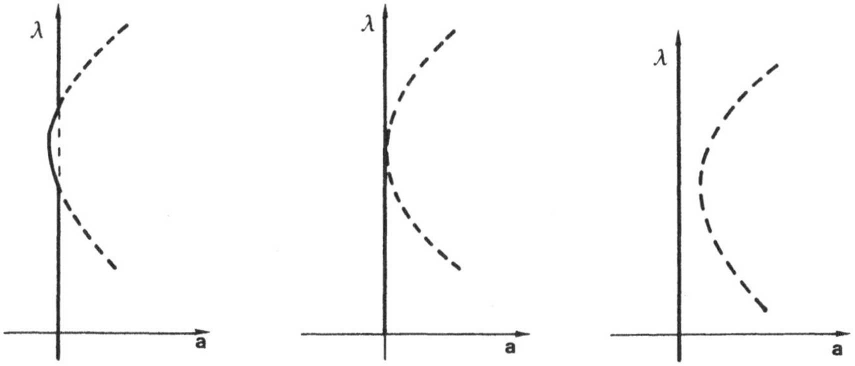

Fig. 10a, $\alpha_{11} < 0$ Fig. 10b, $\alpha_{11} = 0$ Fig. 10c, $\alpha_{11} > 0$

Fig. 10, Tangent bifurcation point (10-b) in the case of a positive α_{12} and of a negative α_{20} and perturbation.

Especially singularities corresponding to a double buckling mode are at least of codimension two. For instance, if the system has two degree of freedom, the matrix P_{11} is in the form :

$$\begin{bmatrix} b_{11} & b_{12} \\ b_{12} & b_{22} \end{bmatrix}$$

and a double buckling mode requires three equations

$$b_{11} = b_{12} = b_{22} = 0 \qquad (3\text{-}53)$$

It does not seem very useful to spend time in the study of singularities of codimension two, because they are not sufficiently common, at least in the present cases of systems with a fundamental path and without additional assumptions. Nevertheless, coincident buckling modes are frequent in application because the engineering structures often are symmetric and because *the symmetries modify the codimension.* In the example of the beam on foundation (Section 2.3), one obtains a double buckling mode if one intents to *optimize the buckling load* (2-30) with respect to the adimensional length L.

In view of these symmetric structures, let us consider the case of n *coincident buckling modes*. It is not very restrictive to assume that the quadratic form F_2^1 (a_i) is negative definite, since in most of the applications, the fundamental path loses its stability at the critical load. Then after rearrangement of the modes, one obtains :

$$F_2' (a_i) = - a_i\, a_i/2 \tag{3-54}$$

We denote by a the Euclidean norm of the amplitude :

$$a_i = a\, A_i \,, \qquad A_i\, A_i = 1 \tag{3-55}$$

Assume that the load increment can be expanded as (this can be proved, see /41/)

$$\varepsilon = \lambda - \lambda_c = a\, \lambda_1 + a^2\, \lambda_2 + \ldots \tag{3-56}$$

Inserting (3-55) (3-56) into the bifurcation equations (3-37), next dividing by a^2, we get :

$$\frac{\partial F_3}{\partial A_i} (A_i) - \lambda_1\, A_i + O(a) = 0 \tag{3-57}$$

When a bifurcating solution goes to the critical state, a tends to zero and the direction A_i tends towards a direction A_i^o such that :

$$\frac{\partial F_3}{\partial A_i} (A_i^o) - \lambda_1\, A_i^o = 0 \tag{3-58}$$

This means that the direction A_i^o is an extremum of the cubic term F_3 on the unit ball defined by (3-55), λ_1 being the extremal value. The existence of a bifurcating branch $A_i(a)$ can be established by using the Inverse Function Theorem /41/. As we have seen in many previous cases, this requires additional conditions, that are here :

$$\lambda_1 \neq 0 \tag{3-59}$$

The extremum A_i^o is not degenerate (3-60)

Furthermore, the bifurcating branch is stable only if :

$\lambda_1 > 0$ and A_i^o is a minimum of F_3 (3-61)

Remark that the parabolic umbilic /24/ does not satisfy (3-60).

A similar analysis can be done when there is no cubic term F_3. Eqs(3-56) (3-58) have to be replaced by :

$$\lambda - \lambda_c = \lambda_c\, a^2 + \lambda_3\, a^3 + \ldots \tag{3-62}$$

$$\frac{\partial F_4}{\partial A_i}(A_i^o) - \lambda_2 A_i^o = 0 \tag{3-63}$$

The additional conditions are (3-60) and $\lambda_2 \neq 0$. The stability condition is

$$\lambda_2 > 0 \text{ and } A_i^o \text{ is a minimum of } F_4 \tag{3-64}$$

The condition (3-60) is necessary, as shown by a counterexample in [41], p. 597.

It is easy to discuss the number of bifurcating equilibrium paths in the case of *two coincident modes*. One lets :

$$A_1 = \cos\Theta \ , \ A_2 = \sin\Theta$$

Hence, F_3 and F_4 are trigonometric polynomials. Because of oddness or evenness, $\Theta + \pi$ is a solution of (3-58) or (3-63) if Θ is. Two such solutions correspond to a single solution path. If F_3 is not zero, we may only have :

- one unstable branch,
- or three unstable branches,
- or three branches, one of which is stable for $\lambda > \lambda_c$,

if we exclude the degenerate cases where (3-59) or (3-60) do not hold. If F_3 is zero and F_4 positive definite, all bifurcating solutions are supercritical ($\lambda_2 > 0$).
We may only have, except in degenerate cases :

- two branches, one stable and one unstable,
- or four branches, two stable and two unstable.

3.4 Alternative method

In a number of cases, we have established that the load λ can be expanded into powers of the amplitude a, the tangent bifurcation point being an exception (see (3-52)). An alternative method is to anticipate such a result and to seek a priori asymptotic expansions is the form (see [9]) :

$$\varepsilon = \lambda - \lambda_c = a\lambda_1 + a^2 \lambda_2 + \dots \tag{3-65}$$

$$u = a\, U_1 + a^2\, U_2 + a^3\, U_3 + \dots \tag{3-66}$$

Formulae (3-65) (3-66) introduce a supplementary parameter a, because, in the initial problem, one only seeks the possible displacements as functions of the load. Hence, one has to require additional conditions for the U_i 's. For instance, in the case of a single buckling mode, one may require :

$$\begin{aligned} &\text{Max } |U_1| = 1 \\ &T_{11}(U_1, U_j) = 0 \quad \text{for } j = 2, 3 \dots \end{aligned} \tag{3-67}$$

With account of (3-17) (3-18), our equation $P_1(\lambda_c + \varepsilon, u ; \delta u) = 0$ may be written as :

$$P_{11}(u,\delta u) + \varepsilon P'_{11}(u,\delta u) + P_{21}(u,\delta u) + P_{31}(u,\delta u) + h.o.t. = 0$$

$$h.o.t. = O(u^4 + \varepsilon u^2 + \varepsilon^2 u) \qquad (3\text{-}68)$$

After having inserted (3-65) (3-66) into (3-68), one gets equations at the order a, a^2 :

$$P_{11}(U_1,\delta u) = 0 \qquad (3\text{-}69)$$

$$P_{11}(U_2,\delta u) + \lambda_1 P'_{11}(U_1,\delta u) + P_{21}(U_1,\delta u) = 0 \qquad (3\text{-}70)$$

First, consider the case of a single buckling mode. With account of the normalization (3-67), the solution of (3-69) is :

$$U_1 = u_1$$

prodided that the buckling mode has been normalized as in (3-67). Eq. (3-70) cannot be solved with respect to U_2 for any λ_1 because the operator P_{11} is singular (see (3-69)). Hence, there is one - and only one - solvability condition for U_2. Because of (3-69) and of the symmetry of the bilinear form $P_{11}(.,.)$, one has

$$P_{11}(U_2, u_1) = 0 \qquad (3\text{-}71)$$

Then, if one chooses $\delta u = u_1$ in (3-70), one gets the

$$\lambda_1 = - P_{21}(u_1,u_1)/P'_{11}(u_1,u_1)$$

such that Eq. (3-70) is solvable in U_2. The latter formula may be rewritten as :

$$\lambda_1 = - 3 P_3(u_1)/2 P'_2(u_1) \qquad (3\text{-}72)$$

Often, the structure is symmetric and this λ_1 is found to be equal to zero. To obtain the next term λ_2, one needs Eq. (3-68) at the order a^3 :

$$P_{11}(U_3,\delta u) + \lambda_1 P'_{11}(U_2,\delta u) + P_{111}(u_1,U_2,\delta u) + P_{31}(U_1,\delta u)$$

$$+ \lambda_2 P'_{11}(U_1,\delta u) + \lambda_1^2 P''_{11}(U_1,\delta u) + \lambda_1 P'_{21}(U_1,\delta u) = 0$$

$$(3\text{-}73)$$

Choosing $\delta u = u_1$ in (3-73), one gets (if $\lambda_1 = 0$) :

$$\lambda_2 = - (P_{21}(u_1, U_2) + 2P_4(u_1))/P_2'(u_1) \tag{3-74}$$

U_2 being the solution of the variational problem (3-70) with λ_1 equal to zero.

Formulae (3-72) (3-74) corroborate previous results (3-41) (3-43) (3-45) that have been obtained by using the Lyapunov-Schmidt reduction. If λ_1 is not zero, the singularity is the asymmetric bifurcation point. If λ_1 is zero and λ_2 not, it is the symmetric bifurcation point which may be stable ($\lambda_2 > 0$) or unstable ($\lambda_2 < 0$). Remark that one can expand the displacement with respect to ε if $\lambda_1 \neq 0$, or with respect to $\varepsilon^{1/2}$ if $\lambda_1 = 0$, or with respect to $(-\varepsilon)^{1/2}$ if $\lambda_1 = 0$, $\lambda_2 < 0$.

A similar procedure can be used with a multiple buckling mode. One has only to change the additional conditions (3-67). The solutions of (3-69) are

$$U_1 = A_i\ u_i$$

The normalized amplitude A_i may be assigned to satisfy :

$$A_i\ A_i = 1 \tag{3-75}$$

With the same assumption and the same rearrangement as in (3-54), the modes satisfy the condition

$$P_{11}'\ (u_i, u_j) = -\ \delta_{ij}$$

where δ_{ij} is the symbol of Kronecker. Then, choosing δu equal to u_i in Eq.(3-70), we get the n equations

$$P_{21}(A_j\ u_j,\ u_i) - \lambda_1\ A_i = 0 \tag{3-76}$$

that are identical to (3-58). This determines the possible values of the A_j 's and of

$$\lambda_1 = 3\ P_3(U_1) \tag{3-77}$$

Eq.(3-70) only gives the projection V_2 of U_2 on the supplementary space $\mathcal{V}$

$$U_2 = V_2 + B_i\ u_i \tag{3-78}$$

where the n numbers B_i are not yet determined. To do this, one chooses $\delta u = C_i\ u_i$ into the third order equation (3-73) :

$$C_1 \{P_{111}(U_1,U_2,u_i) + \lambda_1 P'_{11}(U_2,u_i)\} = - \lambda_2 P'_{11}(U_1,C_i u_i) + C_i f_i(A_j,\lambda_1)$$

which may be rewritten as :

$$\{P_{111}(u_i,u_j,u_k)\ A_k - \lambda_1\ \delta_{ij}\}\ C_i\ B_j = - \lambda_2 P'_{11}(U_1,C_i u_i) + C_i g_i(A_j,\lambda_1) \tag{3-79}$$

Now one has an additional condition for U_2, i.e. for B_i, at one's disposal We propose :

$$P'_{11}(U_1,B_i u_i) = - A_i\ B_i\ = 0 \tag{3-80}$$

and, in (3-79), we limit ourselves to C_i's that satisfy the same condition as the B_i's. This rules out λ_2 from Eq. (3-79). To get B_i from (3-79), one has to invert the restriction to $\{ A_i \}^{\perp}$ of the matrix of the left hand side, i.e. :

$$P_{111}(u_1,u_j,u_k)\ A_k - \lambda_1\ \delta_{ij}$$

This invertibility condition is exactly the condition (3-60) that followed from the Inverse Function Theorem in the previous analysis.

These few examples show that the two methods give the same results. The choice of either of these methods often is a question of personal taste. The present method is a more straight way to practical formulae as (3-72) (3-74). But the Lyapunov - Schmidt is preferable if one wants a quite rigorous proof or if the nature of the studied singularity is not foreseeenable.

3.5 Second case : perturbation of a singular equilibrium

In this Section, we assume no longer the existence of a fundamental equilibrium path on both sides of the critical load but *only the existence of a singular equilibrium* $u = 0$ at a load λ_c. Indeed the imperfect model of Section 2.2 has shown that, generally, there are equilibrium states on a single side of the critical load. Roughly, the matter of this Section is contained in the main references of bifurcation theory, for instance that of Vainberg and Trenogin /67/. More precisely, we follow the analysis of Ref. /42/ Section III.

We limit ourselves to the case of a *single buckling mode* u_1, for the reason given in Section 3.3 : so we shall find all singularities of codimension zero or one. Thus, instead of (3-17) (3-18), we only assume

$$P_1(\lambda_c,0\ ;\ \delta u) = P_1(\delta u) = 0 \tag{3-81}$$

$$P_{11}(\lambda_c,0\ ;\ u_1,\delta u) = 0 \tag{3-82}$$

Because Eq. (3-17) is no longer satisfied, the expansion (3-21) must be completed by terms in εu, $\varepsilon^2 u \ldots$

$$P(\lambda,u) = \varepsilon P_1'(u) + \varepsilon^2 P_1''(u)/2 + P_2(u) + \varepsilon P_2'(u) + P_3(u) + P_4(u) + \text{h.o.t.}$$

$$\text{h.o.t.} = O(\varepsilon^3 u + \varepsilon^2 u^2 + \varepsilon u^3 + u^5) \tag{3-83}$$

The Lyapunov-Schmidt reduction is defined by the same equations (3-22) (3-24) as before. With account of (3-82) (3-83) corresponding variational equation becomes :

$$P_{11}(v,\delta v) + \varepsilon P_1'\,\delta v) + \varepsilon P_{11}'(au_1 + v,\delta v) + P_{21}(au_1 + v,\delta v) + \text{h.o.t.} = 0$$

$$\text{h.o.t.} = O(\varepsilon^2 + \varepsilon u + u^3) \tag{3-84}$$

As previously, Eq. (3-84) has a unique solution $\hat{v}(a,\varepsilon)$ in the form :

$$\hat{v}(a,\varepsilon) = \varepsilon \hat{v}_1 + \varepsilon a \hat{v}_2 + a^2 \hat{v}_3 + \text{h.o.t.}$$

$$\text{h.o.t.} = O(\varepsilon^2 + \varepsilon a^2 + a^3) \tag{3-85}$$

where the $\hat{v}_i$'s solve variational equations

$$P_{11}(\hat{v}_1,\delta v) + P_1'(\delta v) = 0 \tag{3-86}$$

$$P_{11}(\hat{v}_2,\delta v) + P_{11}'(u_1,\delta v) + P_{111}(u_1,\hat{v}_1,\delta v) = 0 \tag{3-87}$$

$$P_{11}(\hat{v}_3,\delta v) + P_{21}(u_1,\delta v) = 0 \tag{3-88}$$

Remark that $\hat{v}_3$ is identical to the $\hat{v}_{11}$ in (3-29). Let us insert (3-85) into the potential energy (3-83). After some simple algebra, one gets :

$$F(a,\varepsilon) = \alpha_{o1}\, a\varepsilon + \alpha_{o2}\, a\varepsilon^2 + \alpha_{11}\, a^2/2 + \alpha_{2o}\, a^3/3 + \alpha_{3o} a^4/4 + \text{h.o.t.} \tag{3-89}$$

where the coefficients α_{ij} are given by the equations :

$$\alpha_{o1} = P_1'(u_1) \tag{3-90}$$

$$\alpha_{o2} = P_1''(u_1)/2 + P_{11}'(u_1,\hat{v}_1) + P_{21}(\hat{v}_1, u_1) \tag{3-91}$$

$$2\alpha_{11} = P_2'(u_1) + P_{21}(u_1,\hat{v}_1) \tag{3-92}$$

$$3\alpha_{2o} = P_3\,(u_1) \tag{3-93}$$

$$4\alpha_{3o} = P_4(u_1) - P_2(\hat{v}_3) \tag{3-94}$$

Eq. (3-91), (3-92) and (3-94) have been simplified by accounting for (3-86) with $\delta u = \hat{v}_2, \delta u = \hat{v}_3$ and for (3-88) with $\delta u = \hat{v}_3$. Remark that $\hat{v}_2$ has no effect on the leading coefficients of the potential energy.

Thus the bifurcation equation is :

$$\alpha_{o1}\,\alpha + \alpha_{2o}\,a^2 + \alpha_{11}\,a\varepsilon + \alpha_{o2}\,\varepsilon^2 + \alpha_{3o}\,a^3 = \text{h.o.t.} = 0 \tag{3-95}$$

If the lowest order coefficient is not zero, i.e.

$$\alpha_{o1} \neq 0 \tag{3-96}$$

the bifurcation equation (3-95) can be solved in ε as follows :

$$\lambda - \lambda_c = \varepsilon = - a^2\alpha_{2o}/\alpha_{o1} + a^3(\alpha_{11}\,\alpha_{2o}/\alpha^2_{01} - \alpha_{30}/\alpha_{o1}) + O(a^4) \tag{3-97}$$

If, in addition to (3-96), we assume :

$$\alpha_{2o} \neq 0 \tag{3-98}$$

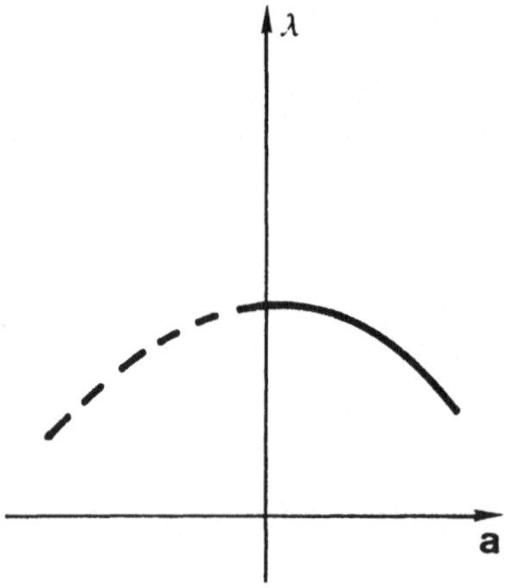

Fig. 11, Limit point when $\alpha_{20} > 0$, $\alpha_{01} > 0$

the singular point is the limit point, as seen in Section 2.2. It is a maximum or a minimum load point, according to signs of α_{2o} and α_{o1} (Figure 11). The conditions for a limit point (3-38) (3-96) (3-98) persist after perturbations of the potential energy. *The limit point is the only robust singularity.*

One finds the singularities of codimension one by ruling out either conditions (3-96) or (3-98), as in Section 3.3. First, assume (3-96) and

$$\alpha_{2o} = 0 \quad , \quad \alpha_{30} \neq 0 \tag{3-99}$$

In this case, there is only one branch that is given by Eq. (3-97). Because of its behaviour after perturbation, i.e. for small values of α_{2o}, this singularity is called *hysteresis point*, according to the termino-logy of Ref. /22/ Indeed, this singularity marks the appearance of a small hysteresis loop, which means that the way is different during loading and unloading. For a small α_{2o} and $\alpha_{2o}\alpha_{3o}$ negative, the hysteresis point splits into two limit points, a maximum and a minimum (Figure 12).

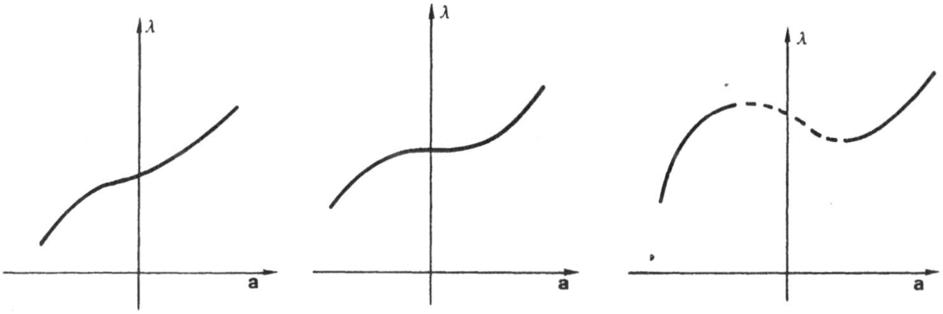

Fig. 12-a, $\alpha_{20} < 0$ Fig. 12-b, $\alpha_{20} = 0$ Fig. 12-c, $\alpha_{20} > 0$

Fig. 12, The hysteresis point (12-b) gives rise to a hysteresis loop (12-c) $\alpha_{01} < 0$, $\alpha_{30} > 0$

Now, one forsakes (3-96) and postulates :

$$\alpha_{o1} = 0 \tag{3-100}$$

The bifurcation equation begins with quadratic terms in both variables and a. Suppose that this quadratic polynomial is positive definite or negative definite :

$$\alpha_{11}^2 < 4\,\alpha_{o2}\,\alpha_{2o} \qquad (3\text{-}101)$$

Thus, the singular point $(\lambda_c, 0)$ is an *isola center*, which means that there is no other equilibrium in the neighbourhood. How can we perturb this isola ? It is not sufficient to reintroduce the eliminated term α_{o1} and one has to consider a constant term (i.e. a small α_{oo}) in the bifurcation equation :

$$\alpha_{oo} + \alpha_{2o}\,a^2 + \alpha_{11}\,a\,\varepsilon + \alpha_{o2}\,\varepsilon^2 + \text{h.o.t.} = 0 \qquad (3\text{-}102)$$

Hence, the isola corresponds to the shrinking of a closed quasi-elliptical loop (Figure 13)

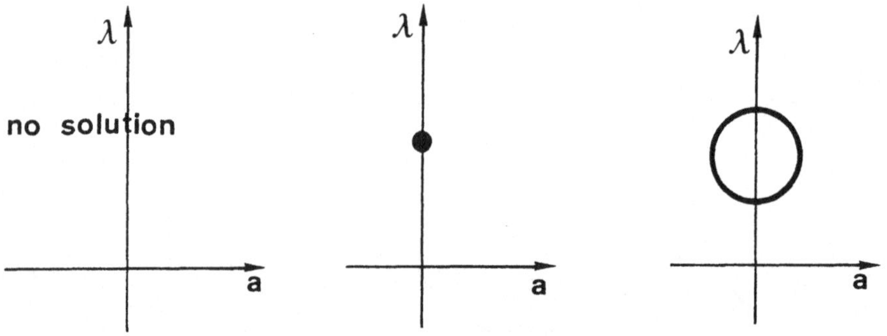

Fig. 13-a, $\alpha_{00} > 0$ Fig. 13-b, $\alpha_{00} = 0$ Fig. 13-c, $\alpha_{00} < 0$

Fig. 13, Isola center (13-b) is the limit of a closed loop $\alpha_{20} > 0$

Now, assume that the quadratic form is not definite, (3-100) being always valid :

$$\alpha_{11}^2 > 4\alpha_{o2}\,\alpha_{2o}\ , \quad \alpha_{o2} \neq 0 \qquad (3\text{-}103)$$

Thus, Eq. (3-95) can be solved twice in ε/a as a function of a. There are two equilibrium paths that cross at the critical point. Hence, it is an *asymmetric bifurcation point*. The perturbed equation (3-102) yields its behaviour after perturbation, which is in accordance with that of Figure 6

We established that the limit point is the only robust singularity. The hysteresis point, the isola center and the asymmetric bifurcation point are the singularities of codimension one, i.e. those characterized by only one supplementary equation (3-99) or (3-100).

Remark that the two latter singularities have raised the question of *the nature of the perturbations* that may be introduced to perturb a non-robust singularity. So, we remarked that it is not always adequate to reintroduce the term that had just been ruled out. For instance, the hysteresis point should be perturbed as :

$$\alpha_{o1}\,\varepsilon + \alpha_{3o}\,a^{3} + \text{h.o.t.} = -\,(\alpha_{oo} + \alpha_{1o}\,a + \alpha_{2o}\,a^{2}) \qquad (3\text{-}104)$$

where the coefficients in the right hand side are small. But the obvious solution of (3-104) shows that α_{oo} and α_{1o} only induce a translation of the inflexion point of Figure 12 and that they do not alter the shape of the path. That is why they have been neglected, as also the linear terms in (3-102) that only translate the center of the ellipse or of the hyperbola. *The mathematical theories, known as catastrophe theory or singularity theory, give qualitative answers to this question,* what will be discussed in Section 5.2.

3.6 Symmetric structure. Symmetry breaking

Hitherto, we followed the classical accounts. But most of the engineering structures are symmetric with respect to a plane, an axis or a point. That is why modern statements of bifurcation theory take into account additional symmetry properties /23/ /51/. To modelize such a symmetry, one assumes that there exists a linear involutive operator S from U into itself, which means that :

$$S^{2} = \text{Id}\ , \qquad S \neq \text{Id} \qquad (3\text{-}105)$$

Furthermore, this symmetry operator leaves the potential energy invariant:

$$P(\lambda, Su) = P(\lambda, u) \qquad (3\text{-}106)$$

In the simple model of Section 2.2, this symmetry operator is $Sa = -a$. For the beam on elastic fondation (Section 2.3), there are three symmetry operators S_1, S_2 and $S_1 S_2$:

$$\begin{aligned} (S_1 u)(x) &= -\,u(x) \\ (S_2 u)(x) &= u(L-x) \end{aligned} \qquad (3\text{-}107)$$

The basic property (3-105) of a symmetry operator allows one to split the space U into a subspace S of symmetric displacements and into a subspace S_- of antisymmetric displacements, that are defined by :

$$S = \{ u \in U \mid S u = u \}$$

$$S_- = \{ u \in U \mid S u = -u \}$$

The splitting property follows from this obvious identity :

$$u = (u + S\, u)/2 + (u - S\, u)/2$$

By taking successive derivatives of Formula (3-106), one gets :

$$P_1(S\, u\, ;\, S\, w) = P_1(u\, ;\, w)$$
$$P_{11}(S\, u\, ;\, S\, w_1, S\, w_2) = P_{11}\, (u\, ;\, w_1, w_2), \text{ etc } \ldots \tag{3-108}$$

From the identities (3-108) and (3-3), one sees that, for any symmetric u, w and any antisymmetric $w\_$, one has :

$$P_1(u\, ;\, w\_) = 0 \tag{3-109}$$

$$P_{11}\, (u\, ;\, w, w\_) = 0 \tag{3-110}$$

$$P_3\, (u\, ;\, w\_) = 0, \text{ etc } \ldots \tag{3-111}$$

We now seek the singularities of the structures that satisfy (3-106). As previously, we limit ourselves to singularities whose codimension is at least one. First, by virtue of (3-109), *a symmetric equilibrium may be defined by a variational problem on* S *only*. Indeed, it is sufficient to consider the δu of S in Formula (3-6). Next, because of (3-110), the second variation P_{11} has no cross term between symmetric and antisymmetric states : the "matrix" P_{11} sends S on S, S - on S -. Hence, *a simple mode is either symmetric or antisymmetric*. With several independent modes, one may rearrange in such a way that some modes are symmetric and the other antisymmetric. Now, begin the classification with the case of a simple symmetric mode :

$$u_1 \in S \tag{3-112}$$

In this case, any solution in a neighbourhood is symmetric. The identities of the type (3-109) to (3-111) *do not imply any change in the analysis of Section* 3.5, because they involve at least an antisymmetric w. Hence, we get only the limit point as a robust singularity, the hysteresis point, the isola center and the asymmetric bifurcation point as singularities of codimension one.

Next, consider the case of a simple antisymmetric mode :

$$u_1 \in S- \tag{3-113}$$

Then, the restriction to S of the matrix P_{11} is invertible. The symmetric problem may be solved by the Inverse Function Theorem. *So there exists an unique path of symmetric equilibria* $u_0(\lambda)$ for a load on both sides of the critical load. Hence, we may apply the analysis of Sections 3.1, 3.3, because of the existence of this fundamental path. But, due to (3-111) (3-113), one has an identity :

$$\alpha_{2o} = 3\ P_3\ (u_o\ ;\ u_1) = 0 \qquad (3\text{-}114)$$

This implies that, in this case, *the robust singularity is the symmetric bifurcation point*, the additional conditions being of course

$$\alpha_{11} \neq 0 \qquad (3\text{-}115)$$

$$\alpha_{3o} \neq 0 \qquad (3\text{-}116)$$

Since the bifurcating solutions are sums of a symmetric part u_o (λ) and of a perturbation which is roughly antisymmetric, the bifurcation goes with a *symmetry breaking*. This means that the normal way of buckling is the symmetric bifurcation if there is symmetry breaking and the limit point if not.

To go further, one selects a supplementary space V that can be splitted on S and S -, which is always possible and can be also formulated as :

$$SV = V \qquad (3\text{-}117)$$

Come back to the auxiliary equation (3-24), that implies :

$$P_1\ (u_o + a\ u_1 + v\ ;\ \delta v) = 0 \qquad (3\text{-}118)$$

$$P_1\ (u_o - a\ u_1 + Sv\ ;\ \delta v) = 0 \qquad (3\text{-}119)$$

To establish (3-119), we have used (3-113), (3-108) and (3-117). Because the solution $\hat{v}$ (a,ε) is unique, the comparison of (3-118) and (3-119) leads to the identity :

$$\hat{v}(-a,\varepsilon) = S\ \hat{v}(a,\varepsilon) \qquad (3\text{-}120)$$

Then, *the reduced potential energy is even with respect to the amplitude* a, because of (3-106), (3-113), (3-120) and of the symmetry of the equilibrium $u_o(\lambda)$:

$$F(a,\varepsilon) = F(-a,\varepsilon) \qquad (3\text{-}121)$$

Indeed, (3-121) can be proved as follows :

$$F(-a,\varepsilon) = P(u_o - a\ u_1 + \hat{v}(-a,\varepsilon)) = P(S(u_o + au_1 + \hat{v}(a,\varepsilon))$$
$$= P(u_o + au_1 + \hat{v}(a,\varepsilon)) = F(a,\varepsilon)$$

Identity (3-121) suppressed the even coefficients in the bifurcation equation :

$$\alpha_{ij} = 0 \quad \text{for } i \text{ even} \qquad (3\text{-}122)$$

which corroborates (3-114).

One gets the singularities of codimension one by ruling out (3-115)or (3-116). First, assume

$$\alpha_{3o} = 0, \quad \alpha_{5o} \neq 0, \quad \alpha_{11} \neq 0 \tag{3-123}$$

Because of (3-122), the bifurcation equation has the form :

$$\partial F/\partial a = a(\alpha_{5o}\, a^4 + \alpha_{11}\, \varepsilon + \text{h.o.t.})$$
$$\text{h.o.t.} = O(\varepsilon^2 + \varepsilon a^2 + a^6) \tag{3-124}$$

Non-symmetric solutions are :

$$\varepsilon = -\, a^4\, \alpha_{5o}/\alpha_{11} + O(a^6) \tag{3-125}$$

The singularity is pictured in Figure 14 and it corresponds to transition between a stable and an unstable symmetric bifurcation. The reduced poten tial begins with terms in a^6 and Thom /58/ calls it *butterfly* for grounds that are not apparent here. There is no good motive to refuse this poetic name.

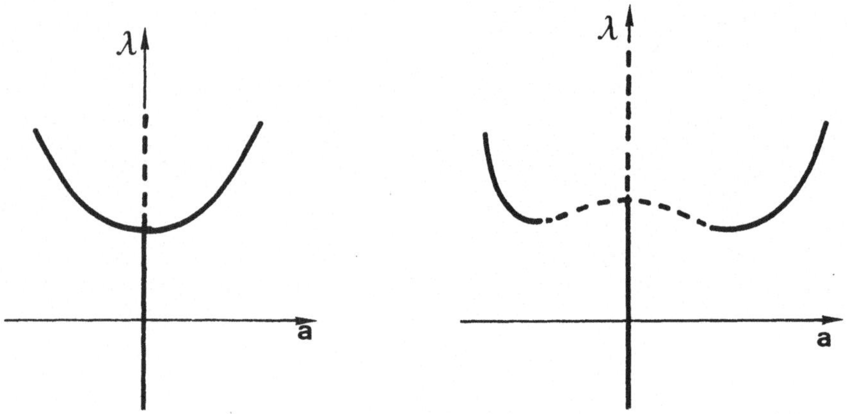

Fig. 14-a, $\alpha_{30} \geqq 0$ Fig. 14-b, $\alpha_{30} < 0$

Fig. 14, The butterfly singularity in the case $\alpha_{50} > 0$, $\alpha_{11} < 0$ and perturbation.

Now, one rules out (3-115) and assume :

$$\alpha_{11} = 0, \quad \alpha_{3o} \neq 0, \quad \alpha_{12} \neq 0 \tag{3-126}$$

The bifurcation equation is :

$$a(\alpha_{30}\, a^2 + \alpha_{12}\, \varepsilon^2 + \text{h.o.t.}) = 0 \tag{3-127}$$

The paths of non-symmetric equilibria have the same shape as in Figure 6 or 13, but there is also a fundamental equilibrium path which is connected with the latter. If the coefficients α_{30} and α_{12} have the same sign, there is only an *isola on the symmetric path*. After perturbation, this isola gives rise to a closed loop (Figure 15). If those coefficients have diffe rent signs, *two branches cross the fundamental path*. After perturbation, one obtains two successive bifurcations or three distinct branches without bifurcation (Figure 16). As for the tangent bifurcation, these singularities occur when the fundamental path is restabilized after the critical load.

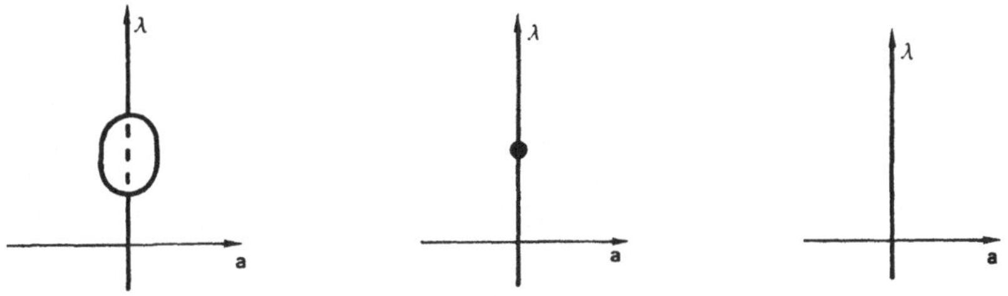

Fig. 15-a, $\alpha_{01} < 0$ Fig. 15-b, $\alpha_{01} = 0$ Fig. 15-c, $\alpha_{01} > 0$

Fig. 15, Isola on a symmetric path ($\alpha_{12} > 0$, $\alpha_{30} > 0$) and perturbation.

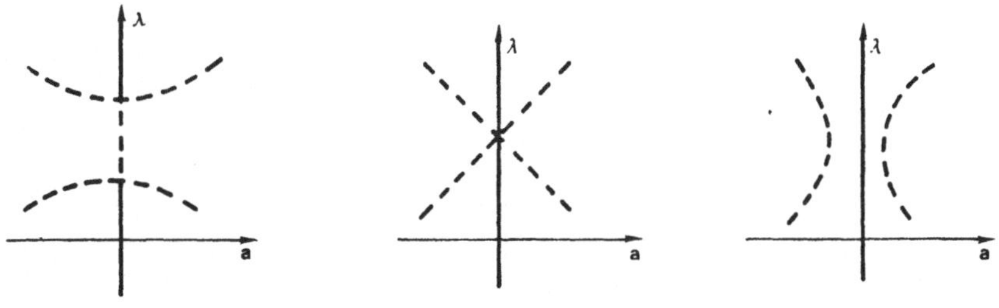

Fig. 16-a, $\alpha_{01} < 0$ Fig. 16-b, $\alpha_{01} = 0$ Fig. 16-c, $\alpha_{01} > 0$

Fig. 16, A singularity of codimension one with two unstable bifurcation branches ($\alpha_{12} > 0$, $\alpha_{30} < 0$) and perturbation.

In Section 3.3, we concluded that a multiple buckling mode leads to singularities whose codimension is at least equal to two, since they are characterized by the three equations (3-53). This no longer holds with symmetric systems, because the coefficient b_{12} is identically zero at the coincidence of a symmetric mode and of an anti-symmetric mode(see Eq. (3-110)).

$$u_1 \in S \quad , u_2 \in S_- \tag{3-128}$$

and only in that case. If one considers the symmetric problem, there is a single buckling mode u_1. The analysis of Section 3.5 still holds and there is only one path of symmetric equilibria which presents a limit point, provide that

$$P_1'(u_1) \neq 0, \quad P_3(u_1) \neq 0 \tag{3-129}$$

these equilibria being given by :

$$\begin{aligned} u &= a_1 u_1 + O(a_1^2) \\ \lambda - \lambda_c &= -a_1^2 P_3(u_1)/3 P_1'(u_1) + O(a_1^3) \end{aligned} \tag{3-130}$$

Symmetry breaking solutions cannot be analyzed as previously. Indeed, the fundamental symmetric path exists on a single side of the critical load. Thus, we compute the bifurcating path as a function of the "symmetric amplitude" a_1 instead of the load λ. After that change, the previous bifurcation analysis may be applied and we get, as in (3-41) (3-45) :

$$\begin{aligned} a_1 &= - a_2^2 \alpha'_{30}/\alpha'_{11} + O(a_2^2) \\ \alpha'_{11} &= 2 \frac{d}{da_1} \{P_2(a_1u_1 + \ldots; u_2)\} \quad a_1 = 0 = P_{111}(u_1,u_2,u_2) \neq 0 \\ \alpha'_{30} &= 4(P_4(u_2) - P_2(\hat{v})) \neq 0 \end{aligned} \tag{3-131}$$

where $\hat{v}$ is given by (3-46) with u_1 replaced by u_2. The bifurcation plot is given in Figure 17, the non-symmetric equilibria being always unstable.

More generally, the *symmetry groups* are the mathematical tools to modelize the symmetry of a structure. In this Section, we have studied systems whose symmetry can be modelized by the simplest group with only two elements : + 1, - 1. This group is called Z_2. We have represented it by the two linear operators Id and S, (3-105) being the only non-trivial group relation. If a *rotational symmetry* (azimutal angle Θ) is broken, the robust singularity looks like a symmetric bifurcation point, due to the invariance with respect to the change of Θ into $-\Theta$. More precisely, the bifurcating solutions lie on an one-parameter sheet generated by azimutal shift /50/. Later on, we shall analyse the postbuckling of the beam on foundation, which is invariant with respect to two independent symmetries by reflexion. A different example of symmetry breaking is the Rivlin's cube /52/, which is invariant with respect to any exchange of the three orthogonal axes. In the latter case, the robust singularity corresponds to two coincident buckling modes, as studied in section 3.3. Of course, there are three bifurcating branches and, as we have established, they are unstable. So the analysis in this Section is only an introduction to the important topics of the influence of symmetries on postbuckling. We refer, for instance, to the references /23/, /51/ to /54/for additional informations. We underline that *there is an appropriate classification for each type of symmetry.*

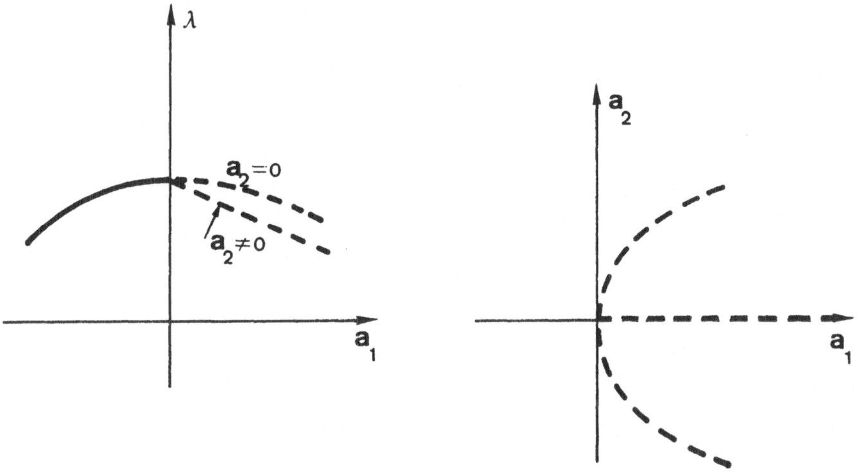

Fig. 17, Coincidence of a limit point and of a symmetric bifurcation point in the case of negative α_{01}, α_{20}, α'_{11} and of a positive α'_{30}.

3.7 Imperfection sensitivity

Real structures cannot be quite perfect, which means that the expected fundamental equilibrium path no longer exists or that the initial state of the structure does not satisfy exactly the wished symmetry. The general analysis of the previous sections and the simple model of Section 2.2 show that *the imperfections change a bifurcation point into one or several limit points*. One generally admits that, as in chapter 2, the initial displacements are the main cause of this change. It was Koiter's contribution /31/ to recognize the influence of these imperfections for explaining the experimental reductions of the critical load and to build up a general theory for this purpose. His theory is presented in this Section.

The singularity theory gives an accurate definition to this idea of perturbation : one says that a functional $P^*(\lambda,u,\mu)$ is an N-parameter *unfolding* of $P(\lambda,u)$ if :

$$P^*(\lambda, u, o) = P(\lambda,u) \qquad (3\text{-}132)$$

where μ designates the N-parameters $\mu_1, \mu_2, \ldots, \mu_N$. Catastroph theory is still more general, by introducing more general unfoldings.

The classical postbuckling theory assumes a *single perturbation parameter*, also designated by μ. Thus, the perturbed potential energy is as follows :

$$P^*(\lambda,u,\mu) = P(\lambda,u) + \mu\, Q(u) + \text{h.o.t.} \qquad (3\text{-}133)$$

$$\text{h.o.t.} = O(\mu^2 u + \lambda\, \mu u)$$

The Lyapunov and Schmidt reduction can be applied with the perturbed potential energy, as in (3-22) (3-24). The unique solution $\hat{v}^*$ to the modified equation (3-24) is given by

$$\hat{v}^*(a_i,\varepsilon,\mu) = \hat{v}(a_i,\varepsilon) + O(\mu) \qquad (3\text{-}134)$$

where $\hat{v}(a_i,\varepsilon)$ is the solution (3-27) of the problem without imperfections. After insertion of (3-134) into the potential energy, one gets a reduced potential energy in the form :

$$F^*(a_i,\varepsilon,\mu) = F(a_i,\varepsilon) + \mu\, Q_1(u_i)a_i + \text{h.o.t.}$$

$$\text{h.o.t.} = O(\mu^2 + \mu\varepsilon a + \mu a^2) \qquad (3\text{-}135)$$

$F(a_i,\varepsilon)$ being the same as in Section 3.2. At the first order of approximation, the effect of the imperfections is taken into account by the single term

$$\mu\, Q_1(u_i)\; a_i \qquad (3\text{-}136)$$

The higher order terms are not useful to compute the reduction of the reduction of the critical load, as in Section 2.2. It is worth noting that the perturbation term (3-136) is nothing but the lower order term $\mu\, Q_1(u)$ in the Taylor expansion of the perturbation $P^* - P$ of the potential energy, where the displacement u has been replaced by its in-mode part. Especially, it will be sufficient to compute the orthogonal part of the displacement (3-134) on the perfect structure (i.e. $\mu = 0$).

We now compute the reduction of the critical load in case the perfect structure has a fundamental equilibrium path and there is a single buckling mode. From the results of the sections 3.2, 3.3 and from the formula(3-130) one gets the reduced potential energy in the form :

$$F^*(\varepsilon,a,\mu) = \alpha_{20}\, a^3/3 + \alpha_{30}\, a^4 + \alpha_{11}\varepsilon\, a^2/2 + \beta\mu a = \text{h.o.t.}$$

$$(3\text{-}137)$$

where the coefficients α_{ij} are given in (3-41) and β is given by :

$$\beta = Q_1(u_1) \qquad (3\text{-}138)$$

Hence, the bifurcation equation is :

$$\partial F^*/\partial a = \alpha_{20}\, a^2 + \alpha_{30}\, a^3 + \alpha_{11}\, a\,\varepsilon + \beta\mu + \text{h.o.t.} \qquad (3\text{-}139)$$

If one neglects the higher order terms - including the cubic term in a if there is a quadratic term -, this bifurcation equation can be identified to the equation (2-12) or (2-13) that governs the imperfection sensitivity of the simple model of Section 2.2. The formulae for the reduction of the critical load are obtained as in that case. Assuming that α_{11} is negative, one finds :

$$\lambda_c - \lambda_m = 3(\beta\mu/2)^{2/3}|\alpha_{30}|^{1/3}/|\alpha_{11}| \qquad (3\text{-}140)$$

$$\lambda_c - \lambda_m = 2(\beta\alpha_{20}\,\mu)^{1/2}/|\alpha_{11}| \qquad (3\text{-}141)$$

respectively for an unstable symmetric bifurcation ($\alpha_{30} < 0$, $\alpha_{20} = 0$) and for an asymmetric bifurcation.

Next, we consider the case of n coincident buckling modes u_i. We assume that there is a cubic term F_3 in the reduced potential energy. After having neglected the higher order terms and rearranged the modes as in (3-54), we get the bifurcation equations as :

$$\frac{\partial F_3}{\partial a_i} - \varepsilon\, a_i + \mu\beta_i = 0 \qquad (3\text{-}142)$$

where the coefficients β_i represent the direction of the imperfections in the space of the buckling modes :

$$\beta_i = Q_1(u_i) \qquad (3\text{-}143)$$

If the cubic term F_3 is zero, it must be replaced by the quartic term F_4 in the bifurcation equations. The n algebraic equations (3-142) do not lead to simple explicit formulae as in (3-140) (3-141). For further discussion of these equations, we refer to Koiter in ref. /10/ to /42/ and to Section 4.4.

How can we justify the elimination of the h.o.t. ? Recently, Catastrophe Theory has put forward qualitative equivalence rules, that will be explained in Section 5.2. Let us sketch a proof using the tools of mathematical analysis /12/ /42/ in the case of a single mode of a symmetric bifurcation and of a negative α_{11}. The considered bifurcation equation is :

$$\alpha_{30}\, a^3 - |\alpha_{11}|\, a\,\varepsilon + \mu\beta + O(a^4 + \varepsilon a^2 + \varepsilon^2 a + \mu^2 + \mu\varepsilon + \mu a) = 0 \tag{3-144}$$

Next, one *scales* by letting :

$$\eta^3 = \mu\,\beta/|\alpha_{11}|$$
$$a = \eta\, A, \qquad \varepsilon = \eta^2\, E \tag{3-145}$$

After division by η^3, one finds :

$$A^3\, \alpha_{30}/|\alpha_{11}| - EA + 1 + O(\eta) = 0 \tag{3-146}$$

Away from the limit point, the Implicit Function Theorem permits us to solve the latter equation with respect to the scaled load E :

$$E = A^2\, \alpha_{30}\,/|\alpha_{11}| + 1/A + O(\eta) \tag{3-147}$$

so that the higher order terms induce only a slight perturbation of the equilibrium path through the term of order η. The solution (3-147) does not hold close to the limit point, but in the latter case, the scaled equation (3-146) can be solved with respect to the amplitude A.

3.8 Classifications of singularities

Here, we present three classifications of singularities according to their codimension. We adopt the rough definition, that was used throughout this chapter : *the codimension of a singularity is the number of equations* that are to be satisfied by the potential energy to define the considered singularity, the common condition (3-14) being not counted. If the codimension is zero, the singularity is said to be *robust*. This notion of codimension seems to coincide with that of Golubitsky and Schaeffer [23] and we shall present the correct definition in Section 5.2. Remark that Thom [58] counts in addition the common condition (3-14).

As we have already pointed out, there are several classifications, depending on the assumed properties of the structure. If the potential energy does not satisfy any special condition, we get the classification of Section 3.5. This general classification is relevant essentially for imperfect structures. We have also established two alternative classifications that hold, respectively, for structures with a known fundamental equilibrium path (Section 3.3) and for structures that are symmetric with respect to a plane, an axis or a point (Section 3.6). Different classifications could be obtained for structures invariant by a bigger symmetry group. Note that in Thom's initial work, there is a single "universal" classification. Two of our three classifications can be found, with more details, in the book by Golubitsky and Schaeffer [23]. *We limit our lists to singularities of codimension zero or one, but all of them are included.*

The *general classification* of Section 3.5 is summarized in the following table.

Terminology	Codimension	Characteristic equations	Figures
Limit point	0	none	11
Hysteresis point	1	$\alpha_{20} = 0$	12
Isola center	1	3-100, 3-101	13
Asymmetric bifurcation	1	3-100, 3-103	1-c, 6-c, 6-d

Table 1

If one a priori assumes that *the structure has a fundamental equilibrium path* for any load, the classification is that discussed in Section 3.3. The next table distinguishes the present codimension (i.e. the codimension with that additional assumption) and the general codimension, as in Table 1.

Terminology	Codim.	General codim.	Characteristic equations	Figures
Asymmetric bifurcation	0	1	none	1 - c
Symmetric bifurcation	1	2	$\alpha_{20} = 0$	1-a or 1-b
Tangent bifurcation	1	2	$\alpha_{11} = 0$	10

Table 2

For structures that are symmetric in the sense of Section 3.6, the general classification holds if the buckling modes are symmetric. Thus, the codimension is the same as in Table 1. If there is an antisymmetric buckling mode, there exist non-symmetric equilibrium paths. We have found five singularities of this type, with a codimension zero or one. Remark that the corresponding general codimension may be much higher.

Terminology	Codim.	General codim.	Nature of modes	Characteristic equations	Figure
Limit point	0	0	sym.	none	11
Symmetric bifurcation	0	2	antisymm.	none	1-a, 1-b
Hysteresis point	1	1	sym.	$\alpha_{20} = 0$	12
Isola center	1	1	sym.	3-100, 3-101	13
Asymmetric bifurcation	1	1	sym.	3-100, 3-103	1-c, 6
Butterfly	1	4	antisymm.	$\alpha_{30} = 0$	14
Isola on a path	1	5	antisymm.	3-126	15
"Trifurcation"	1	5	antisymm.	3-126	16
?	1	?	sym. antisymm.	2 modes	17

Table 3

The singularities of codimension zero give the normal-or "generic" - ways of buckling. *According to Table 1, generally, imperfect structures lose their stability at a limit point. According to Table 2, generally, structures with a fundamental equilibrium path lose their stability at an asymmetric bifurcation point. According to Table 3, generally, symmetric structures lose their stability at a limit point if there is no symmetry breaking and at a symmetric bifurcation point in the presence of symmetry breaking.* Remember that we only considered symmetry by reflexion, cf. (3-103) and (3-105).

The singularities of codimension one are not so interesting because they occur quite rarely. The three lists give all the generic ways such that several generic singularities coincide.

4. APPLICATIONS

In this short chapter, we do not try to present applications of post-buckling theory to problems of practical interest. In this respect, we refer in particular to the lectures by Professor Arcbosz and by Professor

Koiter. We consider only applications to simple illustrative models so that at least one example is given for each singularity that appears in our three classifications.

4.1 A lumped arch

According to the third classification of Section 3.8, a symmetric structure buckles at a symmetric bifurcation point if the buckling occurs with symmetry breaking and at a limit point if not. These two ways to lose stability may be observed with the lumped model for an arch (from Galbe /21/), that is pictured in Figure 18.

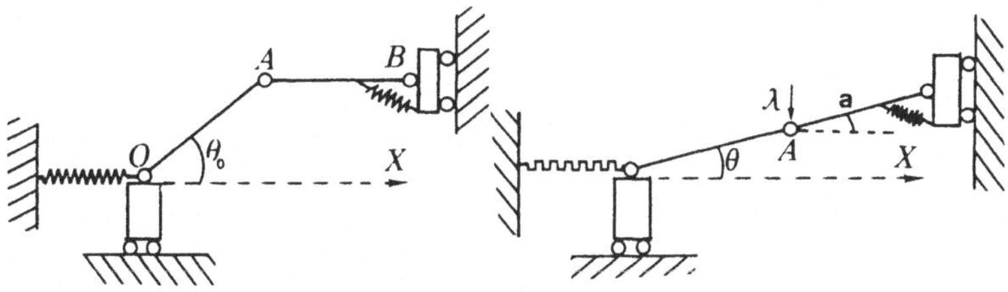

Fig. 18

The arch is modelized by a structure made of two rigid bars OA = ℓ and AB = L. The elasticity is accounted by a spring of stiffness k and a bending spring of stiffness K. The potential energy is a function of the generalized displacement u = (a,Θ) and of the force λ :

$$P(\lambda;a,\Theta) = K\ a^2/2 + k\ \Delta\ell^2/2 + \lambda\ell\ \sin\Theta \qquad (4\text{-}1)$$

$$\Delta\ell(a,\Theta) = \ell(\cos\Theta_0 - \cos\Theta) + L(1 - \cos a) \qquad (4\text{-}2)$$

where $\Delta\ell$ is the length variation of the spring. The structure is symmetric with respect to the axis Ax. Thus, the following symmetry operator :

$$S(a,\Theta) = (-\ a,\Theta)$$

leaves the potential energy invariant. The symmetric equilibrium states (a = 0) are obtained by taking the extremal values of P(λ;0,Θ) with respect to Θ :

$$\partial P/\partial\Theta = k\,\Delta\ell\,\ell\sin\Theta + \lambda\ell\cos\Theta = 0 \qquad (4\text{-}3)$$

$$\Delta\ell = \ell(\cos\Theta_o - \cos\Theta) \qquad (4\text{-}4)$$

The equations (4-3) (4-4) lead to :

$$\lambda = k\,\ell\,(\sin\Theta - \cos\Theta_o\,\mathrm{tg}\Theta) \qquad (4\text{-}5)$$

From (4-3) (4-4), one computes the second derivative :

$$(\partial^2P/\partial\Theta^2)(\lambda;0,\Theta) = k\ell^2(\cos\Theta_o - \cos^3\Theta)/\cos\Theta \qquad (4\text{-}6)$$

The angle-load path is given in closed form by Formula (4-5). This path of symmetric equilibria has two limit points, a maximum and a minimum as pictured in Figure 19. The maximum load point (λ_m,Θ_m) is reached when the second derivative (4-6) becomes zero :

$$\cos\Theta_m = (\cos\Theta_o)^{1/3}, \quad \Delta\ell_m = -\,\ell(\cos\Theta_o^{\,1/3} - \cos\Theta_o) \quad (4\text{-}7)$$

Let us compute the other second derivative :

$$\partial^2P/\partial a^2 = K + k\,\Delta\ell\,L\cos a + k\,L^2\sin^2 a \qquad (4\text{-}8)$$

Remark that the second derivative with respect to a and θ is zero for a=0, in accordance with the general formula (3-110). Symmetry breaking bifurcation occurs when :

$$\partial^2P/\partial a^2(\lambda,0,\Theta) = K + k\,L\,\Delta\ell = 0 \qquad (4\text{-}9)$$

which gives the length variation at the bifurcation point :

$$\Delta\ell_b = -\,K/k\,L \qquad (4\text{-}10)$$

The structure buckles at the limit point if

$$|\Delta\ell_m| < |\Delta\ell|_b$$

and at the symmetric bifurcation point if not. Especially bifurcation buckling is obtained if the bending stiffness K is sufficiently small, which is the case for a continuous slender arch. If the two second derivatives become zero together, we get the last singularity of the Table 3.

As an application of Section 3.2 and 3.3, we now discuss the postbuckling behaviour in the case of a bifurcation. As shown in Section 3.6, the bifurcation is symmetric and the postbuckling behaviour follows from the signs of the coefficients α_{11}, α_{30} given by Formula (3-41). Because the buckling mode is $u_1 = (1,0)$, a derivative in this direction is a derivative with respect to a. Thus, (3-41) leads to :

$$\alpha_{11} = P_2'(u_1) = \frac{1}{2}\frac{d}{d\lambda}\frac{\partial^2 P}{\partial a^2}(0,\Theta(\lambda),\lambda)$$

because the notation P_2' implies a total derivative with respect to λ, as given by (3-16). After derivation of (4-8), we find :

$$\alpha_{11} = k\,\ell\,L\,\sin\Theta(d\Theta/d\lambda)/2$$

The derivative of Θ with respect to the load being computed from (4-5), we obtain

$$\alpha_{11} = -\,L\,\sin\Theta\,\cos\Theta/2\ell(\cos\Theta_o - \cos^3\Theta) \qquad (4\text{-}11)$$

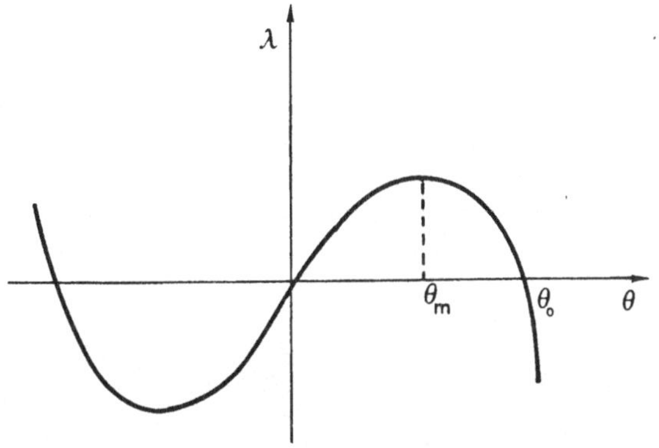

Fig. 19

So α_{11} is negative if bifurcation buckling occurs first. This was expected because this means that the symmetric path is initially stable and becomes unstable at the bifurcation point.

In accordance with the requirement (3-117), we choose $\delta v = (0,1)$ so that a derivative in this direction is a derivative with respect to Θ. Thus (3-41) and (3-36) may be rewritten as :

$$\alpha_{30} = (\partial^4 P/\partial a^4)/6 - 2(\partial^2 P/\partial\Theta^2)v_{11}^2$$

$$(\partial^2 P/\partial\Theta^2)v_{11} + (\partial^3 P/\partial a^2\,\partial\Theta)/2 = 0$$

From that, we get :

$$\alpha_{30} = (\partial^4 P/\partial a^4)/6 - (\partial^3 P/\partial a^2\,\partial\Theta)^2/2(\partial^2 P/\partial\Theta^2)$$

Taking derivatives of (4-8), we obtain :

$$\partial^4 P/\partial a^4 = 3\,k\,L^2 + k\,L\,\ell(\cos\Theta - \cos\Theta_o)$$

$$\partial^3 P/\partial a^2\,\partial\Theta = k\,L\,\ell\,\sin\Theta$$

and finally :

$$\alpha_{30} = -\frac{k\,L^2}{2}(\cos\Theta - \cos\Theta_o)\left(\frac{L}{\cos\Theta_o - \cos^3\Theta} - \frac{\ell}{3}\right)$$

Hence, α_{30} is negative for a shallow arch (i.e. Θ_o and Θ small). In that case, the bifurcation point is unstable-symmetric and thus the structure is imperfection sensitive. The bifurcation point may be stable-symmetric if the ratio ℓ/L is sufficiently large.

4.2 Examples with a restabilization of the fundamental path

In this section, I give examples of buckling where the fundamental path is restabilized after a second bifurcation. What does it happen if the destabilized point and the restabilization point coincide ? *According to the classifications of Section 3.8, one expects that this leads to a tangent bifurcation load (Figure 10) in the cases without symmetry and to an isola on the fundamental path (Figure 15) or to a "trifurcation" (Figure 16) if the buckling occurs with a symmetry breaking.*

The considered structure is similar to that of Section 2.2, but the members are compressible, with stiffness k (Figure 20). It has two degrees of freedom that are the angle a and the length ℓ of the members. The angle moment law is again given by(2-6). In the symmetric case $K_2 = 0$, this

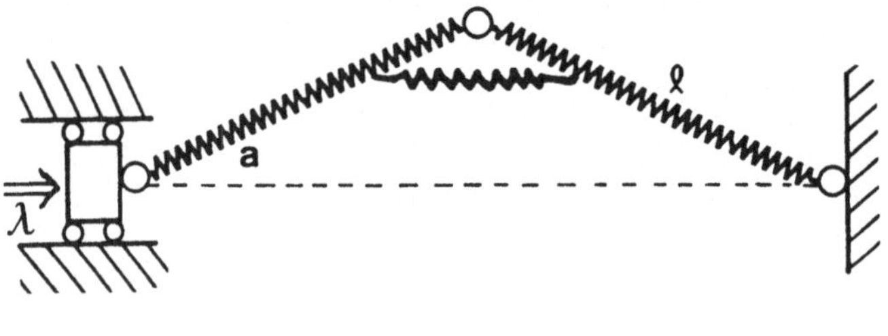

Fig. 20

model is that discussed in the book /23/ p. 245. The potential energy is :

$$P(\lambda,\ell,a) = k(\ell-\ell_o)^2 + K_1 a^2/2 + K_2 a^3/3 + K_4 a^4/4 + 2\lambda\ell \cos a \qquad (4\text{-}12)$$

If $K_2 = 0$, the structure is symmetric with respect to the operator

$$S(\ell,a) = (\ell,-a)$$

The equilibrium equations are :

$$\partial P/\partial a = K_1 a + K_2 a^2 + K_3 a^3 - 2\lambda\ell \cos a = 0$$

$$\partial P/\partial \ell = 2 k(\ell - \ell_o) + 2\lambda \cos a = 0$$

The second one permits us to eliminate the length and to get a single non linear equation for the angle :

$$dP/da = K_1 a + K_2 a^2 + K_3 a^3 - 2\lambda(\ell_o - \lambda \cos a/k) \sin a = 0$$

There is a bifurcation from the symmetric path $a = 0$ when the linear term in a becomes zero. This leads to a quadratic equation for the critical load :

$$2\lambda_c^2 - 2\lambda_c k \ell_o + K_1 k = 0 \qquad (4\text{-}13)$$

For $k\ell_o^2 > 2K_1$, there are two bifurcation points. The smaller one corresponds to a destabilization of the fundamental path, the larger one to a restabilization (look at the sign of d^2P/da^2). These two bifurcation points coincide for :

$$\ell_o^2 = 2K_1/k \qquad \lambda_c = k\ell_o/2 \qquad (4\text{-}14)$$

Thus, the bifurcation equation may be written as :

$$dP/da = a\{K_2 a + (K_3 - k\ell_o^2/6)a^2 + 2(\lambda - \lambda_c)^2/k + \text{h.o.t.}\} = 0$$

Hence, for a non-symmetric structure ($K_2 \neq 0$), the bifurcation path is tangent to the path $a = 0$, which had been pictured in Figure 10. Respectively for $k\ell_o^2$ bigger, equal to or lower than $2K_1$, the equilibrium paths close to the bifurcation point have the same shape as in Figure 10-a, 10-b and 10-c.

For a symmetric structure, one gets an isola - and thus no bifurcation if

$$K_3 > k\ell_o^2/6 \qquad (4\text{-}15)$$

If the opposite inequality holds, there are two unstable bifurcating branches. The plots are similar to those of the figures 15, 16.

The restabilization is easily understood from the formula (2-8) for the critical load of the simple model :

$$\lambda_c = K_1/L \tag{4-16}$$

Of course (4-16) is no longer valid and we use it only for a qualitative explanation. If the members are very compressible, their length decreases rapidly with the load. So the "critical load" (4-16) increases with the load, which may induce a restabilization of the symmetric state. Remark that, because of (4-13) or (4-14), this does not occur before that the length has been reduced by half.

4.3 Bifurcation from a nearly double buckling mode

Now we discuss a simple example of double buckling mode in order to show that *the nonlinear interactions induce considerable changes in the postbuckling paths*. Let us come back to the compressed beam on an elastic bifurcation that has been presented in Section 2.3. We consider only an hardening nonlinearity. After having modified the definitions of the non-dimensional quantities, we get a differential equation and boundary conditions in the form :

$$u^{(4)} + \lambda u^{(2)} + k\,u + u^3 = 0 \quad \text{for} \quad |x| < \pi/2 \tag{4-17}$$

$$u = u'' = 0 \quad \text{for} \quad x = \pm \pi/2 \tag{4-18}$$

The number k is $(\bar{L}/\pi L_o)^4$, where $\bar{L}$ is the true length of the beam and L_o the characteristic length introduced in Section 2.3. If k = 4, then $\lambda_c = 5$ and there are two coincident buckling modes, that are the modes 1 and 2. We study directly the case where the buckling modes are nearly coincident In this respect, we define two small parameters :

$$\varepsilon = \lambda - 5, \qquad \mu_1 = k - 4 \tag{4-19}$$

Then the potential energy may be written as :

$$P(\varepsilon,\mu_1,u) = \int \{ (u''^2 - (5+\varepsilon)\,u'^2 + (4+\mu_1)u^2)/2 + u^4/4 \}\,dx \tag{4-20}$$

The two modes are normalized as follows :

$$u_1(x) = (\cos x)\,(2/\pi)^{1/2}, \quad u_2(x) = (\sin 2x)/(2\pi)^{1/2} \tag{4-21}$$

The structure is invariant with respect to the two symmetry operators S_1, S_2 (and of course $S_1\,S_2 = -I$) :

$$(S_1\,u)\,(x) = u(-x)$$
$$(S_2\,u)\,(x) = -\,u(-x)$$

The two perturbation parameters (4-19) have a similar part : they neither change the fundamental path, nor the symmetry properties of the structure. Hence, the Lyapunov and Schmidt reduction of Section 3.2 is not really modified by the presence of a second perturbation parameter. So we get a reduced potential energy as in (3-31) :

$$F = F_4(a_1,a_2) + \varepsilon F_2'(a_1,a_2) + \mu_1(\partial F_2/\partial\mu_1)(a_1,a_2) + \text{h.o.t.}$$

with : (4-22)

$$F_4 = \int (a_1u_1(x) + a_2u_2(x))^4\, dx/4 = (\beta_1 a_1^4 + 2\beta_2 a_1^2 a_2^2 + \beta_3 a_2^4)/4$$

$$F_2' = \int (a_1u_1'(x) + a_2u_2'(x))^2\, dx/2 = -(a_1^2 + a_2^2)/2$$

$$\partial F_2/\partial\mu_1 = \int (a_1u_1(x) + a_2u_2(x)^2 dx/2 = (a_1^2 + a_2^2/4)/2$$

Straightforward computations give the numerical values :

$$\beta_1 = 3/2\ \pi,\ \beta_2 = 3/4\ \pi,\quad \beta_3 = 3/32\ \pi$$

Now remark that it is equivalent to apply the symmetry S_1 (resp. S_2) or to change the sign of the amplitude a_1 (resp. a_2) :

$$S_1(a_1\, u_1 + a_2\, u_2) = -\, a_1\, u_1 + a_2\, u_2$$

(4-23)

$$S_2(a_1\, u_1 + a_2\, u_2) = a_1\, u_1 - a_2\, u_2$$

In Section 3.6, with a single symmetry operator S, we have established that the reduced potential energy is invariant under S and therefore is an even function of the amplitude. By the same argument the identities (4-23) imply that the reduced potential energy is even with respect to the two amplitudes separetely :

$$F(a_1,\ a_2) = F(-a_1,\ a_2) = F(a_1,\ -a_2) \qquad (4\text{-}24)$$

As a consequence of (4-24), a_1 can be factorized in the first bifurcation equation and a_2 in the second one :

$$\partial F/\partial a_1 = a_1(\beta_1\, a_1^2 + \beta_2\, a_2^2 - \varepsilon + \mu_1 + \text{h.o.t.}) = 0$$

(4-25)

$$\partial F/\partial a_2 = a_2(\beta_2\, a_1^2 + \beta_3\, a_2^2 - \varepsilon + \mu_1/4 + \text{h.o.t.}) = 0$$

In addition to the fundamental solution $a_1 = a_2 = 0$, there is a symmetric (i.e. with respect to S_1) and an antisymmetrical equilibrium branch :

$$a_2 = 0, \ \varepsilon = \mu_1 + \beta_1 \ a_1^2 = \text{h.o.t.} \qquad (4\text{-}26)$$

$$a_1 = 0, \ \varepsilon = \mu_1/4 + \beta_3 \ a_2^2 + \text{h.o.t.} \qquad (4\text{-}27)$$

These two paths bifurcate from the fundamental one for a load increment close to μ_1 and $\mu_1/4$, respectively. The solutions (4-26) - resp.(4-27)- remain symmetric with respect to S_1 - resp. S_2 - and break the alternative symmetry. Lastly, it may happen that there are equilibria with neither of these symmetries. If one neglects the h.o.t., these asymmetric solutions can be expressed as :

$$\begin{aligned} a_1^2(\beta_2^2 - \beta_1 \beta_3) &= \varepsilon(\beta_2 - \beta_3) - \mu_1(\beta_2/4 - \beta_3) \\ a_2^2(\beta_2^2 - \beta_1 \beta_3) &= - \varepsilon(\beta_1 - \beta_3) - \mu_1(\beta_2 - \beta_1/4) \end{aligned} \qquad (4\text{-}28)$$

One finds a possible bifurcation between (4-28) and the symmetric branch (4-26) by letting $a_2 = 0$ in (4-28) :

$$\varepsilon_1 = - \mu_1(\beta_2 - \beta_1/4) \ / \ (\beta_1 - \beta_2) = - \mu_1/2 \qquad (4\text{-}29)$$

Likewise, the possible bifurcation load between (4-28) and (4-27) is :

$$\varepsilon_2 = \mu_1(\beta_2/4 - \beta_3) \ / \ (\beta_2 - \beta_3) = \mu_1/7$$

After having dropped the load in (4-28), one sees that the projection of the asymmetric path on the $a_1 - a_2$ plane has an elliptical shape :

$$(\beta_1 - \beta_2) \ a_1^2 + (\beta_2 - \beta_3) \ a_2^2 = - 3 \ \mu_1/4 \qquad (4\text{-}31)$$

Therefore, these asymmetric equilibria exist only for μ_1 negative.

The equilibrium paths and their stability are pictured in Figure 21. When the load is sufficiently large, the symmetric branch (mode 1) and the antisymmetric one (mode 2) are stable, whatever the first buckling mode is. If the first buckling mode is the mode 1 (i.e. $\mu_1 < 0$), the symmetric path is stable for λ between $5 + \mu_1$ and $5 - \mu_1/2$ and it becomes unstable by crossing the asymmetric branch. In the latter case, *the mode interaction gives rise to snap, to hysteresis and to a close loop of asymmetric equilibria that connects the two primary bifurcating branches.*

The coefficients $\beta_1, \beta_2, \beta_3$ of the quartic term of the reduced energy have a great influence in the previous postbuckling analysis, especially β_2 which accounts for the nonlinear interaction between the two modes. With other numerical values of those coefficients, it may happen that ε_1

is lower than ε_2 and that the close loop of asymmetric solutions is stable. In the latter case, there is no snap and no hysteresis.

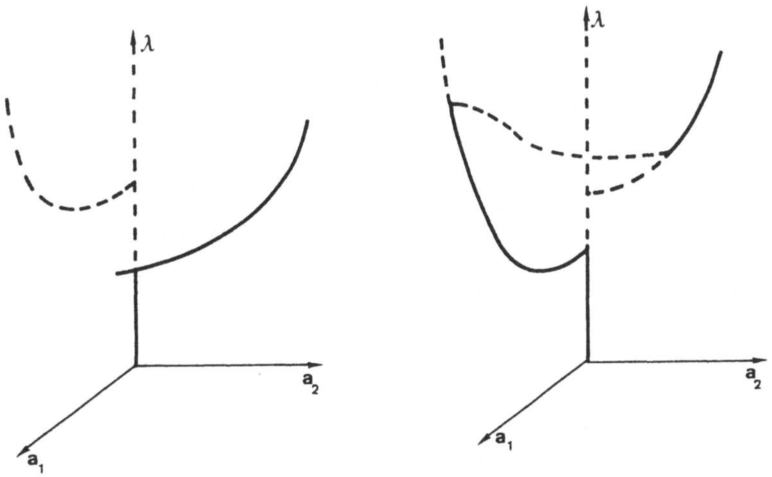

a : $\mu_1 > 0$ b : $\mu_1 < 0$

Fig. 21, The equilibrium paths are to be completed by symmetry

If the nonlinearities are such that :

$$- (\beta_1 \beta_3)^{1/2} < \beta_2 < \beta_3 \qquad \beta_1 > \beta_3 > 0,$$

only the asymmetric solutions are stable for sufficiently large loads, which follows easily from (4-22) (4-25). On the contrary, the symmetric and antisymmetric solutions are stable in the large if

$$\beta_2 > \beta_1 > \beta_3 > 0, \tag{4-32}$$

which occurs in the buckling of a rectangular plate /41/. For more details and different examples of this type, we refer to /23/ /45/ /53/.

It is questionable to neglect the h.o.t., especially close to the secondary bifurcation point. But this can be justified and we shall explain it in the next section.

4.4 Imperfection sensitivity close to a nearly double buckling mode

In this section, we intend to illustrate some concepts introduced in the third chapter, namely the notions of robustness, of a singularity of codimensions one, of symmetry breaking, of a classification that depends on the assumed properties of the structure.

In this respect, we study the effect of imperfections on the buckling of a beam that rests on an elastic foundation. The assumptions are the same as in the previous section, except that we account for an initial displacement. Especially the mode 1 and the mode 2 are nearly coincident. We denote by $\mu_2\ \mu_o(x)$ the initial displacement, $u_o(x)$ being the shape of the imperfections and μ_2 a measure of their size that will be specified later on. With the notations of the previous section, we get the differential equation and the perturbed potential energy as :

$$u^{(4)} + (5+\varepsilon)\,u'' + (4+\mu_1)\,u + u^3 = -\,(5+\varepsilon)\mu_2\,u_o''$$

$$P^{*} = P(\varepsilon,\mu_1,u) - \mu_2(5+\varepsilon)\int u_o'(x)\,u'(x)\,dx$$

where the perfect potential energy P is that given in (4-20). After a few computations, the Lyapunov-Schmidt reduction leads to reduced potential energy :

$$F^{*}(\varepsilon,\mu_1,\mu_2,a_1,a_2) = F + \mu_2(b_1\,a_1 + b_2\,a_2) + \text{h.o.t.} \qquad (4\text{-}33)$$

$$-\,b_i/5 = \int u_o'(x)\,u_i'(x) \qquad i = 1,2 \qquad (4\text{-}34)$$

where F has been given in (4-22) and the u_i's are the buckling modes (4-21). For the sake of convenience, we normalize the size of the imperfections μ_2 by requiring

$$b_1^2 + b_2^2 = 1$$

This allows us to define an angle γ

$$b_1 = \cos\gamma, \qquad b_2 = \sin\gamma \qquad (4\text{-}35)$$

Thus γ represents *the direction of the projection of the imperfections on the space generated by the buckling modes and* μ_2 *the size of this projection* (the projection is defined by the energy scalar product (4-34)).

In a first step, we neglect the higher order terms. Then from (4-22) (4-33) (4-35), we get the bifurcation equations :

$$\partial F^{*}/\partial a_1 = a_1(\beta_1\,a_1^2 + \beta_2\,a_2^2 - \varepsilon + \mu_1) + \mu_2\cos\gamma = 0$$

$$\partial F^{*}/\partial a_2 = a_2(\beta_2\,a_1^2 + \beta_3\,a_2^2 - \varepsilon + \mu_1/4) + \mu_2\sin\gamma = 0 \qquad (4\text{-}36)$$

Our aim is to *discuss the shape of the equilibrium paths* in the amplitude-load space (a_1, a_2, ε) *according to the three perturbation parameters* μ_1, μ_2, γ, that represent, respectively, the deviation from the case of modes coincidence, the size and the direction of the imperfections. Because of similitude, this shape depends only on the angle γ and on

$$M = \mu_1 \ \mu_2^{-2/3} \tag{4-37}$$

First, we analyze the case of a S_1 - *symmetric imperfections*, i.e. $\gamma = 0$. Because the structure remains symmetric with respect to S_1, the expected singularities are those of the third classification of Section 3.8. We can factorize a_2 in the second bifurcation equation. Therefore, there are symmetric equilibrium paths given by

$$\varepsilon = \mu_1 + \beta_1 \ a_1^2 + \mu_2/a_1, \quad a_2 = 0 \tag{4-38}$$

and asymmetric equilibrium paths given by

$$\varepsilon = \mu_1/4 + \beta_2 \ a_1^2 + \beta_3 \ a_2^2 = \mu_1 + \beta_1 \ a_1^2 + \beta_2 \ a_2^2 + \mu_2/a_1 \tag{4-39}$$

The equation (4-39) may be rewritten as

$$(\beta_2 - \beta_3)\varepsilon = (\beta_2^2 - \beta_1 \ \beta_3) \ a_1^2 - \beta_3 \ \mu_2/a_1 - \mu_1(\beta_3 - \beta_2/4) = g(a_1) \tag{4-40}$$

$$h(a_1) = (\beta_1 - \beta_2)a_1^2 + \mu_2/a_1 \tag{4-41}$$

$$a_2^2(\beta_2 - \beta_3) + h(a_1) + 3 \ \mu_1/4 = 0 \tag{4-42}$$

The obtained solutions are real if and only if a_2^2 is positive, which is equivalent to

$$h(a_1) + 3 \ \mu_1/4 \leqslant 0$$

In view of this discussion, let us plot the function $h(a_1)$.
In the case pictured in Figure 22, the asymmetric branches bifurcate three times from the symmetric ones, at the values a_I, a_{II}, a_{III} of the amplitude a_1. If μ_1 increases, the closed loop between a_{II} and a_{III} coalesces. The corresponding singular point is *the isola on a path*, that appears in the theory (Figure 15). The coalescence occurs when M reaches a value characterized by

$$h(a_1) + 3 \ \mu_1/4 = 0 \qquad h'(a_1) = 0$$

Whatever the value of μ_1 may be, there is an asymmetric branch in the half-space $a_1 < 0$ (see Figure 22). Of course, the bifurcation point in a_I is symmetric, but it may be stable or unstable. In order to discuss that, let us compute :

$$\frac{d\varepsilon}{da_2^2} = \frac{d\varepsilon}{da_1} \frac{da_1}{da_2^2} = - g'(a_1) \frac{\beta_2 - \beta_3}{h'(a_1)}$$

where $g(a_1)$ is defined in (4-40). Hence, the direction of the bifurcating branch is inverted for the value of M given by

$$h(a_1) + 3 \mu_1/4 = 0, \qquad g'(a_1) = 0$$

The singular point is the *butterfly* of Figure 14. The equilibrium paths are plotted in Figure 23.

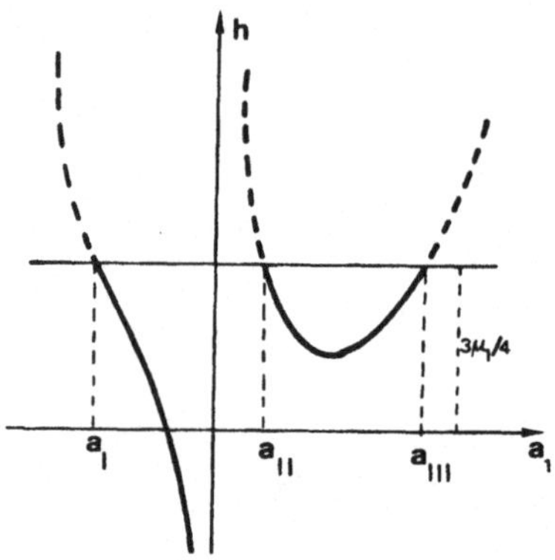

Fig. 22, The hatched part of the curve corresponds to negative a_2^2

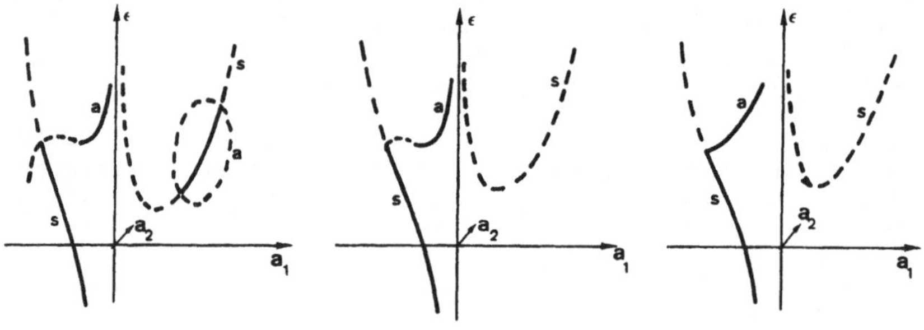

Fig. 23, With symmetric imperfections, there are three possible shapes for the equilibrium path, according the value of M. The s indicates symmetric equilibria and the a asymmetric ones.

If the imperfections are antisymmetric, one finds similar equilibrium paths by the same way. If the *imperfections are general,* it becomes difficult to get solutions of (4-36) in closed form. In a first step, let us seek the nature of the singularities according to the values of the parameters M and γ. From the first classification of Section 3.8., the expected singularities are generally limit points, exceptionally hysteresis points, isola or asymmetric bifurcation points. The singular points are characterized by an eigenvalue problem :

$$\exists A = (A_1, A_2) \; : \quad \frac{\partial^2 F^*}{\partial a_i \partial a_j}(\varepsilon, \mu_1, \mu_2, a_1, a_2)\, A_j = 0$$

From Section 3.5, the isola and the bifurcation point are characterized by

$$\alpha_{01} = 0 \qquad (4\text{-}43)$$

where α_{01} is given by (3-90) with u_1 and P replaced by A and F^*. The hysteresis points are characterized by :

$$\alpha_{20} = 0 \qquad (4\text{-}44)$$

where α_{20} is defined in (3-41). The additional condition (4-43) - or (4-44) - is satisfied if the parameters (M,γ) lie on a curve, which can be computed numerically. These two curves are, respectively, the curves BD and AC of Figure 24-a.

In a second step, the shape of equilibrium paths can be discussed by an argument of continuity. Indeed, these paths are known in special cases ($\gamma = 0$ or $\pi/2$) and we have found the two curves AC and BD where the shape may change. Note that we have also informations in the case of very large $|M|$, which means that the imperfections are small : the obtained curves (especially those of Figure 23) must be in agreement with those found without imperfections (Figure 21).

In Figure 24-a, we have also plotted the shape of the fundamental path on each side of the separatrix AC. On the left, there is hysteresis.

A similar picture is given for the buckling of a rectangular plate in Figure 24-b (see /41/ /45/ for additional details). The gap between the separating curves AC, BD is very small so that the hysteresis loop exists almost never. This gap has been magnified in Figure 24-b to have a more readable schema.

Figure 24 illustrates the interest to classify the singularities by their codimension and to have several classifications. With symmetric ($\gamma = 0$) or antisymmetric ($\gamma = \pi/2$) imperfections, the structure keeps a symmetry : the classification of Table 3 holds. Generally, the singularities are symmetric bifurcation points or limit points (cf Fig. 23). In isolated points of the lines $\gamma = 0$, $\gamma = \pi/2$, we have got butterfly singularities (points A and C), isola on a path (B and D of Fig. 24-a) or trifurcation (B and D of Figure 24-b), that are of codimension one in Table 3.

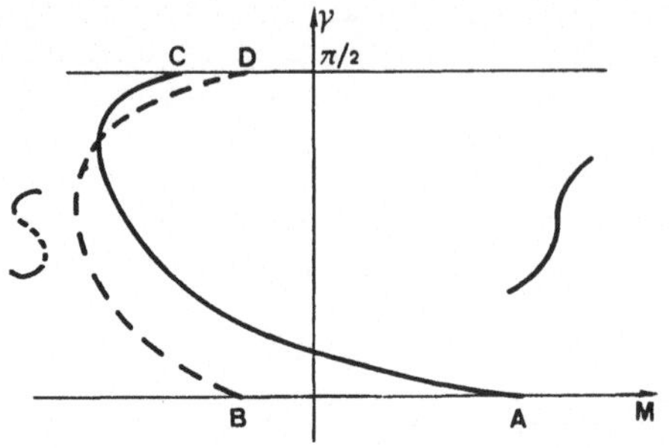

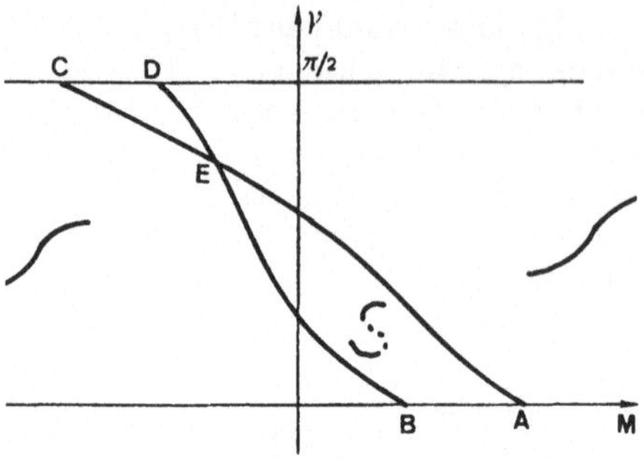

24-a : beam on foundation 24-b : plate buckling

Fig. 24, Shape of the equilibrium paths according to M and to the direction of imperfections γ. When the parameters are on the curve AC, there is an hysteresis point on the fundamental path. If they are on BD, there is an isola (24-a) or an asymmetric bifurcation (24-b). At the point E, the bifurcation point becomes symmetric. Outside these curves, there are only limit points. The shape of the fundamental path is given on each side of the separating curve AC (AC and BD in 24-b)

With general imperfections ($0 < \gamma < \pi/2$), the general classification of Table 1 holds. *Except in some curves, one has got only limit points, that are the only robust singularity. On some curves, we have got* hysteresis points (curve AC), isola (curve BD of Fig. 24-a) or asymmetric bifurcations (curve BD of Fig. 24-b), that are *the singularities of codimension one. In isolated points, one may get singularities of codimension 2* : that is the case for the point E, which corresponds to a symmetric bifurcation point.

Last question : *what is the influence of the higher order terms* on the previous results ? As it is classical /12/ /42/ /45/, let us scale all the unknowns and parameters with respect to a single small parameter η :

$$\mu_2 = \eta^3, \quad \mu_1 = M\,\eta^2, \quad a_i = \eta\,A_i, \quad \varepsilon = \eta^2\,E \qquad (4\text{-}45)$$

Let us insert (4-45) into the potential energy

$$\eta^{-4}\,F^* = F_4(A_i) - E(A_1^2 + A_2^2)/2 + M(A_1^2/2 + A_2^2/8)$$
$$+ A_1 \cos\gamma + A_2 \sin\gamma + O(\eta)$$

where the higher order terms are accounted by the perturbation of order Thus, *the question of the influence of the h.o.t. is equivalent to the problem of the behaviour of equilibrium paths with respect to any perturbation.* In this respect, we need theorems of the following type.

THEOREM

Let $P^*(\lambda,\mu_1,\mu_2,u)$ be a functional that satisfies the symmetry assumption (3-105) (3-106).

(i) Suppose that, for $\mu = (\mu_1,\mu_2) = (0,0)$, there exists a symmetric bifurcation point. Then for any small μ, there exists a unique symmetric bifurcation point $\lambda(\mu),u(\mu)$ in the neighbourhood.

(ii) Suppose that, for $\mu = (0,0)$, there is a butterfly singularity. Then, with suitable additional conditions on $\partial P/\partial\mu_1$, there exists for any small μ_1, a unique butterfly singularity for $\lambda=\lambda(\mu_1)$, $\mu_2 = m(\mu_1)$.

Such a theorem is easily established : see for instance the theorems 5 and 6 of [42], in special cases. The property (i) holds for any robust singularity and the property (ii) for any singularity of codimension one. *This theorem and its extension in the non-symmetric case imply that the higher order terms involve only slight perturbations in the pictures of Figure 24.*

5. MATHEMATICAL FOUNDATIONS

In this chapter, we shortly present some contributions of mathematics to postbuckling theory. It is now very well known that the Lyapunov-Schmidt reduction may be fully justified by the classical tools of mathematical analysis (5.1) The important and difficult question of the dynamical foundation of the energy criterion can also be studied by mathematical analysis (5.3). Lastly, we shall discuss how the quite different approach of catastrophe theory or of singularity theory may improve our view of postbuckling theory (5.2).

5.1 Lyapunov and Schmidt reduction (continued)

In this section, we establish that the arguments involving the Inverse Function Theorem (or Implicit Function Theorem : I.F.T.) can be extended to continuous systems. For the sake of simplicity, we limit ourselves to the beam buckling probem of Section 4.3. Indeed, in that case, the proofs may be achieved by using rather elementary mathematics.

The first consequence of the I.F.T. is the principle of exchange of stability that was stated in Section 3.1. According to this principle, there is no bifurcation so long as the equilibrium path remains stable. In other words :

$$\text{energy criterion stability} \Rightarrow \text{no bifurcation} \qquad (5\text{-}1)$$

First, let us consider discrete systems, which means that the displacement u is represented by n generalized coordinates $(q_1, q_2, \ldots, q_n)$

The equilibrium states solve a system of n equations :

$$\partial P/\partial q_i = g_i(\mu_1,\mu_2, \ldots,\mu_k; q_1, q_2, \ldots, q_n) = 0 \qquad (5\text{-}2)$$

where $\mu = (\mu_1, \mu_2, \ldots, \mu_k)$ designates all the perturbation parameters, including the load. To be brief, the equation (5-2) will be written as :

$$g(\mu;\ u) = 0 \qquad (5\text{-}3)$$

where $g(.,.)$ is a function from R^{k+n} into R^n.
Furthermore, for $\mu = 0$, there is a solution $u = 0$ of (5-3) :

$$g(o;\ o) = 0 \qquad (5\text{-}4)$$

INVERSE FUNCTION THEOREM (discrete case)
Let us assume that g is continuously differentiable close to (o,o) and that its differential in (o,o) satisfies :

$$\det(dg) \neq 0 \qquad (5\text{-}5)$$

Then, locally, the equation (5-3) has a unique equilibrium path $u(\mu)$.
Furthermore, if g is analytic, $u(\mu)$ is also analytic.

The components of the differential dg are $\partial^2 P/\partial q_i \partial q_j$, so that the condition (5-5) of local uniqueness (or non-bifurcation) is equivalent to the fact that the second variation $P_{11}(.,.)$ is non-degenerate. This establishes (5-1).

To extend this argument in the continuous case, one introduces two Banach spaces B_1 and B_2 (i.e. vectorial spaces whose topology is defined by a norm) so that :

$$g(\mu,u) \text{ is continuous from } R^k \times B_1 \text{ into } B_2 \qquad (5\text{-}6)$$

Furthermore, one assumes that, close to (o,o), g has a differential $dg_u(.)$ in the sense of Frechet, i.e. :

$$\| g(u+v) - g(u) - dg_u(v) \|_{B_2} = o(\| v \|_{B_1}) \qquad (5\text{-}7)$$

INVERSE FUNCTION THEOREM (continuous case, from [67] [23])
In addition, let us assume that $dg_o(.)$ has a *bounded inverse*, which means that dg^{-1} is continuous from B_2 into B_1.
Then, the same conclusions as in the discrete case hold.

Now, *we apply this theorem to the simply supported beam* studied in Section 4.3. The differential equation can be written as :

$$g(\lambda;u) = u^{(4)} + \lambda u'' + ku + u^3 = dg_o(u) + u^3 = 0 \qquad (5\text{-}8)$$

Since we deal with a continuous system, the displacement $u(x)$ cannot be represented by an finite number of generalized coordinates. Because of the simplicity of the simply supported boundary conditions, $u(x)$ may be represented by its Fourier series. Limiting ourselves to seek even solutions, one can set :

$$u(x) = \sum_{n\ \mathrm{odd}} q_n (\cos nx)\ (2/\pi)^{1/2} \qquad (5\text{-}9)$$

Thus, it is equivalent to know u or its generalized coordinates q_n. Let us define the norms (m is an integer)

$$\|u\|_o^2 = \int_{-\pi/2}^{+\pi/2} (u(x))^2 dx = \sum_{n\ \mathrm{odd}} q_n^2 \qquad (5\text{-}10)$$

$$\|u\|_m = \int_{-\pi/2}^{+\pi/2} (u^{(m)}(x))^2\ dx = \sum_{n\ \mathrm{odd}} (n^m\ q_n)^2 \qquad (5\text{-}11)$$

These norms are those of the Sobolev spaces L^2 and H^m. After having chosen m = 4, and included the boundary condition, we have defined our Banach spaces :

$$B_2 = L^2(-\pi/2, +\pi/2)$$

$$B_1 = \left\{ u(x) \in H^4(-\pi/2, +\pi/2) \text{ s.t. (4-18) holds} \right\}$$

Let us remark that :

$$m \geqslant m' \geqslant 0 \Rightarrow \qquad \|u\|_m \geqslant \|u\|_{m'} \qquad (5\text{-}12)$$

We want to establish that there is no bifurcation for a load λ lower than λ_c or, in other words, that the fundamental solution $u = 0$ is locally unique up to λ_c. By the I.F.T., this will be done if one proves, first that g is analytical fonction from B_1 into B_2, second that dg has a bounded inverse.

In a first step, let us prove that the linear part of g is continuous from B_1 into B_2. Because of (5-8) (5-9), this linear part is given by

$$dg_o(u) = \sum r_n (\cos nx)\ (2/\pi)^{1/2} \qquad (5\text{-}13).$$

$$r_n = (n^4 - \lambda n^2 + k) q_n \qquad (5\text{-}14)$$

Using (5-11) (5-14) (5-10), we get estimates :

$$\| dg_o(u)\|_o^2 = \sum r_n^2 \leqslant \sum (q_n n^4)^2 \ (1 + \lambda + k)$$

$$\| dg_o(u)\|_o \leqslant \|u\|_4 \ (1 + \lambda + k)^{1/2} \qquad (5\text{-}15)$$

The latter establishes that the linear part satisfies the continuity condition (5-6).

Next, let us consider the nonlinear part u^3. Since u is zero at the boundary and by the Schwarz inequality, we obtain that a u in B_1 is bounded (remark that u in B_2 has only its square integrable and does not need to be bounded) :

$$|u(x)|^2 = \left(\int_{-\pi/2}^{x} u'(y)dy\right)^2 \leqslant \int_{-\pi/2}^{x} (u'(y))^2 \, dy \int_{-\pi/2}^{x} dy$$

Doing the same with the other boundary, we find :

$$|u(x)| \leqslant \|u\|_1 (\pi/2)^{1/2} \leqslant \|u\|_4 \ (\pi/2)^{1/2} \qquad (5\text{-}16)$$

An inequality like (5-16) leads to the so-called Sobolev embedding theorems. This one states that a u in B_1 is bounded. Then, u^3 is also bounded and therefore u is in B_2. It is not necessary to show that g is differentiable : indeed, because of the identity :

$$(u + v)^3 = u^3 + 3 u^2 v + 3 u v^2 + v^3$$

where every term may be estimated by means of (5-16), the function u^3 (and thus g) is analytic from B_1 into B_2.

Last, we have to prove that dg has a bounded inverse. For simplicity, we assume that the beam buckles in mode 1, which occurs when :

$$k < 4, \quad \lambda_c = 1 + k \qquad (5\text{-}17)$$

By (5-14) it is obvious that dg has an inverse for a load lower than the critical one. This inverse is bounded from B_2 into B_1 if there exists a positive constant C such that :

$$\|u\|_4 < C \, \|dg(u)\|_o$$

which is equivalent to :

$$\sum (q_n n^4)^2 < C^2 \sum r_n^2 \qquad (5\text{-}18)$$

Because of (5-17), one has an estimate :

$$n^4 - \lambda n^2 + k \geqslant n^4/2 \qquad \text{for } n \geqslant 2$$

From (5-14), we get two estimates :

$$|q_n| \leqslant 2|r_n|/n^4 \qquad \text{for } n \geqslant 3$$
$$|q_1| = |r_1| \;/\; (\lambda_c - \lambda) \qquad (5\text{-}19)$$

Then, from (5-19), we get the wished estimate (5-18) with :

$$C^2 = \text{Min } \{2,\ 1/(\lambda_c - \lambda)\}$$

The I.F.T. has also been applied for the Lyapunov-Schmidt reduction. Here, the displacement and the out-of-mode part may be easily defined in terms of Fourier coefficients (the first mode is again the mode 1).

$$u = (q_1,\ q_3,\ q_5\ \ldots)$$
$$v = (o,\ q_3,\ q_5,\ \ldots) \qquad (5\text{-}20)$$

Choosing as δv the cos nx $(n \geqslant 3)$, we change the equation for v (3-24) into an infinite number of algebraic equations :

$$h_n(\lambda, q_1\ ; q_3, q_5\ \ldots) = q_n\ (n^4 - \lambda n^2 + k)$$
$$+ \frac{2}{\Pi} \int (\sum_m q_m \cos(mx))^3 \cos(nx)\ dx = 0 \qquad (5\text{-}21)$$

The I.F.T. is applied to (5-21), as it has been done for the complete equation (5-8), the nonlinear term being estimated by Sobolev's inequality (5-16). One obtains easily the condition of a bounded inverse :

$$\| v \|_4^2 \leqslant 2\ \| dh(v) \|^2$$

because the first estimate (5-19) remains valid for λ close to λ_c.

We have established the :

Theorem

(i) For $\lambda < \lambda_c$, there is no bifurcation from the fundamental path $u = 0$
(ii) The Lyapunov-Schmidt reduction can be applied for λ close to λ_c

In the present case, we have not used any advanced tool of functional analysis, because we deal with an ordinary differential equation and because the linear part is diagonal (see (5-14)). For more intricate models, one has also to establish that the number of buckling modes is finite. *The same results may be stated for all models of practical interest,* such as Von Karman plates, Donnell shells. This can be done also within nonlinear tridimensional elasticity provided that the domain of the body is bounded, that the elastic coefficients are strongly elliptic and that there are no stress concentration (cracks, corners ...). We refer to /28/ /41/ /49/ for applications to partial differential equations and to /67/ for the proof of the I.F.T. and historical comments.

5.2 Catastrophe theory : progress or digression

For this section, I have chosen again the title of a recent paper by J.M.T. Thompson [59]. According to its authors Thom and Zeeman, Catastrophe Theory (C.T.) is a new mathematical theory for describing the evolution of *forms* in nature. Its purpose is *qualitative*, almost only qualitative. In fact, such a point of view is present in Postbuckling Theory (P.T.) : when one seeks if a structure is imperfection sensitive or not, this is a qualitative question. But this is only a small part of our interests. C.T. gave rise to polemics that are unusual in scientific circles. Among the most courteous critics, the prominent mathematician S. Smale [57] do not think that the contribution of C.T. is very interesting, but his criticism is not extended to the underlying mathematics, that are called Singularity Theory (S.T.). Here, I shall not try to distinguish the contributions of these two theories, but only to extract what can be of interest in elastic stability.

Since ten years, Thompson and Hunt tried also to explain the interest of C.T. for P.T., but up to now, their efforts were not successful, at least for two reasons. First, the qualitative and mathematical point of view of C.T. is too different from the state of mind in structural mechanics. Second, in its original form, C.T. classifies singularities that do not coincide with those of P.T. For instance, there is no parameter like our loading parameter λ in Thom's primitive analysis [58]. For that reason, the asymmetric bifurcation point is lacking in his classification [61]. Fortunately, Golubitsky and Schaeffer [22] [23] have recently developed S.T. in a way that is in a better agreement with the point of view of P.T.

Let us recall the basic notions of their theory [23], especially the exact definition of the codimension. For the sake of simplicity, we limit ourselves to the perturbation of a single bifurcation equation

$$g(a,\varepsilon) = 0 \tag{5-22}$$

One says that the function $g^*(a,\varepsilon,\mu_1,\mu_2, \ldots,\mu_n)$ is an *n-parameter unfolding* if

$$g^*(a,\varepsilon,o,o, \ldots, o) = g(a,\varepsilon) \tag{5-23}$$

Of course, *the control parameters* $\mu_1,\mu_2 \ldots \mu_n$ may be identified with our perturbation parameters of Section 4.4. Next one says that the two bifurcation equations g and G are *equivalent* if there are smooth changes of variable (= local diffeomorphisms):

$$A = A(a,\varepsilon), \qquad E = E(\varepsilon) \tag{5-24}$$

and a positive function $S(a,\varepsilon)$ such that :

$$g(a,\varepsilon) = S(a,\varepsilon)\, G(A(a,\varepsilon),\, E(\varepsilon)) \tag{5-25}$$

Furthermore, one requires monotony conditions

$$\partial_a A > 0, \quad E' > 0$$

Note that the transformation (5-24) may shift and deform the equilibrium paths, but the vertical directions in the plane (ε,a) remain vertical. So a symmetric bifurcation point is transformed into another symmetric bifurcation point. The number of solutions is not modified by such a transformation, as well as their stability because :

$$\partial_a g = (\partial_a S)\, G + S(\partial_A G)(\partial_a A) = S(\partial_a A)(\partial_A G)$$

One says that an *unfolding* g is *universal* if any other unfolding is equivalent to g for some value of the parameters $\mu_1, \mu_2, \ldots \mu_n$. If the number n of perturbation parameters is minimal to have a universal unfolding, it is called the *codimension* of the *singularity*.

In order to illustrate the contributions of Singularity Theory, let us consider the symmetric bifurcation point with $\alpha_{30} > 0$, $\alpha_{11} < 0$:

$$\alpha_{30}\, a^3 + \alpha_{11}\, \varepsilon\, a + a(\text{h.o.t.}) = 0 \tag{5-26}$$

S.T. establishes that the bifurcation equation (5-26) is equivalent to :

$$A^3 - A\, E = 0,$$

where it is obvious to drop the coefficients, but not the h.o.t. It is also established that this bifurcation point is of codimension two and that a universal unfolding is :

$$A^3 - A\, E + \mu_1 + \mu_2\, A^2 = 0 \tag{5-27}$$

It is straightforward to draw the equilibrium paths of (5-27). Of course, one gets only limit points for almost every value of the parameters. For $\mu_1 = 0$, there is a bifurcation point and for

$$\mu_1 = \mu_c = \mu_2^3/27,$$

there is an hysteresis point. The equilibrium paths are pictured in Fig.25 As in figure 24, one may picture the shape of the equilibrium paths according to the parameters (Fig. 26). As previously, there may exist singularities of codimension one if (μ_1,μ_2) lies on some curves and singularities of codimension two for (μ_1,μ_2) on isolated points (here only the origin). Remark that P.T. introduces only the first perturbation parameter (the exception is the famous analysis of Rourda, quoted in /60/). The figure 26 shows that this point of view is valid except for very small values of the parameter μ_1, which can be identified with the size of the imperfections. Curiously, Thom /58/ obtains the same perturbed bifurcation equation as P.T. ((5-27) with $\mu_2 = 0$), but for bad reasons. Indeed, his *equivalence rule* is not the same as (5-24) (5-25) and he allows changes of variable that do not preserve the vertical lines. So he gets the liberty of identifying a symmetric and an asymmetric bifurcation and to drop the quadratic term of (5-27).

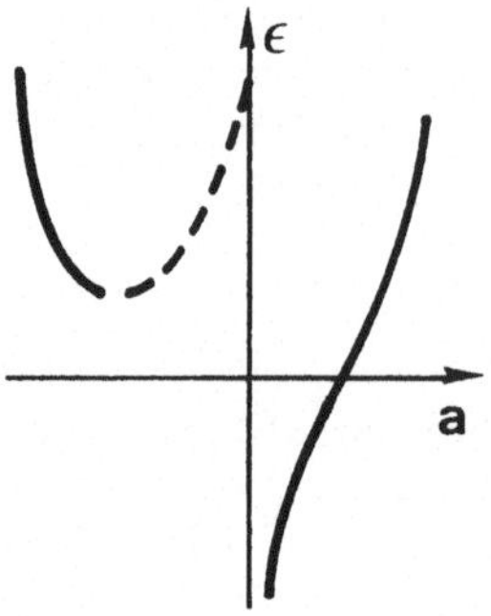

Fig. 25a, $\mu_1 < 0$

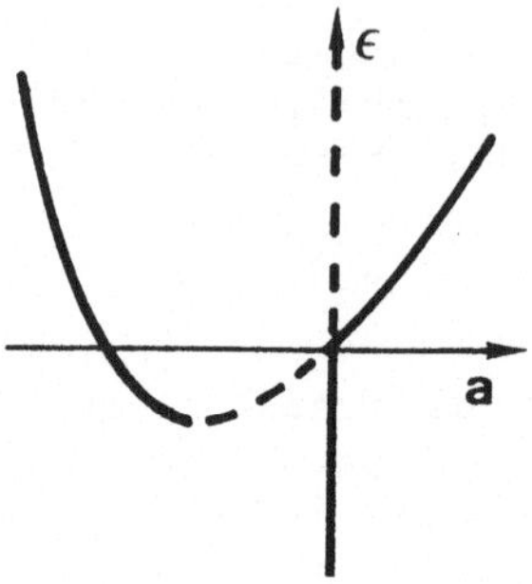

Fig. 25b, $\mu_1 = 0$

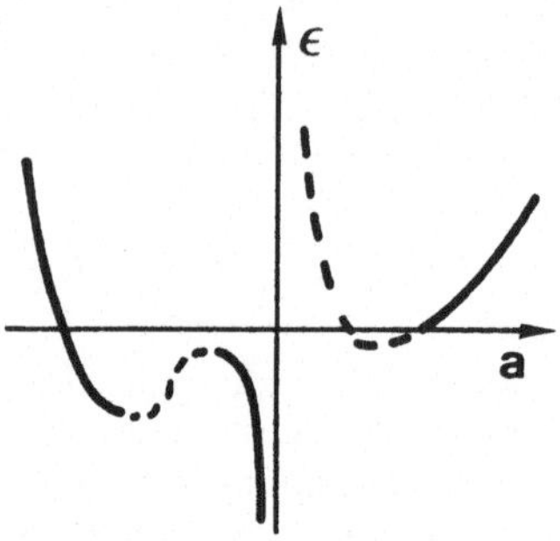

Fig. 25c, $0 < \mu_1 < \mu_c$

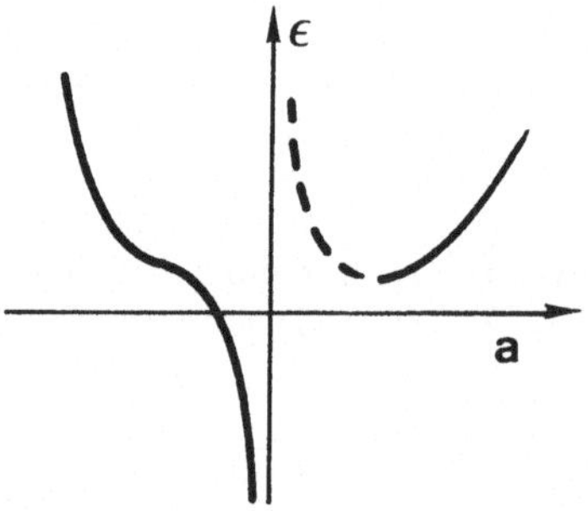

Fig. 25d, $\mu_1 = \mu_c$

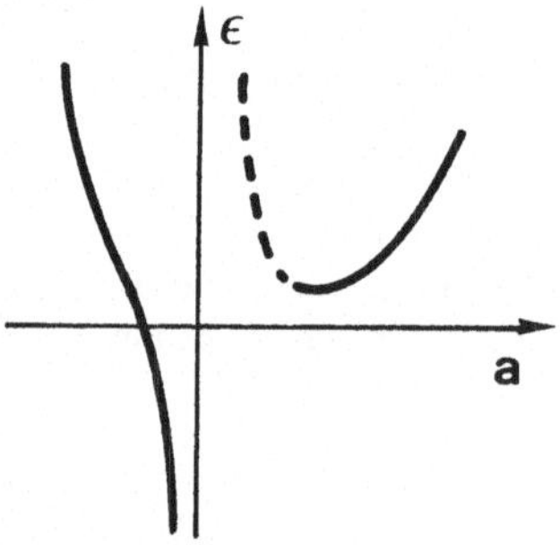

Fig. 25e, $\mu_1 > \mu_c$

Fig. 25, Equilibrium paths for (5-27)

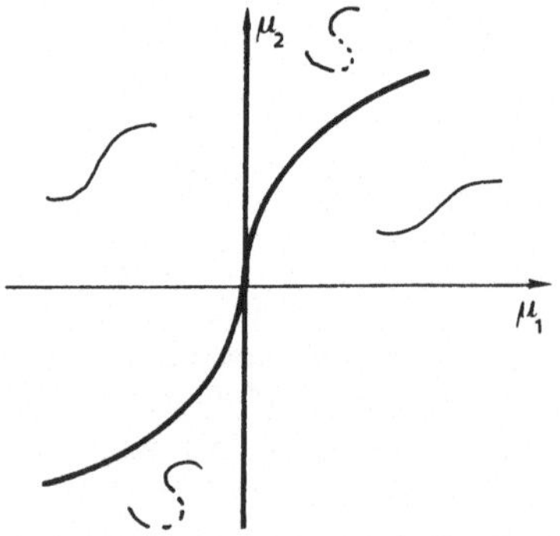

Fig. 26, Fundamental path of (5-27) according to the perturbation parameters

To my opinion, there are two main contributions of C.T. and S.T. The first one is the idea of *simple classifications of singularities according to their robustness, i.e. to their codimension*. This idea is due to René Thom, even if he initially proposes a single classification. This can modify the classical statement of P.T., as we have done in Chapter 3. For singularities of low codimension, these classifications may be achieved in a simple way, by counting the number of additional equations. But this method is not rigorous and perhaps could be misleading for singularities of high codimension. I underline that there are several classification according to the assumed properties of the structure. In this respect, the contribution of group theory may become as important as that of S.T.

The second contribution is *the treatment of controls parameters*, that are introduced in a natural way by means of rather simple algebraic rules. Thus a universal unfolding like (5-27) gives the following information : the 2-parameters perturbation (5-27) provides all possible ways to perturb the bifurcation equation (5-26). More exactly, any other perturbation would be *equivalent*, in the sense defined by (5-24) (5-25), to (5-27) for some values of the parameters. Of course, when one intends to apply such a result for example to the imperfection sensitivity of a symmetric structure, it is possible to consider that some perturbations are not significant in some practical problems (like $\mu_2\ A^2$).

The contribution of C.T. and S.T. would be severely truncated if one does not claim that :

1) generally, any singularity is in the classification(s),

2) any possible perturbation is accounted by the universal unfolding.

But this should not be misunderstood. In a number of important buckling problems, generic conditions like (3-38) to (3-40) are not satisfied. Especially, one gets often many coincident or nearly coincident modes (see the lecture by Prof. Koiter and chapter 6) so that the classifications are not relevant. In an other respect, the equivalence rules of C.T. or S.T. may be misleading, as we have underlined previously. The conclusions of these theories remain qualitative because the change of variable (5-24) is never specified, which leads to an unspecified shift and deformation of the equilibrium paths.

5.3 On the energy criterion

In this Section, we show that the energy criterion for the stability of an equilibrium state cannot be established without the modern methods of functional analysis. We only present the basic ideas, referring to /43/ /44/ /5/ /6/ for a comprehensive account.

During the fifties and the sixties, several ways have been proposed to define the stability of a continuous system. Nowadays, only the following one is accepted.

Definition :

Suppose that the motion of a continuous system is given by a continuous map

$$\{t,\ p(o)\} \rightarrow \quad p(t,\ p(o))$$

from a Banach space $R \times B$ into the Banach space B. Such a map is called a *dynamical system*. Then, an equilibrium state, say $p = 0$, is said to be *stable with respect to the norm* $\|.\|$ if, for any positive time, $\|p(t)\|$ remains smaller than any given positive number ε, as long as $\|p(o)\|$ is smaller than a positive number $\eta(\varepsilon)$. If not, it is unstable.

Unfortunately, this mathematical definition for dynamical stability has to be handled carefully because it depends, first on the chosen norm, second on constitutive assumptions, i.e. on the partial differential equations that govern any dynamical process.

A satisfactory result has been established within the framework of tri dimensional viscoelasticity of Kelvin-Voigt type /44/. The assumed equations of motion are

$$\begin{aligned} &\rho\partial_t^2 u - \text{div } T(\nabla u, \partial_t \nabla u) = 0 \qquad \forall x \in \Omega \\ &u(x, t) = 0 \qquad \forall x \in \partial\Omega \end{aligned} \tag{5-28}$$

where the constitutive law is in the form

$$\begin{aligned} &T = \partial w/\partial\nabla u + T^v(\nabla u, \partial_t \nabla u) \\ &T^v(\nabla u, o) = 0 \end{aligned} \tag{5-29}$$

With minor additional assumptions about the constitutive functions w(.) and $T^v(.,.)$, it has been established that, if the equilibrium state u = 0 is stable according to the energy criterion (3-7), then the following estimate holds for any positive time and for any small initial data :

$$\|u(t)\|_{2,p} + \|\partial_t u(t)\|_{2,p} \leqslant C\{\|u(o)\|_{2,p} + \|\partial_t u(o)\|_{2,p}\}\exp(-\beta t) \tag{5-30}$$

The constants β,C are positive, p is greater than 3 and the norm 2-p may be defined as

$$\|u\|_{2,p} = \left\{ \int_\Omega (|u|^p + |\nabla u|^p + \sum_{i,j,k} |\partial^2 u_i/\partial x_j \partial x_k|^p) dx \right\}^{1/p} \tag{5-31}$$

Hence, the estimate (5-30) ensures that the energy criterion is a sufficient condition for stability with respect to the norm (5-31), the governing equations being (5-28) (5-29). This is satisfactory, because no signi ficant constitutive restrictions have been needed, especially about the nonlinearities. Nevertheless, this must be extended to more realistic boun dary conditions and to other constitutive assumptions. Remark that the proof of such a result is much more intricate than those needed to justify the Lyapunov-Schmidt reduction.

Although the previous definition for dynamic stability is the best one, its application raises mathematical problems. Indeed, contrary to the discrete case, there are many non-equivalent norms and the notion of dynamic stability actually depends on the choice of a norm. An example of norm-dependance arises within linear isotropic elasticity. Let us consider a spherically converging motion such that the initial data lie in a thin spherical crown. Looking at the solution in closed form, one sees that the wave may focus nearby the center inside a much smaller volume. Since the total energy is constant, the maximal strain becomes arbitrarily large, however small the maximal strain may be. Hence, the rest state is unstable with respect to the norm which measures the maximum of strain and of velocity. Hence, this apparently attractive norm is not suitable to define linear elastic stability. This difficulty may be overcomed while choosing energy norms, at least in cases where the field equations are linear.

The task becomes much more difficult if the field equations are nonlinear. For instance, it is well-known that the nonlinear hyperbolic equation

$$\partial_t^2 u - \partial_x \sigma(\partial_x u) = 0 \tag{5-32}$$

may admit of non smooth solutions for any positive time, even if the linear stability condition "$\sigma'(o)$ positive" is satisfied. More precisely, with some constitutive functions $\sigma(.)$, most of the solutions of (5-32) are such that the quantity :

$$\operatorname{Sup}_x\{|\partial_x^2 u(x,t)| + |\partial_x \partial_t u(x,t)|\} \tag{5-33}$$

becomes infinite after finite time. Hence linear dynamic stability is not equivalent to nonlinear stability with respect to the norm (5-33) and the latter norm is not suitable to define "nonlinear elastic stability". This difficulty no longer exists within Kelvin-Voigt viscoelasticity (see 5-30) or within theory of beams with moderate rotations /2/.

Since a century, the energy criterion has been extensively applied. A discrepancy between theory and experiment has never been explained by removing this criterion. Hence, those mathematical difficulties should not be used to question the validity of this stability test. On the contrary, mathematics are to comply to physical evidence and some norms or some constitutive assumptions must be rejected if the corresponding notion of stability is not equivalent to the energy criterion.

6. STRUCTURES WITH A LARGE ASPECT RATIO

6.1 Limitations of classical postbuckling theory

In chapter III, we have presented the classical postbuckling theory. To apply it, it is necessary that the buckling modes are quite distinct and that their number is finite. In most applications, there are one or two modes, except in the cases of a spherical shell or of an axially compressed cylindrical shells (see Koiter's lecture in this volume). Consider the example of the beam on foundation, where the length is large with respect to wavelength of the buckling modes :

$$\bar{L} \gg L_o = (EI/K_1)^{1/4} \tag{6-1}$$

Whatever the boundary conditions may be, there are many nearly coincident buckling modes that are characterized by their wavenumber q. Hence, the classical theory cannot be used for these structures with a large aspect ratio (i.e. structures such that a condition like (6-1) holds).

In another respect, some experimental results are difficult to explain within the classical theory. For instance, Boucif, Guyon and Wesfreid /5/ have tested long rectangular plate in axial compression. The long sides were simply supported. The observed values of (λ,q) are not in the whole region above the neutral stability curve. With clamped short sides, the selected wavenumbers seem to be inside a cone that is pictured in Figure 27. Furthermore, the postbuckling deflection is not exactly periodic, but its amplitude is modulated close to the short sides, the shape of the amplitude depending on the load. In the supported case, the deflection is periodic and the observed values of (λ,q) are in the region above a parabolic looking curve.

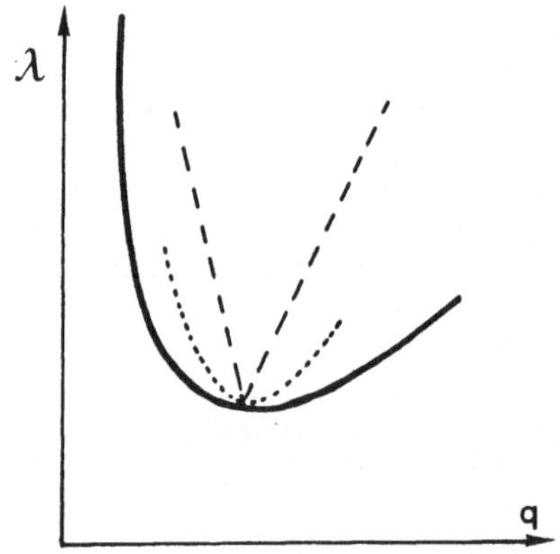

Fig. 27, Experimental wavenumbers for long rectangular plates. The continuous line is the computed neutral stability curve. The two others are lower limits with clamped (dashes) and supported (dots) short sides.

To account for a large number of solutions and for a continuously varying postbuckling pattern, the interaction of a large number of modes would be necessary. Remark that the need of continuously varying patters was early recognized (Koiter /30/and, sometimes, these problems have been solved by pragmatical variational methods /34/ /35/. Here, we follow a more accurate way which allows for the interaction of an infinite number of buckling modes. Thus one is able to explain the amplitude modulation, the existence of many solutions characterized by their wavenumber (what is called the wavelength selection problem) and the localization of buckling patterns that is observed in the buckling of mild steel plates (see /39/

and the lecture by Tvergaard). These methods had been introduced within fluid mechanics and we refer to a recent book /68/ for a lot of theoretical and experimental results on these topics.

6.2 Landau-Ginzburg equation

Let us come back to the beam on foundation with a hardening nonlinearity (K_3 positive) and without imperfections. The aspect ratio is large (i.e. (6-1) holds) and the boundary conditions are not yet specified. Let us denote, respectively, by

$$\lambda_c = 2, \quad q_c = 1, \quad \varepsilon = \lambda - 2$$

the critical load, the critical wavenumber (that correspond to the minimum of the neutral stability curve) and the load increment. The equation (2-21) may be written as :

$$(\partial_x^2 + 1)^2 u + \varepsilon \partial_x^2 u + u^3 = 0 \qquad (6\text{-}2)$$

In the supercritical range (ε positive), there is a continuous band of instability modes, whose width is :

$$|q - q_c| = o(\varepsilon^{1/2}) \qquad (6\text{-}3)$$

If the classical postbuckling theory would be applied, the postbuckling solutions would be linear combinations of buckling modes, because of (3-22) (3-27). It is well-known that a superposition of harmonics with near wavenumbers yields an oscillation of slowly varying amplitude. This suggests to use the *double scale expansion method* and to seek a deflection in the following form :

$$u(x, X), \qquad X = \varepsilon^{1/2} x \qquad (6\text{-}4)$$

According to the rule of the double scale expansion method, the derivative ∂_x in (6-2) must be replaced by

$$\partial_x + \varepsilon^{1/2} \partial_X \qquad (6\text{-}5)$$

As in Section 3.4, the deflection is expanded into powers :

$$u = \varepsilon^{1/2} u_1 + \varepsilon u_2 + \varepsilon^{3/2} u_3 + \ldots \qquad (6\text{-}6)$$

After insertion of (6-5), (6-6) into (6-2), one gets equations at the order $\varepsilon^{1/2}$, ε, $\varepsilon 3/2$:

$$D^2 u_1 = 0 \qquad \text{with } D = \partial_x^2 + 1 \qquad (6\text{-}7)$$

$$D^2 u_2 = -4 \, \partial_x \, \partial_X \, D u_1 \qquad (6\text{-}8)$$

$$D^2 u_3 = - 4 \partial_x \partial_X D u_2 - 2 \partial_x^2 D u_1 - 4 \partial_x^2 \partial_X^2 u_1 - \partial_x^2 u_1 - u_1^3 \tag{6-9}$$

According to the rules of the double expansion method, the rapid variable x and the slow variable X are assumed to be independant. Because the operator D acts only on the rapidly varying functions, two arbitrary complex functions of X appears in the bounded solutions of (6-7), (6-8) :

$$u_1 = (A_1(X) \exp i x + c.c.)/3^{1/2} \tag{6-10}$$

$$u_2 = (A_2(X) \exp i x + c.c.)/3^{1/2}$$

Because the first amplitude A_1 is complex, a solution like (6-10) can account for an amplitude modulation of the buckling patterns as well as a slow variation of the wavelength. To distinguish these two effects, let us introduce the real amplitude r and the phase Θ :

$$A_1(X) = r(X) \exp i \Theta(X) \tag{6-11}$$

The linear equation (6-9) has not a bounded solution u_3, except if there is no term exp ix in the right hand-side. Because one requires a bounded u_3, these terms have to be dropped, which leads to a *differential equation for the complex amplitude*.

$$4 A_1'' + A_1 - A_1 |A_1|^2 = 0 \tag{6-12}$$

The three terms of these equations derive, respectively, from the mode interaction, from the increase of the load and from the nonlinearities (look at (6-9) and (6-2)).

Let us introduce the true amplitude of the buckling pattern :

$$a(x) = \varepsilon^{1/2} A_1(X)$$

which satisfies the differential equation

$$4 a'' + (\lambda - \lambda_c) a - a |a|^2 = 0$$

Then, if one neglects the amplitude modulation and if one seeks only even solutions (a is real), one gets the same equation as (3-42), which would be obtained by applying the classical postbuckling theory to the search of periodic even solutions. Thus, we have introduced a new term in the amplitude equation, which allows for a variation of the amplitude and of the phase.

The amplitude equation (6-12) can be integrated in the following form, Q and E being constants :

$$r^2 \Theta' = Q \tag{6-13}$$

$$2 r'^2 = E - r^2/2 + r^4/4 - 2 Q^2/r^2 \tag{6-14}$$

The fundamental equation (6-10) is as general as Eq. (3-42) for classical postbuckling. It has been discovered in relation with problems of convection /56/ /40/. This equation has been found for Eq. (6-1) in /1/ /36/; for elastic plates /15/for cylindrical shells and for plates with a nonlinear stress-strain law /14/.

6.3 Amplitude modulation of buckling patterns

As it is recalled in Section 6.1, the postbuckling patterns of a long axially compressed rectangular plate are not periodic if the short sides are clamped. This can be explained by the amplitude equation (6-12). Let us consider the simpler model of the previous section, the two ends being clamped :

$$u(o) = \partial_x u(o) = 0 \tag{6-15}$$

$$u(L) = \partial_x u(L) = 0 \tag{6-16}$$

On account of (6-6) (6-10), both boundary conditions (6-15) can be satisfied only if the amplitude is zero at the boundary. The same holds at the other end :

$$r(o) = 0, \qquad r(L\,\varepsilon^{1/2}) = 0 \tag{6-17}$$

By virtue of (6-13) (6-17), the constant Q is zero. But the real amplitude r(X) will be different from zero inside the beam. Because of (6-13), the phase Θ is a constant. Hence, at this stage, *one does not find solutions with different wavelengths*, as it has been experimentally observed.

The amplitude modulation is governed by equation (6-14), where Q is zero, and by boundary conditions (6-17). In addition to the fundamental solution r = 0, there are solution branches that bifurcate from the fundamental solution for the values ε_n of the load increment (Figure 28). The bifurcation values are given by the linearized equations :

$$4\,r'' + r = 0 \qquad \text{and} \quad (6\text{-}17)$$

Then, the bifurcation values and the corresponding modes are :

$$\varepsilon_n = (2n\,\Pi/L)^2, \qquad r_n = \sin(n\,\Pi\,x/L) \tag{6-18}$$

As expected, only the first bifurcation branch is stable. Remark that the first bifurcation value is very small because it is the square of the ratio of the wavelength on the length of the beam. Beyond this value, the shape of the amplitude varies with the load increment (Figure 29). For $\varepsilon/\varepsilon_1$ greater than six, the initial sine shape is completely modified. The amplitude is constant and nearly equal to unity in the center of the beam, while there is a modulation nearby the boundaries. When the load is in this range, the amplitude is nearly equal to the following solution of (6-12) :

$$r(x) = \tanh\,(x/(8\varepsilon)^{1/2}) \tag{6-19}$$

in the left part of the beam. The latter solution corresponds to the saddle-saddle loop of the phase portrait of Figure 29-a.

Hence, shortly after the bifurcation, the postbuckling pattern is periodic in the center and there are boundary layers of which the thickness is $O(\varepsilon^{-1/2})$. Thus, the size of these layers decays when the load increases.

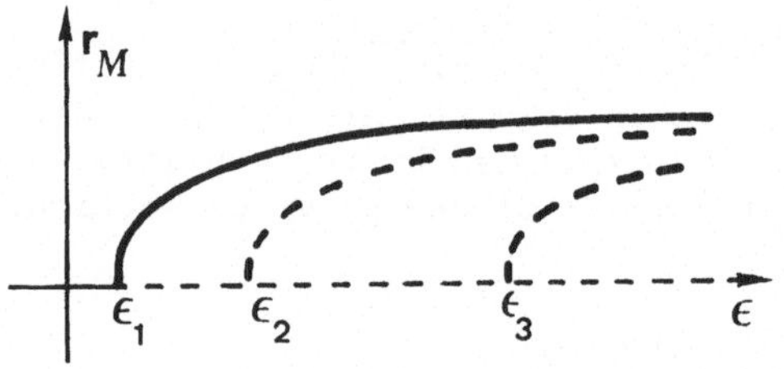

Fig. 28, The bifurcation plot in the plane of the load increment and of the maximal amplitude r_M, r being defined by (6-6) (6-10) (6-11).

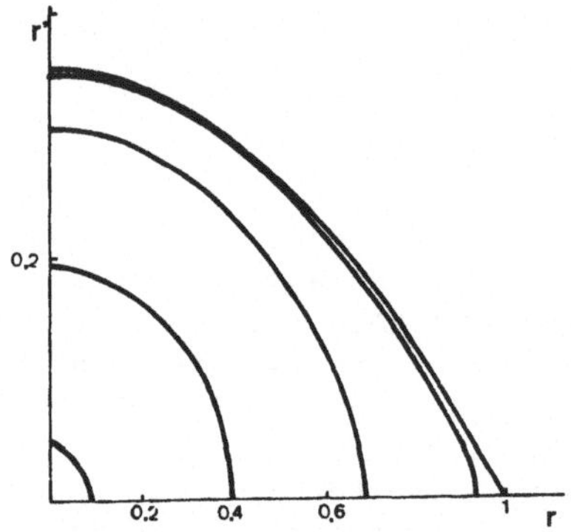

a) Phase portrait in a quarter of plane. The origin is a focus. The point (1,0) is a saddle.

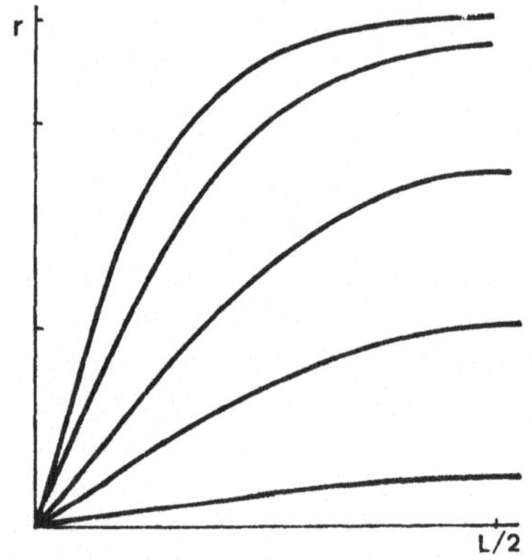

b) Patterns amplitude along the half plate. Appearance of the boundary layer.

Fig. 29, The pictured maximal amplitudes are 0.1,0.4,0.7,0.95,0.995. The corresponding $\varepsilon/\varepsilon_1$'s are 1.008, 1.13, 1.60, 3.95, 9.06.

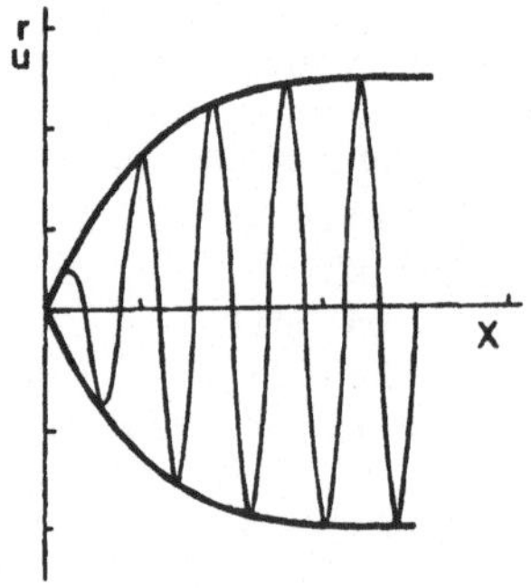

Fig. 30, Amplitude and deflection in the boundary layer.

6.4 Tendency to localization

More or less localized postbuckling patterns have been observed in many experiments carried out on mild steel rectangular plate in axial compression. After a snap, the plate reached an equilibrium state that involved only a few buckles, although the buckling mode is periodic. For additional details, we refer to /39/ and to Tvergaard lecture. It is not surprising that, after a snap, the shape of the postbuckling pattern is very different from the shape of the buckling mode. *Here we establish that the shape of the bifurcation solutions becomes more and more localized if first the bifurcation branch is unstable, and second the aspect ratio is large.*

We again consider the beam on elastic foundation, but with a softening nonlinearity and in the presence of quasimodal imperfections. The initial displacement is such that :

$$- \lambda_c u_o'' = \eta^3 \{\exp i x \beta(\eta x) + c.c.\}/3^{1/2}$$

where the small parameter η is a measure of the size of the imperfections and the unspecified complex function β a measure of their shape. The equilibrium equation is

$$(\partial_x^2 + 1)^2 u + (\lambda - \lambda_c)\partial_x^2 u - u^3 = \eta^3\{\exp ix \beta(\eta x) + c.c.\}/3^{1/2} \tag{6-20}$$

As in Section 6.2, one uses the double expansion method with respect to the two variables x and $X = \eta x$. The deflection is expanded into powers of η, the load increment being scaled by

$$\lambda = \lambda_c - m \eta^2$$

in such a way that m is positive in the subcritical range. The obtained amplitude equation is similar to (6-12), but with different signs and with a new term that accounts for the imperfections :

$$4 A_1'' - m A_1 + A_1 |A_1|^2 = \beta(X)$$

If there is no imperfections ($\beta = 0$) and the boundary conditions are (6-17), one gets the same first bifurcation load as in Section 6.2 :

$$\lambda_{bif} = \lambda_c + 4 \Pi^2/L^2$$

and the same sine mode (6-18). But the bifurcation is symmetric and unstable. The shape of the envelope is modified when one goes away from the threshold. For large m, a good approximation is given by

$$A_1 = (m/2)^{1/2}/\cos h((x - L/2)/2\eta\, m^{1/2}) \tag{6-22}$$

except in the neighbourhood of the boundaries. The solution (6-22) of equation (6-21) is small outside an interval whose width is of order $(\lambda_c - \lambda)^{1/2}$. *When the maximal amplitude growths, there is a shrinking of the interval where the amplitude is not negligible. This behaviour could explain in many problems the tendency of buckling patterns to localization.* Nevertheless, one has to keep on mind that these equilibrium states are unstable and the final collapse mode may be very different from the latter.

Tvergaard and Needleman [65] [66] have proposed an alternative explanation of the localization process. Their analysis relies on several buckling models, whose common properties are softening nonlinearities, large or moderate aspect ratios, and imperfections. They select very special periodic imperfections, very special values of the aspect ratio and they consider simply supported structures in order that there is a branch of periodic solutions. They established that some branches of nonperiodic solutions bifurcate from the periodic branch not too far from the maximum load point. They assert that "the basic mechanism of localization involves a bifurcation subsequent to the maximum load point".

We now show how the same result may be established from the amplitude equation (6-21). In order to be in the same conditions, let us assume imperfections and boundary conditions such that

$$\lambda_c u_o'' = 2\eta^3 \cos x \qquad \Rightarrow \beta = -1$$

$$u(o) = u(L) = \partial_x^2 u(o) = \partial_x^2 u(L) = 0 \tag{6-23}$$

$$L = (2n+1)\pi, \qquad n \text{ integer}$$

Hence, the amplitude equation becomes :

$$4 A_1'' - m A_1 + A_1 |A_1|^2 + 1 = 0 \tag{6-24}$$

From the boundary conditions (6-23), one gets boundary conditions for the phase and the real amplitude :

$$\begin{aligned} &\Theta(o) = 0, \qquad \Theta(L) = k\pi, \qquad k \text{ integer} \\ &r'(o) = 0, \qquad r'(L) = 0, \end{aligned} \tag{6-25}$$

If the equation (6-20) has solutions that are periodic with the critical wavenumber, the corresponding amplitude is real and constant :

$$A_1(X) = R, \qquad R^3 - m R + 1 = 0, \tag{6-26}$$

the latter equation being identical to the one obtained in Section 3.7. There is a maximum load point

$$R = (1/2)^{1/3}, \qquad m = 3(2)^{-2/3}$$

Let us seek bifurcation points from the path (6-26) of constant and real solutions. We set :

$$\rho(X) = r(X) - R \qquad (6\text{-}27)$$

Next, we insert (6-27) in (6-24) and we linearize :

$$4\rho'' + (3R^2 - m)\,\rho = 0$$

$$4\rho'' + (2R^2 - 1/R)\rho = 0$$

On account of the boundary conditions (6-25), the bifurcation modes and the corresponding R's are :

$$\rho = \cos(p\,\pi X/L\eta), \quad p \text{ integer}$$
$$2\,R^2 - 1/R = (p\,\pi/L\eta)^2 \qquad (6\text{-}28)$$

These bifurcation points always lie on the unstable part of the path of periodic equilibria. The largest $L\eta$ is, the closest to the maximum load point the bifurcation points are. Since both (6-22) and the bifurcation analysis provide unstable solutions, the analysis remains as much qualitative as the one of Tvergaard and Needleman. But the mathematical method and the primary cause of localization are different. Here *the softening nonlinearity and the large aspect ratio give rise to localized solutions. The bifurcations subsequent to the maximum load point follow from the latter assumptions.*

6.5 Wavenumber selection

Let us come back to the case of a stable postbuckling and to equation (6-2). As established in Section 6.3, there is no phase modulation at the order $\varepsilon^{1/2}$ if the ends are clamped. This remains true with many boundary conditions, but not if the ends are simply supported. Indeed, because of the boundary conditions (6-25), solutions with a constant amplitude can exist

$$r(X) = R, \qquad \Theta = QX + \psi \qquad (6\text{-}29)$$

in the whole region above the neutral stability curve. Taking the derivative of (6-14), one gets another form of the equation for the real amplitude and a relation between amplitude and wavenumber :

$$4(r'' - r\,\Theta'^2) + r - r^3 = 0 \;\rightarrow\; R^2 = 1 - 4\,Q^2 \qquad (6\text{-}30)$$

so that, in the first approximation, the region above the neutral stability curve is equivalent to :

$$Q^2 < 1/4 \qquad (6\text{-}31)$$

But all the solutions (6-29) (6-30) are not stable in the whole region (6-31). By using a simple argument involving only the amplitude equation (6-12) and the boundary conditions (6-25) /15/, one establishes that those periodic solutions are stable only above a curve of parabolic shape :

$$Q^2 = 1/12$$

or (6-32)

$$\lambda - \lambda_c = 12(q-1)^2 + o(q-1)^3$$

This is in agreement with experimental results (Figure 27). This type of wavelength selection by phase instability is known as Eckhaus instability [18].

Now, we consider a clamped left end or more generally :

$$u(o) = 0 \qquad \partial_x u(o) = k\, \partial_x^2 u(o) \tag{6-33}$$

where k is a positive number. It is sufficient to consider a semi-infinite beam and to seek solutions of (6-2) (6-33) that converge to a periodic solution. But the wavelength of the solution of a nonlinear equation cannot be a priori prescribed. The assumed form of the deflection (6-4) does not permit ones to get exactly a solution in that form. Thus it is not surprising that the expansions (6-4) (6-6) no longer holds for large X [13]. It is more convenient to seek the deflection in the form :

$$u(\xi, X), \qquad X = \varepsilon^{1/2} x$$

$$\xi = q(\varepsilon)x, \qquad q(\varepsilon) = 1 + \varepsilon Q_1 + \ldots \tag{6-34}$$

and to require that u is 2π-periodic with respect to its first argument. In this way, the admissible wavenumbers q are not a priori prescribed.

The expansions may be performed with the modified scales (6-34), as in Section 6.2. We refer to [13] [15] [46] [68] for details. The first terms (6-10) and the first amplitude equation (6-12) remain valid. Because the length of the beam is infinite, the only solution (6-19) is relevant, the phase being constant and arbitrary :

$$A_1(X) = \exp i\varphi \tanh(X/2^{3/2})$$

At the next order, one finds that the second amplitude satisfies an ordinary differential equation with variable coefficients :

$$4A_2'' + A_2 - 2\,|A_1(X)|^2 A_2 - A_1^2(X)\bar{A}_2 = 4iA_1''' + 2iA_1' - 8iQ_1 A_1' \tag{6-35}$$

Furthermore, from (6-33), one may establish boundary conditions for A_2. Last, one requires that the second amplitude is bounded, which leads to an explicit value of Q_1 as a function of the phase φ. Because this phase is arbitrary, one gets an interval of possible values of Q_1. This band of admissible wavenumber may be written as [46]

$$q_- \leqslant q(\varepsilon) \leqslant q_+$$

$$q_{\pm} = 1 + \{4 \pm (1 + 4\,k^2)^{1/2}\}\,\varepsilon/16 + O(\varepsilon^{3/2})$$

Because the right limit is greater than one, there exist solutions whose wavelength is lower than the critical one. If the beam is clamped (k is zero), the left limit is also greater than one and the wavelength decreases with the load (Figure 31-a). The finite domain analysis /13/ establishes that this wavelength diminution comes with a sequence of snap-through. If the flexibility coefficient k is large, the left limit q_- is negative and the wavelength may remain constant when one increases the load. But if one reduces the load from a solution having a wavenumber different from one, then the wavenumber goes to one after a sequence of snap. The latter behaviour has been observed in plate buckling experiment /5/as well as in the corresponding theoretical analysis /15/.

This fine analysis has been first done for problems of convections /13/.

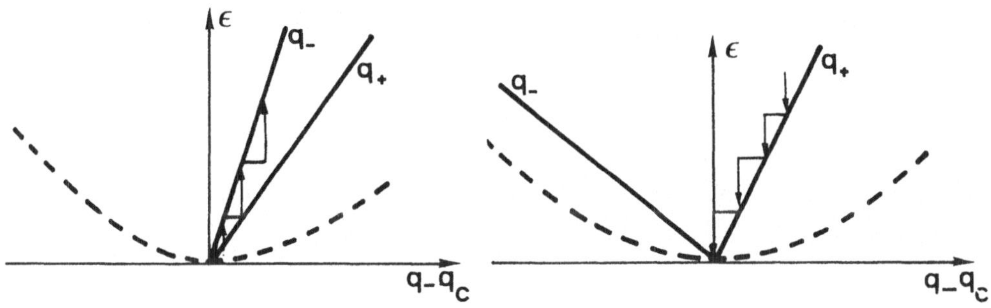

a Clamped end

b $k^2 > 15/4$

Fig. 31, The cone of admissible wavenumbers depends on the boundary conditions

REFERENCES

1. AMAZIGO, J.C., BUDIANSKI, B. and CARRIER, G.F. : Asymptotic analysis of the buckling of imperfect columns on nonlinear elastic foundations, Int. J. Solids Struct., 6 (1970), 1341-1356.
2. BALL, J.M. : Extensible beam stability theory, J. Diff. Equat., 14 (1974), 399-418
3. BAYLAC, G. (Ed.) : Recent Advances in Nuclear Component Testing and Theoretical Studies on Buckling, E.D.F.-S.E.P.T.E.N., Lyon (1984)

4. BOLOTIN, V.V.: Nonconservative Problems of the Theory of Elastic Stability, Pergamon Press, Oxford 1963.
5. BOUCIF, M., WESFREID, J.E., GUYON, E.: Role of boundary conditions on mode selection in a buckling instability, J. Physique Lettres 45 (1984) L 413-418.
6. BRUN, L., POTIER-FERRY, M.: Constitutive inequalities and dynamic stability in the linear theories of elasticity and thermoelasticity, J. Thermal Stresses, 7 (1984), 35-49.
7. BRUN, L., POTIER-FERRY, M.: Thermodynamics, Stability and Constitutive Inequalitites, Symposium Thermoelasticita Finita, Roma 1985, G. Grioli Ed., to appear.
8. BRUSH, D.O. and ALMROTH, B.O.: Buckling of Bars, Plates and Shells, McGraw-Hill, New York 1975.
9. BUDIANSKY, B.: Theory of buckling and postbuckling behaviour of elastic structures, Advances in Appl. Mech., 14 (1974), 1-65.
10. BUDIANSKY, B.(Ed.): Buckling of Structures, IUTAM Symposium, Springer-Verlag, Berlin 1976.
11. BUDIANSKY, B. and HUTCHINSON, J.W.: Dynamic Buckling of Imperfection-Sensitive Structures, Proc. Int. Congr. Appl. Mech. XI, Springer-Verlag Berlin 1966, 636-651.
12. CHOW, S.N., HALE, J.K., MALLET-PARET, J.: Applications of generic bifurcation, I Arch. Rat. Mech. Anal. 59 (1975), 159-188, II id. 62 (1976), 209-235.
13. CROSS, M.C., DANIELS, P.G., HOHENBERG, P.C., SIGGIA, E.D.: Phase-winding solutions in a finite container above the convective threshold J. Fluid Mech., 127 (1983), 155-183.
14. DAMIL, N.: Sélection de Longueur d'Ondes et Localisation du Mode de Flambement, Thèse 3° Cycle, Paris VI 1984.
15. DAMIL, N., POTIER-FERRY, M.: Wavelength selection in the postbuckling of a long rectangular plate, to appear in Int. J. Solids Struct.
16. DOWELL, E.H.: Aeroelasticity of Plates and Shells, Sijthoff and Noordhoff, Aalphen aan den Rijn 1975.
17. DUHEM, P.: Traité d'Energétique Générale, Paris 1911.
18. ECKHAUS, W.:Studies in Nonlinear Stability, Springer-Verlag, Berlin 1965.
19. ERICKSEN, J.L.: A thermokinetic view of elastic stability, Int. J. Solids Struct., 2 (1966), 573-580.
20. EULER, L.: Methodus Inveniendi Lineas Curvas Maximi Minimive Proprietate (Appendix, De Curvis Elasticis), Marcum Michaelem Bousquet, Lausanne and Geneve 1744.
21. GALBE, G.: Bifurcation et analyse non linéaire in réf. 38 (1981),39-50.
22. GOLUBITSKY, M., SCHAEFFER, D.G.: A theory for imperfect bifurcation via singularity theory, Comm. Pure Appl. Match, (1979), 21-98.
23. GOLUBITSKY, M., SCHAEFFER, D.G.: Singularities and Groups in Bifurcation Theory, vol. 1, Springer-Verlag, New-York 1985.
24. HUNT, G.W.: Elastic stability in structural mechanics and applied mathematics, in ref. 62 (1983), 125-147.
25. HUSEYIN, K.: Nonlinear Theory of Elastic Stability, Noordhoff, Leyden 1975.

26. HUTCHINSON, J.W., KOITER, W.T.: Postbuckling theory, Appl. Mech. Rev. 23 (1970), 1353-1366.
27. IOOSS, G., JOSEPH, D.D.: Elementary Stability and Bifurcation Theory Springer-Verlag, New-York 1981.
28. KELLER, J.B., ANTMAN, S. (Eds.): Bifurcation Theory and Nonlinear Eigenvalue Problems, Benjamin, New-York 1969.
29. KNOPS, R.J., WILKES, E.W.: Theory of Elastic Stability, Handbook of Physics VIa, 3, Springer-Verlag, Berlin (1973), 125-302.
30. KOITER, W.T.: The effective Width of Infinitely Long Flat Rectangular Plates under Various Conditions on Edge Restraint, National Lucht-en Ruim, NLR Report 5-287, in Dutsch (1943).
31. KOITER, W.T.: On the Stability of Elastic Equilibrium, Thesis, Delft 1945, English translation NASA Techn. Trans. F-10, 833 (1967).
32. KOITER, W.T.: Elastic Stability and Postbuckling Behaviour, in "Non-linear Problems", R.E. Langer Ed., Univ. Wisconsin Press (1963), 257-275.
33. KOITER, W.T.: Elastic Stability, Buckling and Postbuckling, D.E. Carlson and R.T. Shield Eds., Martinus Nijhoff Publishers, The Hague (1981), 13-24.
34. KOITER, W.T.: Amplitude Modulation of Short Wave Buckling Modes, Delft University Techn. Report n° 740 (1983).
35. KOITER, W.T., KUIKEN, G.D.C.: The Interaction between Local Buckling and Overall Buckling on the Behaviour of Built-up Column, Delft University Techn. Report WTHD-23 (1971).
36. LANGE, C.G., KRIEGSMANN: The axisymmetric behaviour of complete spherical shells, Quart. Appl. Math. 39 (1981), 145-178.
37. LANGE, C.G., NEWELL, A.C.: The postbuckling problem for thin elastic shells, SIAM J. Appl. Math., 21 (1971), 605-629.
38. L'HERMITE, R. (Ed.): Le Flambement des Structures, Editions Bâtiment Travaux Publics, Paris 1981.
39. MOXHAM, K.E.: Cambridge Univ. Engrg. Dept., Reports C.U.E.D./ C-Struct. /TR2 and TR3 (1971).
40. NEWELL, A.C., WHITEHEAD, J.A.: Finite bandwidth, Finite amplitude convection, J. Fluid Mech., 38 (1969), 279-403.
41. POTIER-FERRY, M.: Bifurcation et stabilité pour des systèmes dérivant d'un potentiel, J. Mécanique, 17 (1978), 579-608.
42. POTIER-FERRY, M.: Perturbed bifurcation Theory, J. Diff. Equat., 33 (1979), 112-146.
43. POTIER-FERRY, M.: Critères de l'énergie en élasticité et viscoélasticité in ref. 38 (1981), 23-37.
44. POTIER-FERRY, M.: On the mathematical foundations of elastic stability theory I, Arch. Rat. Mech. Anal., 78 (1982), 55-72.
45. POTIER-FERRY, M.: Imperfection sensitivity of a nearly double bifurcation point in ref. 55 (1982), 201-214.
46. POTIER-FERRY, M.: Amplitude modulation, phase modulation and localization of buckling patterns in ref. 62 (1983), 149-159.
47. RAMM, E. (Ed.) : Buckling of Shells, A State of the Art-Colloquium, Springer-Verlag, Berlin (1982).
48. RHODES, J., WALKER, A.C. (Edts.): Developmemnts in Thin-Walled Struc-

tures, Applied Science Publ., London, Vol. 1 (1982) and Vol. 2 (1984)
49. SATTINGER, D.H.: Topics in Stability and Bifurcation Theory, Lecture Notes in Math 309, Springer-Verlag, Berlin (1973).
50. SATTINGER, D.H.: Transformation Groups and Bifurcation at Multiple Eigenvalue, Bull. A.M.S. 79 (1973), 709-711.
51. SATTINGER, D.H.: Group Theoretic Methods in Bifurcation Theory, Lecture Notes in Math 762, Springer-Verlag, Berlin (1979).
52. SCHAEFFER, D.G.: Topics in Bifurcation Theory, in "Systems of Nonlinear P.D.E." (J.M. Ball Ed.), D. Reidel Publishing Company, Dordrecht (1983), 219-262.
53. SCHAEFFER, D.G., GOLUBITSKY, M.: Boundary conditions and mode jumping in the buckling of a rectangular plate, Comm. Math. Phys., 69 (1979), 209-236.
54. SCHROEDER, F.H. (Ed.): Stability in the Mechanics of Continua, IUTAM Symposium, Springer-Verlag, Berlin 1982.
55. SEGEL, L.A.: Distant Sidewall Cause Slow Amplitude Modulation of Cellular Convection, J. Fluid Mech., 38 (1969), 203-224.
56. SEWELL, M.J.: On the branching of equilibrium paths, Proc. Roy. Soc. London, A 315 (1970), 499-518.
57. SMALE, S.: Review of the book "Catastrophe Theory : Selected Papers", by E.C. ZEEMAN, Bull. A.M.S. 84 (1978), 1360-1367.
58. THOM, R.: Stabilité Structurelle et Morphogénèse, Benjamin, New York (1972), Engl. Translation, id. (1975).
59. THOMPSON, J.M.T.: Catastrophe theory in mechanics : progress or digression, J. Struct. Mech., 10 (1982), 167-175.
60. THOMPSON, J.M.T., HUNT, G.W.: A General Theory for Elastic Stability, Wiley, New-York 1973.
61. THOMPSON, J.M.T., HUNT, G.W.: Towards a unified bifurcation theory, Z. Angew. Math. Phys., 26 (1975), 581-603.
62. THOMPSON, J.M.T., HUNT, G.W. (Eds): Collapse, the Buckling of Structures in Theory and Practice, IUTAM Symposium, Cambridge Univ. Press (1983).
63. TIMOSHENKO, S.P., GERE, J.M.: Théorie de la Stabilité Elastique, Dunod, Paris 1966.
64. TVERGAARD, V.: Buckling Behaviour of Plate and Shell Structures, Int. Congress Theor. Appl. Mech., North-Holland, Amsterdam (1977), 233-247.
65. TVERGAARD, V., NEEDLEMAN, A.: On the localization of buckling patterns, J. Appl. Mech., 47 (1980), 613-619.
66. TVERGAARD, V., NEEDLEMAN, A.: On the Development of Localized Buckling Patterns, in ref. 62 (1983), 1-17.
67. VAINBERG, M.M., TRENOGIN, V.A.: The methods of Lyapunov and Schmidt in the study of nonlinear equations and their further development, Russ. Math. Surveys, 17 (1962), 1-60.
68. WESFREID, J.E., ZALESKI, S. (Eds.): Cellular structures in instability problems, Lecture Notes in Physics 210, Springer-Verlag, Berlin(1984).

POST-BUCKLING BEHAVIOUR OF STRUCTURES

NUMERICAL TECHNIQUES FOR MORE COMPLICATED STRUCTURES

Johann Arbocz
Professor of Aircraft Structures
Department of Aerospace Engineering
Delft University of Technology
Delft, The Netherlands

ABSTRACT

This paper deals with the stability problem of axially compressed imperfect orthotropic cylindrical shells. The initial imperfections are represented by a double Fourier series. Approximate solutions are derived for a single axisymmetric, a single asymmetric, a 2-modes and a multi-mode imperfection model. The effect of boundary conditions is studied by reducing the stability problem to the solution of a 2-point nonlinear boundary value problem. A reliability based stochastic stability approach is described, which makes it possible to include the results of the Imperfection Sensitivity Theory directly into an Improved Shell Design Procedure.

1. INTRODUCTION

It is well known that thin-walled, stiffened or unstiffened shells exhibit very favorable strength over weight ratios. Thus it is not surprising that they play an important role in modern engineering design, especially when it comes to weight sensitive applications. However, as pointed out in the preceding chapters, thin shell structures are often prone to buckling instabilities. In the last few decades, due to the manyfold applications in the aerospace, off-shore and related fields literally thousands of technical papers dealing with shell stability have been published. For extensive reviews the interested readers should consult References [1-3].

When talking about buckling of thin-walled shells one must distinguish between collapse at the maximum point in a load versus deflection curve and bifurcation. To obtain the critical load levels one can carry out an asymptotic analysis or a general nonlinear analysis.

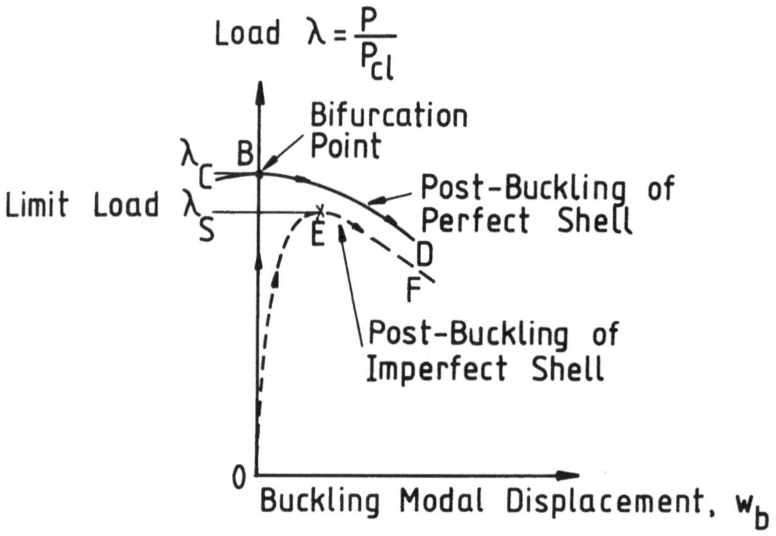

Fig. 1, Bifurcation point and limit point via asymptotic analysis

Applying the asymptotic analysis to an axially compressed perfect orthotropic shell, initially the buckling displacement W_b will be identically zero until the bifurcation load λ_c at point B has been reached (see Fig. 1). Following bifurcation the initial failure of the perfect structure will be characterized by a rapidly growing asymmetric deformation along the path BD with a decreasing axial load λ.

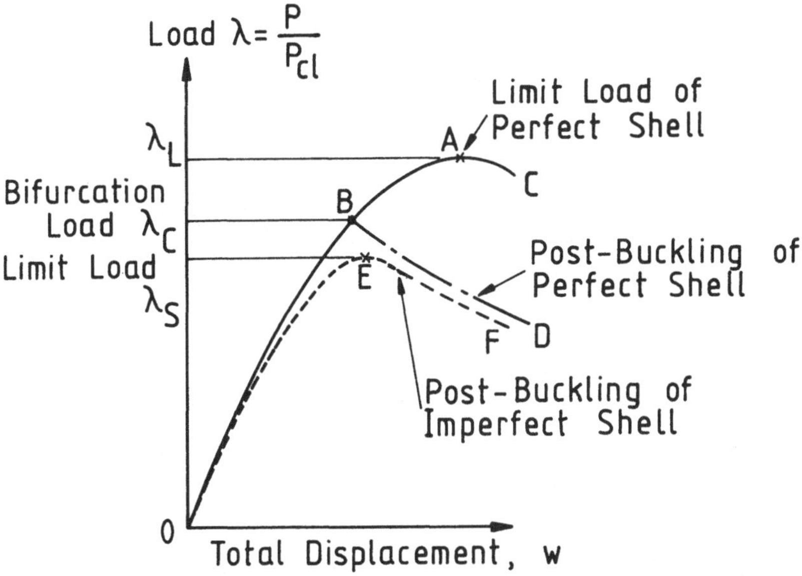

Fig. 2, Bifurcation point and limit point via nonlinear analysis

On the other hand, if one employs the general nonlinear analysis the axially compressed perfect orthotropic shell deforms axisymmetrically along the path OA (see Fig. 2) until a maximum (or limit) load λ_L is reached at point A. However, here a bifurcation point B lies between O and A. Thus, once the bifurcation load λ_c has been reached, the initial failure of the perfect structure will be characterized by a rapidly growing asymmetric deformation along the path BD with a decreasing axial load λ. Thus, in this case, the collapse load of the perfect structure λ_L is of no engineering significance.

In the case of real shells, which contain unavoidable initial imperfections, for both approaches the structural response will follow a fundamental path OEF, with failure occurring as a "snapthrough" at point E at the limit (collapse) load λ_s. In this case there are no bifurcation points between O and E. However, considering Figures 1 and 2, one can state that if there are no significant prebuckling load redistributions then the bifurcation model often leads to a good approximation of the failure load and mode, especially in cases involving significant pre-bifurcation symmetries.

2. MATHEMATICAL FORMULATION OF THE STABILITY PROBLEM

According to the sign convention defined in Fig. 3 the Donnell type equations for imperfect stiffened cylindrical shells are written [4]

$$L_H(F) - L_Q(W) = - W,_{xx}/R - \frac{1}{2} L_{NL} (W, W + 2\bar{W}) \qquad (1)$$

$$L_Q(F) + L_D(W) = F_{,xx}/R + L_{NL}\,(F,\, W + \bar{W}) \tag{2}$$

where the linear operators are

$$L_D(\;) = D_{xx}(\;)_{,xxxx} + D_{xy}(\;)_{,xxyy} + D_{yy}(\;)_{,yyyy} \tag{3}$$

$$L_H(\;) = H_{xx}(\;)_{,xxxx} + H_{xy}(\;)_{,xxyy} + H_{yy}(\;)_{,yyyy} \tag{4}$$

$$L_Q(\;) = Q_{xx}(\;)_{,xxxx} + Q_{xy}(\;)_{,xxyy} + Q_{yy}(\;)_{,yyyy} \tag{5}$$

and the nonlinear operator is

$$L_{NL}(S,T) = S_{,xx}\,T_{,yy} - 2\,S_{,xy}\,T_{,xy} + S_{,yy}\,T_{,xx} \tag{6}$$

Here $\bar{W}$ is the initial radial imperfection, W is the component of displacement normal to the shell midsurface and F is the Airy stress function. The stiffener parameters D_{xx}, H_{xx}, ..., etc. are defined in Appendix A. If we let

$$W = W^* + \hat{W} \tag{7}$$

$$F = F^* + \hat{F}$$

where

W^*, F^* = prebuckling solutions

$\hat{W}$, $\hat{F}$ = small perturbations at buckling

then a direct substitution into Eqs. (1) and (2) and deletion of squares and products of the perturbation quantities yields a set of nonlinear governing equations for the prebuckling quantities,

$$L_H(F^*) - L_Q(W^*) = -\frac{1}{R}\,W^*_{,xx} - \frac{1}{2}\,L_{NL}\,(W^*,\, W^* + 2\bar{W}) \tag{8}$$

$$L_Q(F^*) + L_D(W^*) = \frac{1}{R}\,F^*_{,xx} + L_{NL}\,(F^*,\, W^* + \bar{W}) \tag{9}$$

and a set of linearized stability equations governing the perturbation quantities,

$$L_H(\hat{F}) - L_Q(\hat{W}) = -\frac{1}{R}\,\hat{W}_{,xx} - \frac{1}{2}\,L_{NL}\,(W^*,\, \hat{W}) - \frac{1}{2}\,L_{NL}\,(W^* + 2\bar{W},\, \hat{W}) \tag{10}$$

$$L_Q(\hat{F}) + L_D(\hat{W}) = \frac{1}{R}\,\hat{F}_{,xx} + L_{NL}\,(F^*,\, \hat{W}) + L_{NL}\,(W^* + \bar{W},\, \hat{F}) \tag{11}$$

These equations, together with the appropriate boundary conditions, govern the behaviour of the imperfect cylindrical shells

1. In the prebuckling stress and deformation state.
2. At the bifurcation point or limit point (if there is one).
3. In the postbuckling stress and deformation state.

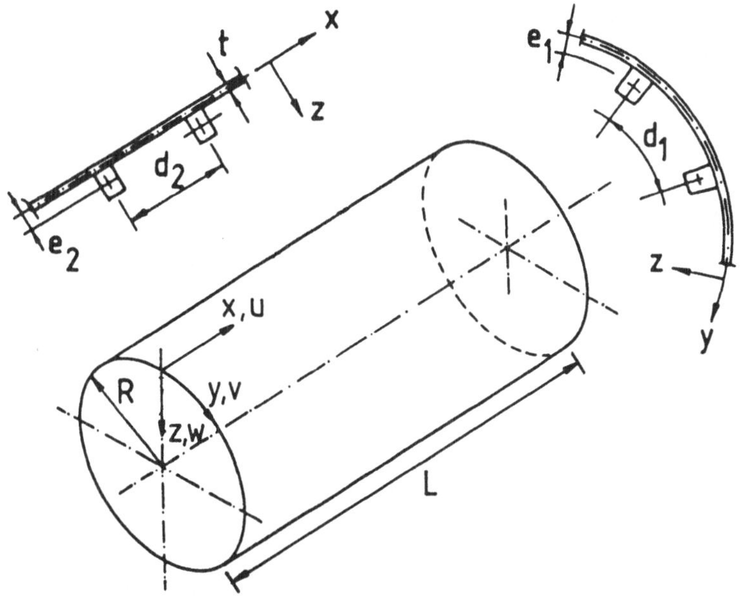

Fig. 3, Notation and sign convention

3. CLASSIC LINEARIZED SMALL DEFLECTION THEORY FOR A PERFECT SHELL

For a perfect shell $\bar{W} = 0$. If one assumes the following "membrane" solution to represent the prebuckling stress and deformation state

$$W^* = -\frac{\upsilon}{c}\frac{\bar{H}_{xx}}{1+\mu_1}\lambda t \tag{12}$$

$$F^* = -\frac{Et^2}{c}\frac{y^2}{2R}\lambda \tag{13}$$

then Eqs. (8) and (9) governing the prebuckling state are identically satisfied and Eqs. (10) and (11), the linearized stability equations, reduce to

$$L_H(\hat{F}) - L_Q(\hat{W}) = -\frac{1}{R}\hat{W}_{,xx} \tag{14}$$

$$L_Q(\hat{F}) + L_D(\hat{W}) = \frac{1}{R}\hat{F}_{,xx} - \frac{Et^2}{cR}\lambda\hat{W}_{,xx} \tag{15}$$

Assuming simply supported boundary conditions ($\hat{W} = \hat{W}_{,xx} = 0$ at $x = 0, L$) these equations admit solutions of the form

$$\hat{W} = A \sin\frac{k\pi x}{L}\cos\frac{\ell y}{R} \tag{16}$$

$$\hat{F} = B \sin\frac{k\pi x}{L}\cos\frac{\ell y}{R} \tag{17}$$

Substitution and regrouping yields a standard eigenvalue problem with the eigenvalues

$$\lambda_{c_{k\ell}} = \frac{1}{2}\left\{\frac{\bar{\gamma}_{D,k,\ell}}{\alpha_k^2} + \frac{(\bar{\gamma}_{Q,k,\ell}+\alpha_k^2)^2}{\alpha_k^2\bar{\gamma}_{H,k,\ell}}\right\} \tag{18}$$

and the eigenfunctions

$$\hat{W} = t \sin\frac{k\pi x}{L}\cos\frac{\ell y}{R} \tag{19}$$

$$\hat{F} = \frac{Et^3}{2c}\,\frac{(\bar{\gamma}_{Q,k,\ell}+\alpha_k^2)}{\bar{\gamma}_{H,k,\ell}}\sin\frac{k\pi x}{L}\cos\frac{\ell y}{R} \tag{20}$$

where

$$\bar{\gamma}_{D,k,\ell} = \bar{D}_{xx}\,\alpha_k^4+\bar{D}_{xy}\,\alpha_k^2\,\beta_\ell^2 + \bar{D}_{yy}\,\beta_\ell^4 \quad , \quad \alpha_k^2 = k^2\,\frac{Rt}{2c}\left(\frac{\pi}{L}\right)^2$$

$$\bar{\gamma}_{H,k,\ell} = \bar{H}_{xx}\,\alpha_k^4+\bar{H}_{xy}\,\alpha_k^2\,\beta_\ell^2 + \bar{H}_{yy}\,\beta_\ell^4 \quad , \quad \beta_\ell^2 = \ell^2\,\frac{Rt}{2c}\left(\frac{1}{R}\right)^2$$

$$\bar{\gamma}_{Q,k,\ell} = \bar{Q}_{xx}\,\alpha_k^4+\bar{Q}_{xy}\,\alpha_k^2\,\beta_\ell^2 + \bar{Q}_{yy}\,\beta_\ell^4 \quad , \quad c = \sqrt{3(1-\nu^2)}$$

$$\lambda = c\,\frac{\sigma}{E}\,\frac{R}{t}$$

Notice that the eigenvalues $\lambda_{c_{k\ell}}$ depend not only on the geometric parameters but also on the axial and circumferential wave numbers k and ℓ.

For isotropic shells Eq. (18) reduces to

$$\lambda_{c_{k\ell}} = \frac{1}{2}\left\{\frac{(\alpha_k^2+\beta_\ell^2)^2}{\alpha_k^2} + \frac{\alpha_k^2}{(\alpha_k^2+\beta_\ell^2)^2}\right\} = \frac{1}{2}\left\{\Lambda + \frac{1}{\Lambda}\right\}$$

Minimizing it

$$\frac{d\lambda_{c_{k\ell}}}{d\Lambda} = 0 = \frac{1}{2}\left\{1 - \frac{1}{\Lambda^2}\right\}$$

yields

$$\Lambda = \frac{(\alpha_k^2+\beta_\ell^2)^2}{\alpha_k^2} = 1 \tag{21}$$

Thus all mode shapes which satisfy the above equation have the same (lowest) eigenvalue of $\lambda_c = 1.0$. Regrouping Eq. (21) one gets the well known Koiter [5] circle

$$\alpha_k^2 + \beta_\ell^2 - \alpha_k = 0 \qquad (22)$$

which is the locus of a family of modes belonging to the lowest eigenvalue $\lambda_c = 1.0$ (see also Fig. 4).

To locate the lowest or critical buckling load Eq. (18) is used in the computer program AXBIF [6] to generate classic buckling load maps. Figures 5, 6 and 7 display such maps for an isotropic shell (A-8), an externally ring stiffened shell (AR-1) and an externally stringer stiffened shell (AS-2). The geometric properties of these shells are given in Table 1.

Table 1: Geometric and Material Properties of the Shells

Shell A-8		Shell AR-1		Shell AS-2		
t	$= 0.01179$	t	$= 0.02360$	t	$= 0.01966$	cm
L	$= 20.32$	L	$= 13.335$	L	$= 13.97$	cm
R	$= 10.16$	R	$= 10.16$	R	$= 10.16$	cm
		d_2	$= 0.635$	d_1	$= 8.03402 \cdot 10^{-1}$	cm
		e_2	$= -2.66192 \cdot 10^{-2}$	e_1	$= -3.36804 \cdot 10^{-2}$	cm
		A_2	$= 3.06406 \cdot 10^{-3}$	A_1	$= 7.98708 \cdot 10^{-3}$	cm^2
		I_{22}	$= 2.40286 \cdot 10^{-7}$	I_{11}	$= 1.50384 \cdot 10^{-6}$	cm^4
		I_{t_2}	$= 7.75647 \cdot 10^{-7}$	I_{t_1}	$= 4.94483 \cdot 10^{-6}$	cm^4
E	$= 10.48 . 10^{+6}$	E	$= 6.89472 \cdot 10^{+6}$	E	$= 6.89472 \cdot 10^{+6}$	N/cm^2
υ	$= 0.3$	υ	$= 0.3$	υ	$= 0.3$	

As mentioned above for the isotropic shell A-8 there is a family of modes belonging to the lowest eigenvalue $\lambda_c = 1.0$. There are also many modes whose eigenvalues are only slightly higher than the lowest eigenvalue $\lambda_c = 1.0$. Thus for the sake of clarity in Fig. 5 only parts of the total eigenvalue map are shown. See also Fig. 4 for the location of the areas displayed on the Koiter circle.

For the externally ring stiffened shell AR-1, the lowest eigenvalue ($\rho_c = 1.0$) is single valued and is associated with the short wavelength axisymmetric mode i = 17. Further it must be mentioned that all the eigenvalues have been normalized by the lowest one $\lambda_{c_{17,0}} = 1.10$. Thus $\rho_{k\ell} = \lambda_{c_{k\ell}}/(\lambda_{c_{k\ell}})_{min}$. Once again there are several modes whose eigenvalues are only slightly higher than the lowest eigenvalue. However, as can been seen from the results of Fig. 6, the ring stiffened shell AR-1 has only a few modes with eigenvalues less than or equal to 1.01 (within 1% of the lowest eigenvalue $\rho_c = 1.0$) and that all these modes have short wavelength in the axial direction.

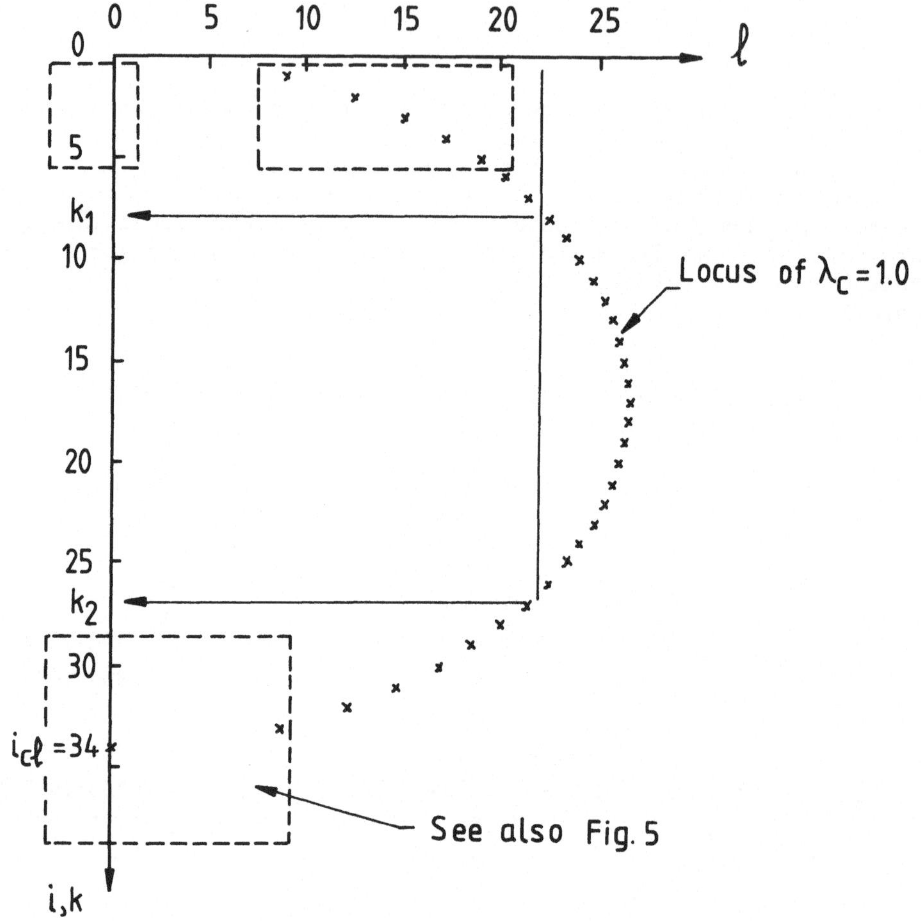

Fig. 4, Basic combination on the Koiter circle

The lowest eigenvalue (ρ_c = 1.0) of the externally stringer stiffened shell AS-2 is also single valued but it is associated with the asymmetric mode that has one half wave in the axial and 10 full waves in the circumferential direction. Once again all eigenvalues have been normalized by the lowest one $\lambda_{c_{1,10}}$= 1.44. As can be seen from Fig. 7 there are only three modes with eigenvalues less than 1.10 (within 10% of the lowest eigenvalue ρ_c = 1.0). As a matter of fact there are only a few modes with eigenvalues less than 1.50.

L=	0		8	9	10	11	12	13	14	15	16	17	18	19	20
K= 1	577.20	··	1.11	1.00	1.08	1.31	1.69	2.21	2.89	3.75	4.80	6.08	7.61	9.42	11.54
K= 2	144.32	··	2.67	1.85	1.38	1.13	1.01	1.01	1.09	1.24	1.46	1.75	2.11	2.55	3.07
K= 3	64.15	··	4.31	3.05	2.24	1.70	1.36	1.15	1.04	1.00	1.02	1.09	1.21	1.38	1.59
K= 4	36.09	··	5.30	3.94	2.98	2.30	1.82	1.49	1.26	1.11	1.03	1.00	1.01	1.06	1.15
K= 5	23.10	··	5.61	4.37	3.44	2.73	2.20	1.80	1.51	1.30	1.16	1.06	1.01	1.00	1.02
...	...									...	...	...	...		
...	...									...	...	...	...		

L=	0	1	2	3	4	5	6	7	8	9
K=29	1.05	1.05	1.05	1.05	1.05	1.04	1.04	1.04	1.03	1.03
K=30	1.03	1.03	1.03	1.03	1.03	1.03	1.02	1.02	1.02	1.02
K=31	1.02	1.02	1.02	1.02	1.01	1.01	1.01	1.01	1.01	1.01
K=32	1.01	1.01	1.01	1.01	1.01	1.01	1.00	1.00	1.00	1.00
K=33	1.00	1.00	1.00	1.00	1.00	1.00	1.00	1.00	1.00	1.00
K=34	1.00	1.00	1.00	1.00	1.00	1.00	1.00	1.00	1.00	1.00
K=35	1.00	1.00	1.00	1.00	1.00	1.00	1.00	1.00	1.01	1.01
K=36	1.01	1.01	1.01	1.01	1.01	1.01	1.01	1.01	1.01	1.01
K=37	1.01	1.01	1.01	1.02	1.02	1.02	1.02	1.02	1.02	1.02
K=38	1.03	1.03	1.03	1.03	1.03	1.03	1.03	1.03	1.03	1.04

Fig. 5, Buckling loads from linear theory for shell A-8
(SS-3 boundary condition: $N_x = v = w = M_x = 0$)

4. EFFECT OF AXISYMMETRIC IMPERFECTION

If the initial imperfection is axisymmetric, e.g.

$$\bar{W} = t\,\bar{\xi}_1 \cos \frac{i\pi x}{L} \tag{23}$$

then the prebuckling solution will also be axisymmetric, namely

$$W^* = -\frac{\upsilon}{c}\lambda t \frac{\bar{H}_{xx}}{1+\mu_1} + w^*(x) \tag{24}$$

$$F^* = -\frac{Et^2}{c}\frac{y^2}{2R}\lambda + f^*(x) \tag{25}$$

A substitution of Eqs. (24) and (25) into Eqs. (8) and (9) yields the ordinary differential equations

$$H_{xx}\, f^*{}_{,xxxx} - Q_{xx}\, w^*{}_{,xxxx} = -\frac{1}{R} w^*{}_{,xx} \tag{26}$$

$$D_{xx}\, w^*{}_{,xxxx} + Q_{xx}\, f^*{}_{,xxxx} =$$

$$= \frac{1}{R} f^*{}_{,xx} - \lambda \frac{Et^2}{cR} w^*{}_{,xx} + \lambda \frac{Et^3}{cR}\bar{\xi}_1 \left(\frac{i\pi}{L}\right)^2 \cos\frac{i\pi x}{L} \tag{27}$$

L=	0	1	2	3	4	5	6	7	8	9	10	11	12	13
K= 1	135.97	92.07	40.76	17.30	7.93	4.05	2.37	1.69	1.56	1.79	2.35	3.21	4.42	6.00
K= 2	34.00	30.58	23.09	15.74	10.31	6.73	4.48	3.08	2.25	1.77	1.55	1.52	1.65	1.93
K= 3	15.12	14.42	12.62	10.32	8.09	6.19	4.70	3.59	2.78	2.21	1.83	1.60	1.49	1.48
K= 4	8.53	8.31	7.70	6.83	5.87	4.93	4.09	3.37	2.79	2.33	1.98	1.72	1.55	1.46
K= 5	5.48	5.40	5.14	4.76	4.31	3.83	3.36	2.92	2.54	2.21	1.95	1.73	1.58	1.47
K= 6	3.84	3.80	3.68	3.49	3.26	3.00	2.73	2.47	2.23	2.01	1.82	1.67	1.54	1.45
K= 7	2.86	2.84	2.78	2.68	2.55	2.41	2.25	2.09	1.94	1.80	1.67	1.56	1.47	1.40
K= 8	2.24	2.23	2.19	2.14	2.07	1.98	1.89	1.79	1.70	1.60	1.52	1.45	1.38	1.33
K= 9	1.83	1.82	1.80	1.77	1.72	1.67	1.62	1.56	1.50	1.44	1.39	1.34	1.30	1.27
K=10	1.54	1.54	1.53	1.51	1.48	1.45	1.42	1.38	1.35	1.31	1.28	1.25	1.23	1.21
K=11	1.35	1.34	1.34	1.32	1.31	1.29	1.27	1.25	1.23	1.21	1.19	1.18	1.17	1.16
K=12	1.21	1.21	1.20	1.20	1.19	1.18	1.17	1.16	1.15	1.14	1.13	1.12	1.12	1.12
K=13	1.11	1.11	1.11	1.11	1.10	1.10	1.10	1.09	1.09	1.09	1.08	1.09	1.09	1.10
K=14	1.05	1.05	1.05	1.05	1.05	1.05	1.05	1.05	1.05	1.05	1.06	1.06	1.07	1.08
K=15	1.02	1.02	1.02	1.02	1.02	1.02	1.02	1.03	1.03	1.04	1.04	1.05	1.07	1.06
K=16	1.00	1.00	1.00	1.00	1.01	1.01	1.01	1.02	1.03	1.03	1.04	1.06	1.07	1.09
K=17	1.00	1.00	1.00	1.00	1.01	1.01	1.02	1.02	1.03	1.04	1.06	1.07	1.09	1.10
K=18	1.01	1.01	1.01	1.02	1.02	1.03	1.03	1.04	1.05	1.06	1.08	1.09	1.11	1.13

Fig. 6, Buckling loads from linear theory for shell AR-1
(SS-3 boundary condition: $N_x = v = w = M_x = 0$)

L=	0	1	2	3	4	5	6	7	8	9	10	11	12	13
K= 1	116.48	79.97	37.43	17.13	8.43	4.53	2.66	1.73	1.26	1.05	1.00	1.07	1.25	1.52
K= 2	28.87	26.07	19.99	14.03	9.58	6.57	4.60	3.31	2.47	1.92	1.56	1.34	1.21	1.16
K= 3	12.72	12.16	10.74	8.94	7.20	5.71	4.52	3.61	2.92	2.41	2.04	1.76	1.57	1.44
K= 4	7.15	6.98	6.52	5.88	5.17	4.47	3.84	3.30	2.86	2.49	2.20	1.96	1.80	1.67
K= 5	4.66	4.60	4.42	4.16	3.84	3.51	3.19	2.89	2.63	2.40	2.20	2.05	1.92	1.82
K= 6	3.40	3.38	3.30	3.18	3.04	2.88	2.72	2.56	2.41	2.27	2.16	2.06	1.98	1.91
K= 7	2.75	2.73	2.70	2.65	2.58	2.50	2.42	2.34	2.26	2.19	2.13	2.07	2.03	1.99
K= 8	2.42	2.42	2.40	2.38	2.35	2.31	2.28	2.24	2.20	2.17	2.14	2.12	2.10	2.09
K= 9	2.30	2.30	2.30	2.29	2.28	2.27	2.25	2.24	2.23	2.22	2.22	2.21	2.22	2.22
K=10	2.33	2.33	2.33	2.33	2.33	2.33	2.33	2.33	2.33	2.34	2.35	2.36	2.37	2.39

Fig. 7, Buckling loads from linear theory for shell AS-2
(SS-3 boundary condition: $N_x = v = w = M_x = 0$)

whose solution is

$$w^*(x) = \frac{\lambda}{\lambda_{c_i} - \lambda}\, t\, \bar{\xi}_1 \cos\frac{i\pi x}{L} \tag{28}$$

$$f^*(x) = \frac{Et^2}{c}\, \frac{(1 + \alpha_i^2\, \bar{Q}_{xx})}{2\,\alpha_i^2\, \bar{H}_{xx}}\, \frac{\lambda}{\lambda_{c_i} - \lambda}\, t\, \bar{\xi}_1 \cos\frac{i\pi x}{L} \tag{29}$$

where

$$\lambda_{c_i} = \frac{1}{2}\left\{\alpha_i^2\, \bar{D}_{xx} + \frac{(1 + \alpha_i^2\, \bar{Q}_{xx})^2}{\alpha_i^2\, \bar{H}_{xx}}\right\} \tag{30}$$

and the effect of the boundary conditions has been neglected. Notice that λ_{c_i} is the classic axisymmetric buckling load.
If now one assumes that the buckling mode is represented by

$$\hat{W} = C_{k\ell} \sin\frac{k\pi x}{L}\, \cos\frac{\ell y}{R} \tag{31}$$

(an expression that satisfies simply supported boundary conditions $\hat{W} = \hat{W}_{,xx} = 0$ at $x = 0$, L) then the linearized stability equations can be solved as follows. First the compatibility equation (10) is solved exactly for $\hat{F}$. This guarantees that a kinematically admissible displacement field is associated with an approximate solution of the equilibrium equation (11) via Galerkin's method. This procedure is then equivalent to an approximate minimization of the second variation of the potential energy by the Rayleigh-Ritz method, which guarantees that the eigenvalue so obtained is an upper bound to the actual buckling load. Straightforward calculation yields the following eigenvalue equation

$$\begin{aligned} \lambda^3 - \lambda^2 (\lambda_{c_{k\ell}} + 2\, \lambda_{c_i} - \underline{C_4\, \bar{\xi}_1}) + \lambda\Big[\lambda_{c_i}^2 + 2\, \lambda_{c_i} \lambda_{c_{k\ell}} \\ - (\underline{C_4 - C_3 - C_6})\, \lambda_{c_i}\, \bar{\xi}_1\Big] - \Big[\lambda_{c_{k\ell}} + (\underline{C_3 + C_6})\, \bar{\xi}_1 \\ + (C_{11} - C_{13}\, \delta_{ik})\, \bar{\xi}_1^2\Big]\, \lambda_{c_i}^2 = 0 \end{aligned} \tag{32}$$

where the underscored coupling terms vanish identically unless the condition $i = 2k$ is satisfied. The coefficients C_1 through C_{13} are listed in Appendix B.

The solution of Eq. (32) was first carried out for an isotropic shell by Koiter[7] in 1963. He used an imperfection in the form of the classic axisymmetric buckling mode

$$\bar{W} = t\, \bar{\xi}_1 \cos i_{c\ell} \frac{\pi x}{L} \qquad \text{where } i_{c\ell} = \frac{L}{\pi}\sqrt{\frac{2c}{Rt}} \tag{33}$$

and found that the minimum buckling load occurred when $k = \frac{1}{2} i_{c\ell}$ for different values of ℓ (which depended on the value of $\bar{\xi}_1$). These results are shown in Fig. 8. Hutchinson and Amazigo [8] extended this analysis to ring- and stringer stiffened shells.

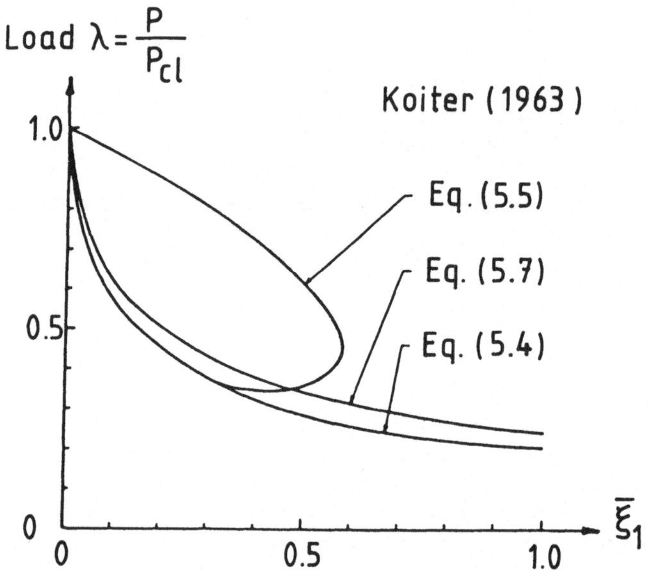

Fig. 8, Effect of axisymmetric imperfection on buckling of cylindrical shells

The previously mentioned computer program AXBIF[6] can also be used to calculate the effect of a given axisymmetric initial imperfection on the critical buckling load. Using

$$\frac{\bar{W}}{t} = - 0.4 \cos 2 \frac{\pi x}{L} \tag{34}$$

Fig. 9 displays the effect of this axisymmetric initial imperfection on the buckling load of the isotropic shell A-8. Comparing the calculated buckling loads with the corresponding values for the perfect shell A-8 shown in Fig. 5, one sees that only if the condition $k = i/2 = 1$ is satisfied are there any significant decreases. Notice that the minimum value $\lambda_{BIF} = 0.86$, occurring at $\ell = 9$, corresponds to one of the multiple eigenvalues $\lambda_c = 1.0$ on the Koiter circle.

It is interesting to observe that since the eigenvalues of the ring stiffened shell AR-1 are isolated and the lowest buckling load of $\rho_c = 1.0$ corresponds to the short wavelength axisymmetric mode $i = 17$, therefore, as can be seen from Fig. 10, the long wavelength axisymmetric

K \ L=	1	2	3	4	5	6	7	8	9	10	11	12	13	14
K= 1	143.72	81.89	25.93	9.65	4.39	2.25	1.34	0.96	0.86	0.94	1.14	1.45	1.88	2.42
K= 2	118.99	73.08	39.49	21.02	11.58	6.72	4.12	2.67	1.85	1.38	1.13	1.02	1.01	1.09
K= 3	58.74	46.06	32.49	21.68	14.21	9.36	6.28	4.32	3.05	2.24	1.70	1.36	1.15	1.04
K= 4	34.33	29.75	23.94	18.28	13.54	9.93	7.22	5.30	3.94	2.98	2.30	1.82	1.49	1.26
K= 5	22.37	20.38	17.60	14.58	11.71	9.23	7.21	5.61	4.37	3.44	2.73	2.20	1.80	1.51
K= 6	15.70	14.70	13.24	11.54	9.79	8.15	6.70	5.46	4.44	3.62	2.96	2.44	2.03	1.72
K= 7	11.61	11.06	10.23	9.22	8.12	7.03	6.01	5.09	4.28	3.60	3.03	2.56	2.17	1.86
K= 8	8.93	8.61	8.10	7.47	6.76	6.02	5.30	4.62	4.00	3.46	2.98	2.57	2.22	1.94
K= 9	7.09	6.89	6.56	6.15	5.67	5.17	4.65	4.15	3.68	3.24	2.86	2.51	2.21	1.95
K=10	5.77	5.63	5.42	5.14	4.81	4.45	4.08	3.71	3.35	3.01	2.70	2.41	2.16	1.93

Fig. 9, Buckling loads for shell A-8 with axisymmetric imperfection ($i = 2$, $\bar{\xi}_1 = -0.4$)

imperfection of Eq. (34) does not produce any ρ_{BIF} less than 1.0.

On the other hand also the eigenvalues of the stringer stiffened shell AS-2 are isolated. However, its lowest buckling load of $\rho_c = 1.0$ corresponds to the asymmetric mode ($k = 1$, $\ell = 10$). Thus, as can be seen from Fig. 11, the long wavelength axisymmetric mode of Eq. (34) produces significant decreases in the buckling loads for $k = i/2 = 1$, with the minimum value $\rho_{BIF} = 0.75$ occurring at $\ell = 10$. As mentioned earlier all the eigenvalues of the ring stiffened shell AR-1 shown in Fig. 10 are normalized by $\lambda_{c_{17,0}} = 1.10$, whereas the eigenvalues of the stringer stiffened shell AS-2 shown in Fig. 11 are normalized by $\lambda_{c_{1,10}} = 1.44$.

5. EFFECT OF ASYMMETRIC IMPERFECTION

Koiter[5] has shown that the imperfection sensitivity of a structure is closely related to its initial postbuckling behaviour. If an axially compressed cylindrical shell prossesses a unique asymmetric buckling mode associated with the lowest buckling load, then its buckling and initial postbuckling behaviour is as shown in Fig. 1. If the initial imperfections are taken affine to the unique asymmetric buckling mode, then Koiter has shown that the buckling load of an imperfect shell λ_s, (defined as the maximum load the structure can support prior to buckling), is related to the imperfection amplitude $\bar{\xi}_2$ and the postbuckling coefficient "b" by

$$\left(1 - \frac{\lambda_s}{\lambda_{c_{k\ell}}}\right)^{3/2} = \frac{3}{2} \sqrt{-3b} \, \frac{\lambda_s}{\lambda_{c_{k\ell}}} |\bar{\xi}_2| \qquad \text{for } b < 0 \tag{35}$$

Imperfection sensitive structures are characterized by negative values of "b".

To calculate the postbuckling coefficient "b" for the case where a unique buckling mode $W^{(1)}$, $F^{(1)}$ corresponds to the classic buckling load

L=	1	2	3	4	5	6	7	8	9	10	11	12	13	14
K= 1	33.79	31.14	15.75	7.09	3.48	1.94	1.35	1.24	1.45	1.89	2.54	3.38	4.40	5.60
K= 2	30.58	23.09	15.74	10.31	6.73	4.48	3.09	2.25	1.77	1.55	1.52	1.66	1.93	2.36
K= 3	14.42	12.62	10.33	8.09	6.19	4.71	3.59	2.78	2.21	1.83	1.60	1.49	1.48	1.57
K= 4	8.31	7.70	6.03	5.87	4.93	4.09	3.37	2.79	2.33	1.98	1.72	1.55	1.46	1.44
K= 5	5.40	5.14	4.76	4.31	3.83	3.36	2.92	2.54	2.21	1.95	1.74	1.58	1.47	1.41
K= 6	3.80	3.68	3.49	3.26	3.00	2.73	2.47	2.23	2.01	1.82	1.67	1.54	1.45	1.39
K= 7	2.84	2.78	2.68	2.55	2.41	2.25	2.09	1.94	1.80	1.67	1.56	1.47	1.40	1.35
K= 8	2.23	2.19	2.14	2.07	1.98	1.89	1.79	1.70	1.60	1.52	1.45	1.38	1.33	1.30
K= 9	1.82	1.80	1.77	1.72	1.67	1.62	1.56	1.50	1.44	1.39	1.34	1.30	1.27	1.25
K=10	1.54	1.53	1.51	1.48	1.45	1.42	1.38	1.35	1.31	1.28	1.25	1.23	1.21	1.20
K=11	1.34	1.34	1.32	1.31	1.29	1.27	1.25	1.23	1.21	1.19	1.18	1.17	1.16	1.16
K=12	1.21	1.20	1.20	1.19	1.18	1.17	1.16	1.15	1.14	1.13	1.12	1.12	1.12	1.13
K=13	1.11	1.11	1.11	1.10	1.10	1.10	1.09	1.09	1.09	1.08	1.09	1.09	1.10	1.11
K=14	1.05	1.05	1.05	1.05	1.05	1.05	1.05	1.05	1.05	1.06	1.06	1.07	1.06	1.10
K=15	1.02	1.02	1.02	1.02	1.02	1.02	1.03	1.03	1.04	1.04	1.05	1.07	1.08	1.10
K=16	1.00	1.00	1.00	1.01	1.01	1.01	1.02	1.03	1.03	1.04	1.06	1.07	1.09	1.11
K=17	1.00	1.00	1.00	1.01	1.01	1.02	1.02	1.03	1.04	1.06	1.07	1.09	1.10	1.13
K=18	1.01	1.01	1.02	1.02	1.03	1.03	1.04	1.05	1.06	1.08	1.09	1.11	1.13	1.15

Fig. 10, Buckling loads for shell AR-1 with axisymmetric imperfection (i = 2, $\bar{\xi}_1$ = - 0.4)

L=	1	2	3	4	5	6	7	8	9	10	11	12	13	14
K= 1	28.73	27.17	15.63	7.62	3.97	2.24	1.39	0.98	0.79	0.75	0.81	0.94	1.13	1.38
K= 2	26.07	19.98	14.03	9.58	6.57	4.60	3.31	2.47	1.92	1.56	1.34	1.21	1.16	1.17
K= 3	12.16	10.74	8.94	7.20	5.71	4.52	3.61	2.92	2.41	2.04	1.76	1.57	1.44	1.36
K= 4	6.98	6.52	5.88	5.17	4.47	3.84	3.30	2.86	2.49	2.20	1.98	1.80	1.67	1.58
K= 5	4.60	4.42	4.16	3.84	3.51	3.19	2.89	2.63	2.40	2.20	2.05	1.92	1.82	1.74
K= 6	3.38	3.30	3.18	3.04	2.88	2.72	2.56	2.41	2.27	2.16	2.06	1.98	1.91	1.86
K= 7	2.73	2.70	2.65	2.58	2.50	2.42	2.34	2.26	2.19	2.13	2.07	2.03	1.99	1.97
K= 8	2.42	2.40	2.38	2.35	2.31	2.28	2.24	2.20	2.17	2.14	2.12	2.10	2.09	2.09
K= 9	2.30	2.30	2.29	2.28	2.27	2.25	2.24	2.23	2.22	2.22	2.21	2.22	2.22	2.23
K=10	2.33	2.33	2.33	2.33	2.33	2.33	2.33	2.33	2.34	2.35	2.36	2.37	2.39	2.41

Fig. 11, Buckling loads for shell AS-2 with axisymmetric imperfection (i = 2, $\bar{\xi}_1$ = - 0.4)

$\lambda_{c_{k\ell}}$, one begins by assuming a solution valid in the initial postbuckling region in the form of an asymptotic expansion

$$W = \lambda W^{(0)} + \xi_2 W^{(1)} + \xi_2^{\,2} W^{(2)} + \ldots$$

$$F = \lambda W^{(0)} + \xi_2 F^{(1)} + \xi_2^{\,2} F^{(2)} + \ldots \tag{36}$$

$$\frac{\lambda}{\lambda_{c_{k\ell}}} = 1 + a\, \xi_2 + b\, \xi_2^{\,2} + \ldots$$

A formal substitution of this expansion into the nonlinear Donnell-type equations (1) and (2) for a perfect shell ($\bar{W} = 0$) generates a sequence of (linear) equations for the functions appearing in the expansion.

The set of equations for $W^{(0)}$ and $F^{(0)}$ is identical to the nonlinear equations (Eqs. 8 and 9) governing the prebuckling states of a perfect shell ($\bar{W} = 0$). If one neglects the edge effects, then the clasic membrane prebuckling solution

$$W^{(0)} = -\frac{\nu t}{c}\frac{\bar{H}_{xx}}{1+\mu_1} \tag{37}$$

$$F^{(0)} = -\frac{Et^2}{cR}\frac{y^2}{2} \tag{38}$$

satisfies the governing equations of the 0th order state identically. In this case the set of equations for $W^{(1)}$ and $F^{(1)}$ become the classic eigenvalue problem (Eqs. 14 and 15), the solution of which are given by Eqs. (18-20). Finally from the next higher order terms in the expansion one gets the governing equations for $W^{(2)}$ and $F^{(2)}$ as:

$$L_H\,(F^{(2)}) - L_Q\,(W^{(2)}) + \frac{1}{R} W^{(2)}{}_{,xx} = W^{(1)}{}_{,xy} W^{(1)}{}_{,xy} - W^{(1)}{}_{,xx} W^{(1)}{}_{,yy} \tag{39}$$

$$L_Q\,(F^{(2)}) + L_D\,(W^{(2)}) - \frac{1}{R} F^{(2)}{}_{,xx} + \frac{Et^2}{cR}\lambda_{c_{k\ell}} W^{(2)}{}_{,xx} = L_{NL}\,(F^{(1)}, W^{(1)}) \tag{40}$$

General expressions for the postbuckling coefficients "a" and "b" have been derived by Budiansky and Hutchinson[9]. For the case under consideration these expressions reduce to

$$a = \frac{\frac{3}{2}\int_S \left(F^{(1)}_{,xx} W^{(1)}_{,y} W^{(1)}_{,y} - 2F^{(1)}_{,xy} W^{(1)}_{,x} W^{(1)}_{,y} + F^{(1)}_{,yy} W^{(1)}_{,x} W^{(1)}_{,x}\right) dS}{\frac{Et^2}{cR}\lambda_{c_{k\ell}} \int_S W^{(1)}_{,x} W^{(1)}_{,x}\, dS} \tag{41}$$

$$b = \frac{2\int_S \{F_{,xx}^{(1)} W_{,y}^{(1)} W_{,y}^{(2)} - F_{,xy}^{(1)} (W_{,x}^{(1)} W_{,y}^{(2)} + W_{,y}^{(1)} W_{,x}^{(2)}) + F_{,yy}^{(1)} W_{,x}^{(1)} W_{,x}^{(2)}\} dS}{\frac{Et^2}{cR} \lambda_{c_{k\ell}} \int_S W_{,x}^{(1)} W_{,x}^{(1)} dS} + \frac{\int_S (F_{,xx}^{(2)} W_{,y}^{(1)} W_{,y}^{(1)} - 2F_{,xy}^{(2)} W_{,x}^{(1)} W_{,y}^{(1)} + F_{,yy}^{(2)} W_{,x}^{(1)} W_{,x}^{(1)}) dS}{\frac{Et^2}{cR} \lambda_{c_{k\ell}} \int_S W_{,x}^{(1)} W_{,x}^{(1)} dS} \quad (42)$$

It is easily verified that the first postbuckling coefficient "a" is identically zero, and, therefore, it is necessary to solve for $W^{(2)}$ and $F^{(2)}$ in order to calculate "b". Equations (39) and (40) admit solutions of the form.

$$W^{(2)} = t \left\{ \sum_{j=1}^{\infty} A_j \sin \frac{j\pi x}{L} + \cos \frac{2\ell y}{R} \sum_{j=1}^{\infty} B_j \sin \frac{j\pi x}{L} \right\} \quad (43)$$

$$F^{(2)} = \frac{Et^2}{cR} \left\{ \sum_{j=1}^{\infty} C_j \sin \frac{j\pi x}{L} + \cos \frac{2\ell y}{R} \sum_{j=1}^{\infty} D_j \sin \frac{j\pi x}{L} \right\} \quad (44)$$

Notice that each individual term in the above series satisfies simply supported boundary conditions ($W^{(2)} = W_{,xx}^{(2)} = F^{(2)} = F_{,xx}^{(2)} = 0$) at $x = 0,L$. The coefficients are readily determined by the Galerkin procedure. Finally the postbuckling coefficient "b" is calculated by evaluating the integrals indicated by Eq. (42) to obtain

$$b = -\frac{4c}{\pi} \frac{\beta_\ell^2}{\lambda_{c_{k\ell}}} \left\{ \sum_{j=1,3,\ldots}^{\infty} (2\bar{F} B_j + D_j) \frac{1}{j} + - 2 \sum_{j=1,3,\ldots}^{\infty} (2\bar{F} A_j + C_j) \frac{j}{j^2 - 4k^2} \right\} \quad (45)$$

The coefficients A_j, B_j, C_j, D_j and $\bar{F}$ are listed in Appendix C. The series can be evaluated numerically to any degree of accuracy desired. This solution was first obtained by Hutchinson and Amazigo[8] using an asymmetric imperfection of the form

$$\bar{W} = t\,\bar{\xi}_2 \sin \frac{k\pi x}{L} \cos \frac{\ell y}{R} \quad (46)$$

for stringer stiffened shells. Knowing "b", one can use Eq. (35) to obtain the $\lambda_s/\lambda_{c_{k\ell}}$ versus $\bar{\xi}_2$ curve of Fig. 12. More recently, Cohen[10] and Hutchinson and Frauenthal[11] have presented solutions that take into account the effect of nonlinear prebuckling deformations and different boundary conditions.

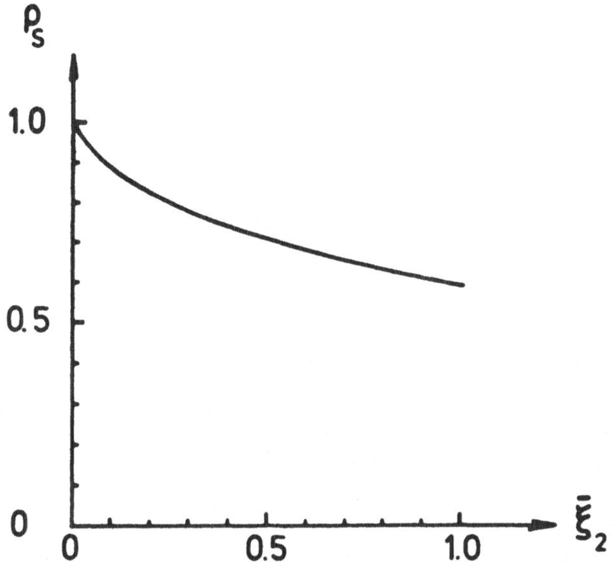

Fig. 12, Imperfection sensitivity for asymmetric imperfection only (Shell AS-2, b = - 0.0308)

Fig. 13 shows that the type of prebuckling analysis employed can influence strongly the value of the "b" factor calculated. The curves labeled linear and nonlinear have been obtained with SRA, a general shell of revolution code developed by Cohen[10] which has the capability of computing the imperfection-sensitivity factors "a" and "b" for general meridional shapes and different boundary conditions.

To estimate the degree of imperfection sensitivity as a function of the magnitude of the postbuckling coefficient "b", one can use the curves shown in Fig. 14. For the sake of calibration a curve taken from Koiter[7], showing the effect of axisymmetric imperfections on the buckling load of axially compressed isotropic cylinders, is also included in Fig. 14.

6. EFFECT OF AXISYMMETRIC AND ASYMMETRIC IMPERFECTIONS (2-MODES SOLUTION)

If one assumes that the initial radial imperfection is given by

$$\bar{W} = t\,\bar{\xi}_1 \cos\frac{i\pi x}{L} + t\,\bar{\xi}_3 \sin\frac{k\pi x}{L} \cos\frac{\ell y}{R} \qquad (47)$$

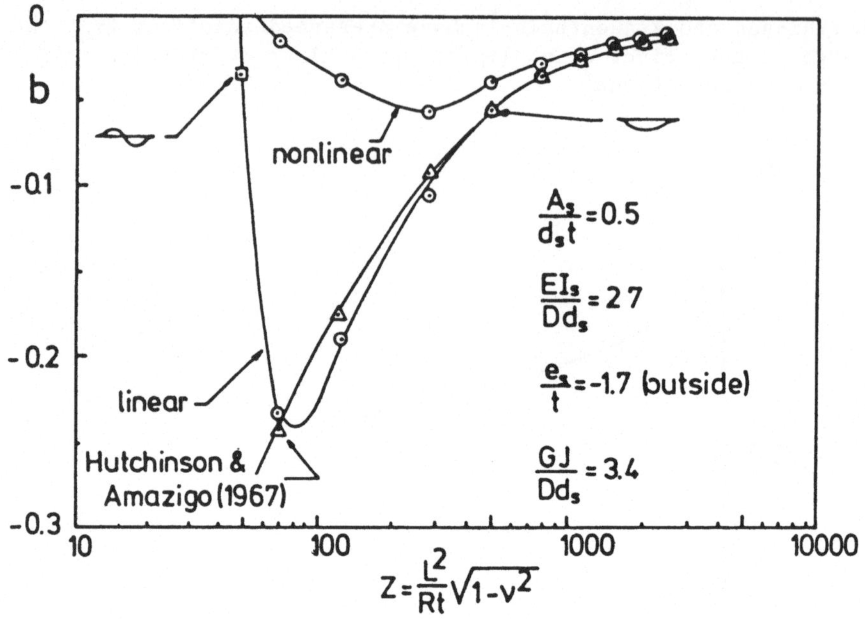

Fig. 13, Imperfection sensitivity of axially compressed stringer stiffened cylinders (AS-shells, $N_x = v = w = M_x = 0$)

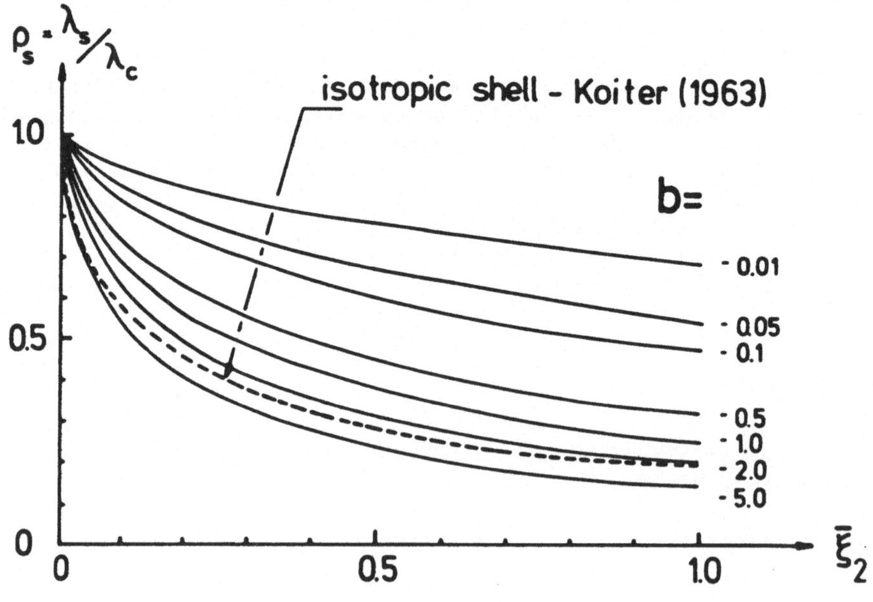

Fig. 14, Variation of buckling load with imperfection amplitude for various values of "b"

then any equilibrium state of the axially loaded cylinder can be represented by

$$W = t\,\xi_0 + t\,\xi_1 \cos\frac{i\pi x}{L} + t\,\xi_3 \sin\frac{k\pi x}{L}\cos\frac{\ell y}{R} \tag{48}$$

$$F = -\frac{Et^2}{c}\frac{y^2}{2R}\lambda + f \tag{49}$$

The value of the term

$$\xi_0 = -\frac{\upsilon}{c}\frac{\bar{H}_{xx}}{1+\mu_1}\lambda + \frac{c}{4}\beta_\ell^2\,(\xi_3 + 2\,\bar{\xi}_3)\,\xi_3 \tag{50}$$

is obtained by enforcing the circumferential periodicity condition (see Ref. 12).

An approximate solution of the nonlinear Donnell type equations is obtained as follows. First, the compatibility equation (1) is solved exactly for the stress function f in terms of the assumed radial displacement W and the measured imperfection $\bar{W}$. In this solution, only the effect of the initial imperfections on the buckling load is of interest. Hence, only a particular solution of Eq. (1) needs to be considered. Second, the equation of equilibrium (2) is solved approximately by substituting therein F, W and $\bar{W}$, and then applying Galerkin's procedure. This approach yields the following set of nonlinear algebraic equations in terms of the unknown amplitudes ξ_1 and ξ_3:

$$(\lambda_{c_i} - \lambda)\,\xi_1 + \underline{C_1\,(\tfrac{1}{2}\,\xi_3^2 + \xi_3\,\bar{\xi}_3) + C_2\,(\xi_3 + \bar{\xi}_3)\,\xi_3}$$
$$+ (C_8 - C_{12}\,\delta_{ik})\,(\xi_3 + \bar{\xi}_3)\,(\xi_1\,\xi_3 + \xi_1\,\bar{\xi}_3 + \bar{\xi}_1\,\xi_3) = \lambda\,\bar{\xi}_1 \tag{51}$$

$$(\lambda_{c_{k\ell}} - \lambda)\,\xi_3 + \underline{C_3\,\left(2\xi_3\,(\xi_1 + \bar{\xi}_1) + \xi_1\,\bar{\xi}_3\right) + C_4\,\xi_1\,(\xi_3 + \bar{\xi}_3)}$$
$$+ (C_9 + C_{10})\,(\xi_3 + \bar{\xi}_3)\,(\tfrac{1}{2}\,\xi_3^2 + \xi_3\,\bar{\xi}_3)$$
$$+ (C_{11} - C_{13}\,\delta_{ik})\,(\xi_1 + \bar{\xi}_1)\,(\xi_1\,\xi_3 + \xi_1\,\bar{\xi}_3 + \bar{\xi}_1\,\xi_3) = \lambda\,\bar{\xi}_3 \tag{52}$$

The coefficients C_1 through C_{13} are listed in Appendix B. The underscored quadratic coupling terms vanish identically unless the condition i = 2k is satisfied. It has been demonstrated in Reference 13 that contrary to earlier assertions the cubic nonlinearities cannot be neglected. Solution of the nonlinear algebraic equations (51) and (52) yields the equilibrium configurations of the finite shell as a function of λ. Regrouping of

Eqs. (51) and (52) yields

$$(\lambda_{c_i} - \lambda)\, \xi_1 + D_1\, \xi_1 + D_2\, \xi_3 + D_3\, \xi_1\, \xi_3 + D_4\, \xi_3^2 + D_5\, \xi_1\, \xi_3^2 = \lambda\, \bar{\xi}_1 \tag{53}$$

$$(\lambda_{c_{k\ell}} - \lambda)\, \xi_3 + D_6\, \xi_3 + D_7\, \xi_1 + D_8\, \xi_1^2 + D_9\, \xi_1\, \xi_3 + D_{10}\, \xi_3^2 + D_{11}\, \xi_1^2\, \xi_3 + D_{12}\, \xi_3^3 = \lambda\, \bar{\xi}_3 \tag{54}$$

where

$$\begin{aligned}
D_1 &= (C_8 - C_{12}\, \delta_{ik})\, \bar{\xi}_3^2 \\
D_2 &= (C_1 + C_2)\, \delta_{i=2k}\, \bar{\xi}_3 + (C_8 - C_{12}\, \delta_{ik})\, \bar{\xi}_1\, \bar{\xi}_3 \\
D_3 &= 2\, (C_8 - C_{12}\, \delta_{ik})\, \bar{\xi}_3 \\
D_4 &= (\tfrac{1}{2}\, C_1 + C_2)\, \delta_{i=2k} + (C_8 - C_{12}\, \delta_{ik})\, \bar{\xi}_1 \\
D_5 &= C_8 - C_{12}\, \delta_{ik} \\
D_6 &= 2\, C_3\, \delta_{i=2k}\, \bar{\xi}_1 + (C_{11} - C_{13}\, \delta_{ik})\, \bar{\xi}_1^2 + (C_9 + C_{10})\, \bar{\xi}_3^2 \\
D_7 &= (C_3 + C_4)\, \delta_{i=2k}\, \bar{\xi}_3 + (C_{11} - C_{13}\, \delta_{ik})\, \bar{\xi}_1\, \bar{\xi}_3 \\
D_8 &= (C_{11} - C_{13}\, \delta_{ik})\, \bar{\xi}_3 \\
D_9 &= (2\, C_3 + C_4)\, \delta_{i=2k} + 2\, (C_{11} - C_{13}\, \delta_{ik})\, \bar{\xi}_1 \\
D_{10} &= \frac{3}{2}\, (C_9 + C_{10})\, \bar{\xi}_3 \\
D_{11} &= C_{11} - C_{13}\, \delta_{ik} \\
D_{12} &= \frac{1}{2}\, (C_9 + C_{10})
\end{aligned} \tag{55}$$

δ_{ik} = KRONECKER delta

$\delta_{i=2k}$ = GENERALIZED KRONECKER delta
= 1 if i=2k
= 0 otherwise

6.1 <u>For a perfect shell $\bar{\xi}_1 = \bar{\xi}_3 = 0$</u>

Then Eqs. (53) and (54) reduce to

$$(\lambda_{c_i} - \lambda)\, \xi_1 + D_4\, \xi_3^2 + D_5\, \xi_1\, \xi_3^2 = 0 \tag{56}$$

$$\left\{\lambda_{c_{k\ell}} - \lambda + D_9\,\xi_1 + D_{11}\,\xi_1^{\,2} + D_{12}\,\xi_3^{\,2}\right\}\xi_3 = 0 \tag{57}$$

If $\lambda_{c_i} < \lambda_{c_{k\ell}}$ then initially $\xi_1 = \xi_3 = 0$ until $\lambda = \lambda_{c_i}$, when bifurcation into the ξ_1-mode occurs. ξ_1 becomes nonzero at constant value of $\lambda = \lambda_{c_i}$ with $\xi_3 = 0$ until

$$\xi_1 = -\frac{D_9}{2\,D_{11}}\left\{1 + \sqrt{1 - \frac{4\,D_{11}}{D_9^{\,2}}\,(\lambda_{c_{k\ell}} - \lambda_{c_i})}\right\} \tag{58}$$

when bifurcation into the ξ_3-mode occurs. Following this bifurcation deformations in both ξ_1 and ξ_3 occur with decreasing values of λ (see Fig. 15). If, however, $\lambda_{c_{k\ell}} < \lambda_{c_i}$ then initially $\xi_1 = \xi_3 = 0$ until $\lambda = \lambda_{c_{k\ell}}$ when bifurcation into the ξ_3-mode occurs. Following bifurcation deformations in both ξ_1 and ξ_3 occur with decreasing values of λ (see Fig. 16).

It should be remarked here that with the modes assumed in Eq. (47) for a perfect shell bifurcation will occur with $\xi_1 < 0$ as shown in Figs. 15 and 16.

6.2 For a shell with axisymmetric imperfection $\bar{\xi}_1 \neq 0$, $\bar{\xi}_3 = 0$

Then Eqs. (53) and (54) reduce to

$$(\lambda_{c_i} - \lambda)\,\xi_1 + D_4\,\xi_3^{\,2} + D_5\,\xi_1\,\xi_3^{\,2} = \lambda\,\bar{\xi}_1 \tag{59}$$

$$\left\{\lambda_{c_{k\ell}} - \lambda + D_6 + D_9\,\xi_1 + D_{11}\,\xi_1^{\,2} + D_{12}\,\xi_3^{\,2}\right\}\xi_3 = 0 \tag{60}$$

Since the imperfection is purely axisymmetric, therefore the prebuckling deformation is also axisymmetric. Hence initially $\xi_3 = 0$, which satisfies Eq. (60) identically. From Eq. (59) initially

$$\xi_1 = \frac{\lambda\,\bar{\xi}_1}{\lambda_{c_i} - \lambda} \tag{61}$$

Bifurcation into the ξ_3-mode (that is $\xi_3 \neq 0$) will occur when from Eq. (60)

$$\lambda_{c_{k\ell}} - \lambda + D_6 + D_9\,\xi_1 + D_{11}\,\xi_1^{\,2} = 0 \tag{62}$$

Following bifurcation deformation in both ξ_1 and ξ_3 occur with decreasing values of λ (see Figures 17 and 18).

Notice that it depends on the sign of the axisymmetric imperfection wether λ_{BIF}, the buckling load of the imperfect shell, is less than (if $\bar{\xi}_1 < 0$) or greater than (if $\bar{\xi}_1 > 0$) $\lambda_{c_{k\ell}}$, the buckling load of the

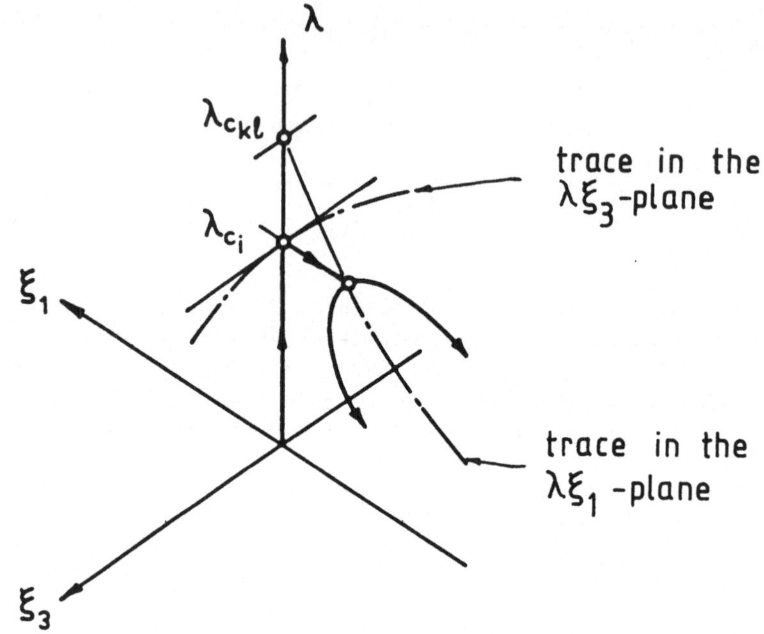

Fig. 15, Bifurcation for a perfect shell if $\lambda_{c_i} < \lambda_{c_{k\ell}}$

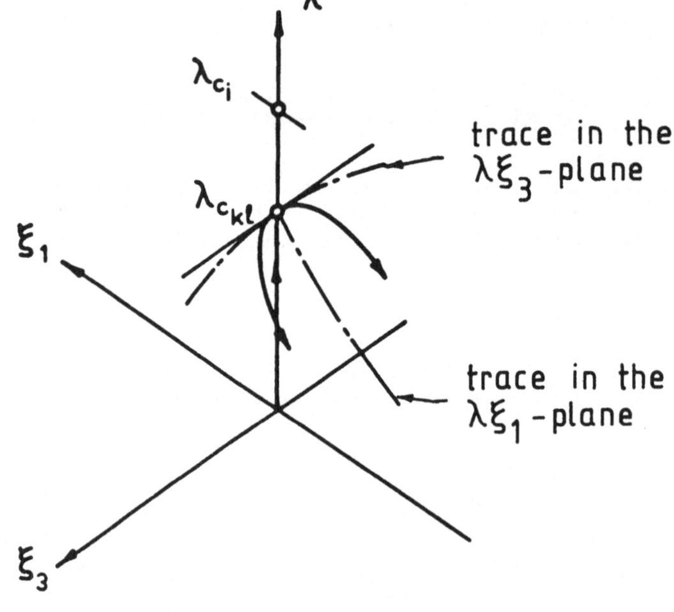

Fig. 16, Bifurcation for a perfect shell if $\lambda_{c_{k\ell}} < \lambda_{c_i}$

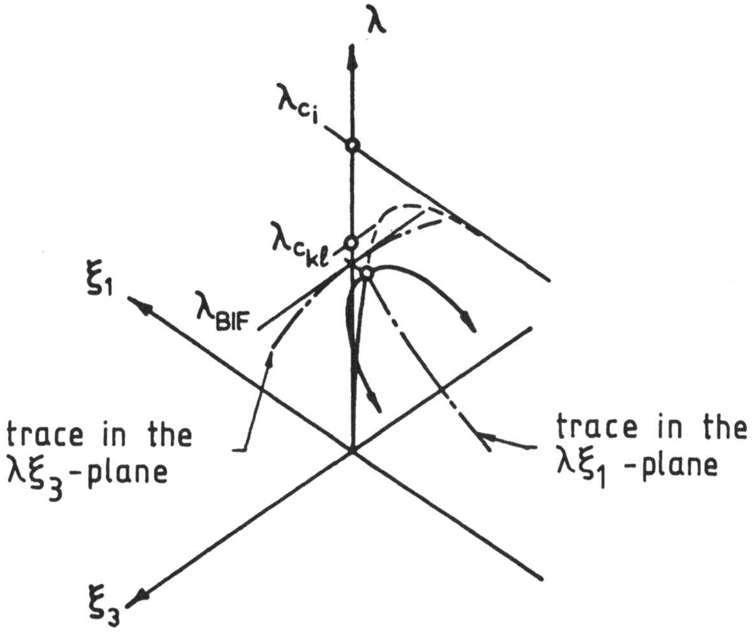

Fig. 17, Bifurcation for a shell with axisymmetric ($\bar{\xi}_1 < 0$) imperfection

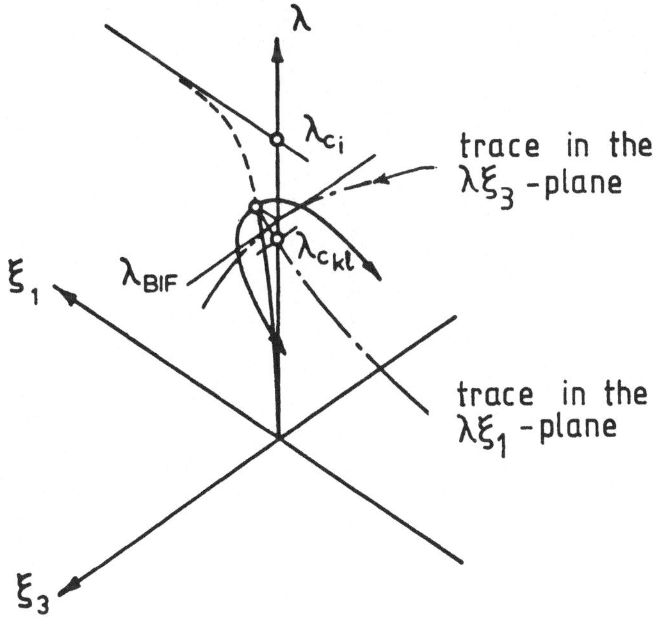

Fig. 18, Bifurcation for a shell with axisymmetric ($\bar{\xi}_1 > 0$) imperfection

perfect shell.

6.3 For a shell with asymmetric imperfection $\bar{\xi}_1 = 0$, $\bar{\xi}_3 \neq 0$

Then Eqs. (53) and (54) reduce to

$$(\lambda_{c_i} - \lambda)\,\xi_1 + D_1\,\xi_1 + D_2\,\xi_3 + D_3\,\xi_1\,\xi_3 + D_4\,\xi_3^{\,2} + D_5\,\xi_1\,\xi_3^{\,2} = 0 \quad (63)$$

$$(\lambda_{c_{k\ell}} - \lambda)\,\xi_3 + D_6\,\xi_3 + D_7\,\xi_1 + D_8\,\xi_1^{\,2} + D_9\,\xi_1\,\xi_3 + D_{10}\,\xi_3^{\,2} + D_{11}\,\xi_1^{\,2}\,\xi_3 + D_{12}\,\xi_3^{\,3} = \lambda\,\bar{\xi}_3 \quad (64)$$

These equations define a limit point, the location of which can be found by plotting the trace of the solution curve in the $\lambda\xi_3$-plane (see Figure 19). Notice that for $\lambda > 0$ one gets immediately nonzero values for both ξ_1 and ξ_3.

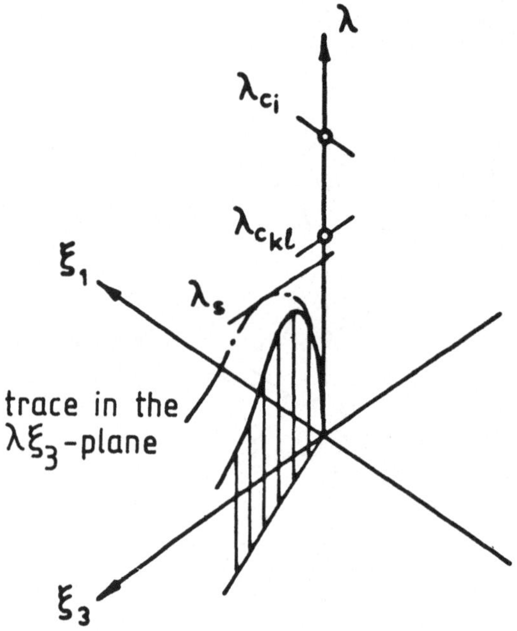

Fig. 19, Location of the limit point for a shell with asymmetric imperfection ($\lambda_s < \lambda_{c_{k\ell}}$ always)

6.4 For a shell with both axisymmetric and asymmetric imperfections $\bar{\xi}_1 \neq 0$, $\bar{\xi}_3 \neq 0$

Then one must solve Eqs, (53) and (54), which define a limit point type behaviour. The location of the limit point is found by plotting the trace of the solution curve in the $\lambda\xi_3$-plane (see Figures 20 and 21). Though Figures 19 and 20 appear nearly identical, it should be remembered that for $\bar{\xi}_1 < 0$ the shell is imperfection sensitive. Thus λ_s in Fig. 20, where both $\bar{\xi}_1$ and $\bar{\xi}_3$ are different from zero, is lower than λ_s in Fig. 19, where only $\bar{\xi}_3$ is different from zero. Further, as can be seen from Fig. 21, for $\bar{\xi}_1 > 0$ it can happen that $\lambda_s > \lambda_{c_{k\ell}}$, that is, the shell appears to be insensitive to the specified imperfections.

The 2-modes solution was first employed by Hutchinson [14] in 1965 and was extended by Arbocz [6] in 1973 to orthotropic shells. It is available as a computer code called TWOMOD [6]. Hutchinson has restricted his imperfections to the form of classic buckling modes. In the present analysis the imperfections are quite general, with the one restriction that $i=2k$. As mentioned earlier, without this restriction all the quadratic terms in ξ_1 and ξ_3 vanish identically and the resulting equations would describe a system with stable postbuckling states which are known to be insensitive to imperfections.

It must be mentioned here that the form of the 2-modes simplified imperfection model (Eq. 47) is dictated by the results of a 1974 paper by Arbocz and Sechler [13], in which the effects of boundary conditions are treated. For further details the reader is referred to Section 8 of this paper.

7. EFFECT OF GENERAL IMPERFECTIONS - MULTIMODE SOLUTION

Up to now the effect of initial geometric imperfections has been studied using simplified imperfection models. However, if one considers the measured initial imperfections of the isotropic shell A-8, the ring stiffened shell AR-1 and the stringer stiffened shell AS-2 shown in Figures 22-24 one must conclude, that it is unlikely that the full effect of the very general shapes can be reliably represented by one or two trigonometric functions.

Thus in this section an analytical solution of the imperfect shell equations is derived that incorporates general imperfection shapes. Initially the nonlinear Donnell type shell equations are reduced to a set of linear partial differential equations by Newton's method of quasilinearization [15]. Next a combination of Fourier expansion and Galerkin's procedure is used to obtain a set of algebraic equations in terms of the Fourier components of the correction terms. Finally, the system of algebraic equations is solved by a standard iterative procedure.

7.1 Newton's Method of Quasilinearization

Let us represent the $(m+1)^{th}$ approximation to a solution of Eqs. (1) and (2) in the following form

$$\begin{aligned} W_{m+1} &= W_m + \delta W_m \\ F_{m+1} &= F_m + \delta F_m \end{aligned} \tag{65}$$

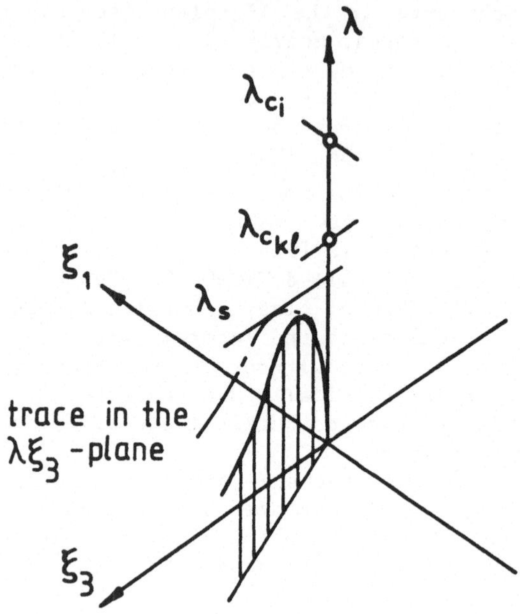

Fig. 20, Location of the limit point for a shell with both axisymmetric ($\bar{\xi}_1 < 0$) and asymmetric ($\bar{\xi}_3 > 0$) imperfections ($\lambda_s < \lambda_{c_{k\ell}}$ always)

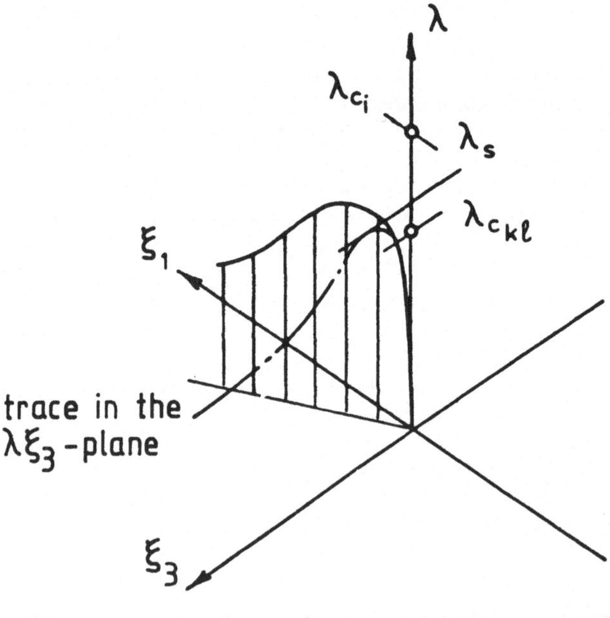

Fig. 21, Location of the limit point for a shell with both axisymmetric ($\bar{\xi}_1 > 0$) and asymmetric ($\bar{\xi}_3 > 0$) imperfections ($\lambda_s > \lambda_{c_{k\ell}}$)

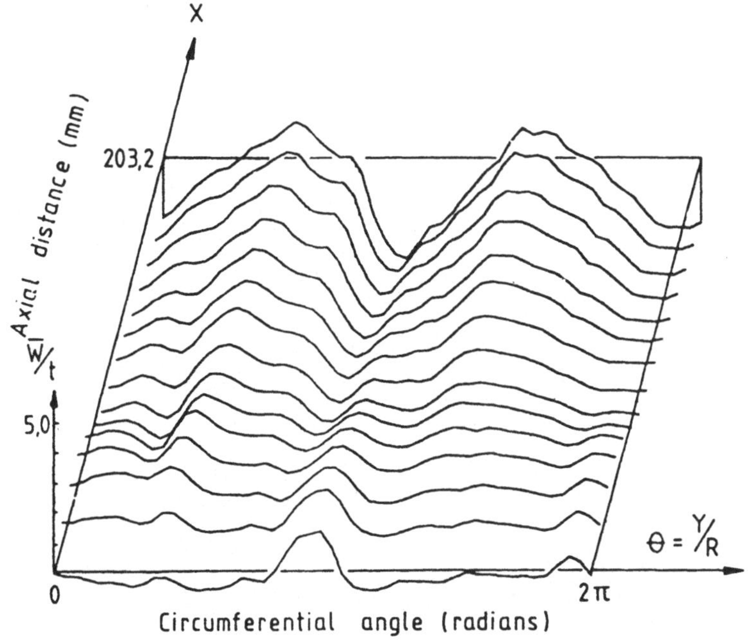

Fig. 22, Measured initial imperfections of the isotropic shell A-8

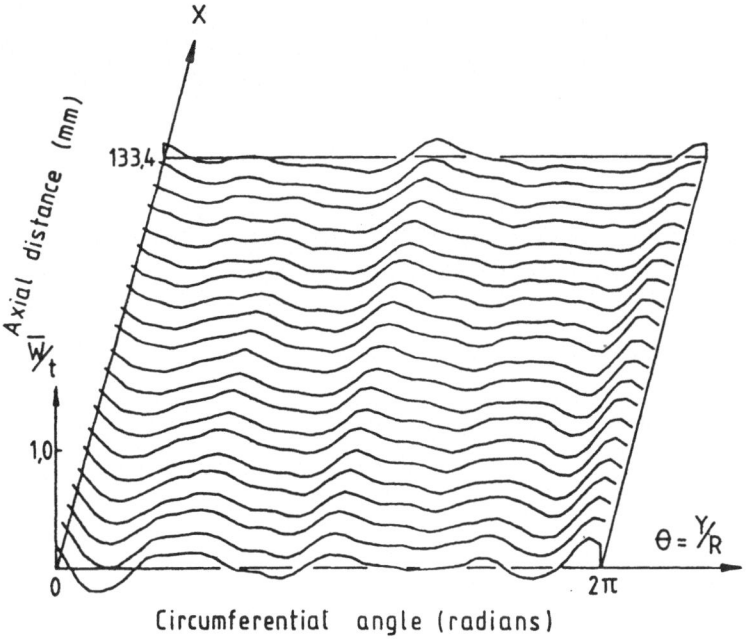

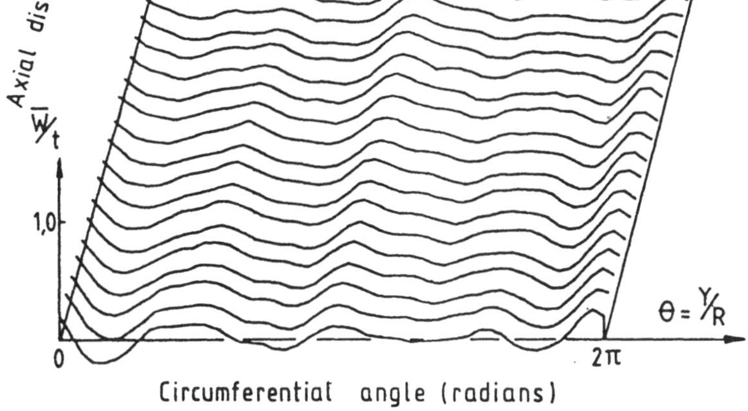

Fig. 23, Measured initial imperfections of the ring stiffened shell AR-1

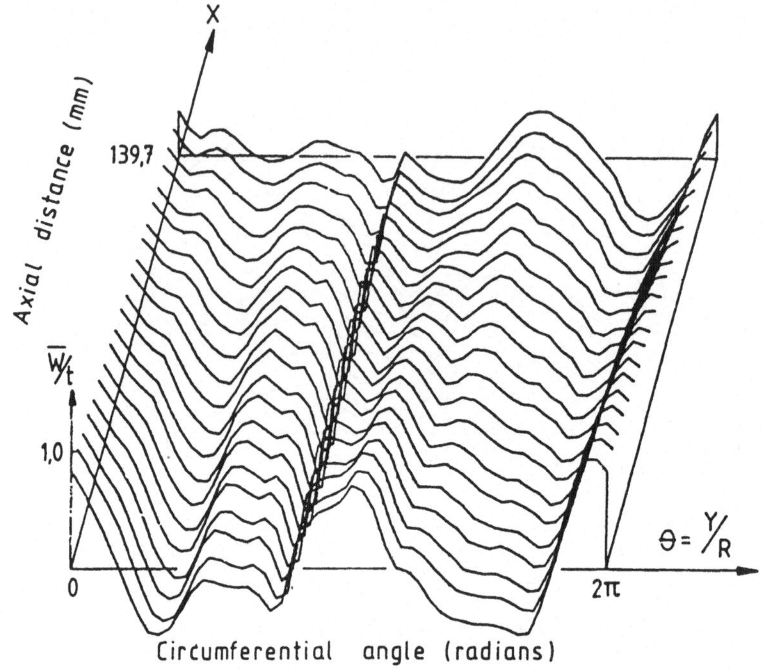

Fig. 24, Measured initial imperfections of the stringer stiffened shell AS-2

where W_m, F_m = m^{th} approximation to the solution

δW_m, δF_m = correction to the mth approximation

Substituting into Eqs. (1) and (2) and neglecting products of the correction quantities yields the following set of linearized equations for determining the correction terms

$$L_H(\delta F_m) - L_Q(\delta W_m) + \delta W_{m,xx}/R + L_{NL}\,(W_m + \bar{W},\ \delta W_m) = -\,E_m^{(1)} \tag{66}$$

$$L_Q(\delta F_m) + L_D(\delta W_m) - \delta F_{m,xx}/R - L_{NL}\,(F_m,\ \delta W_m) + \\ - L_{NL}(\delta F_m,\ W_m + \bar{W}) = -\,E_m^{(2)} \tag{67}$$

where:

$$E_m^{(1)} = L_H(F_m) - L_Q(W_m) + W_{m,xx}/R + \frac{1}{2}\,L_{NL}(W_m,\ W_m + 2\bar{W})$$

$$E_m^{(2)} = L_Q(F_m) + L_D(W_m) - F_{m,xx}/R - L_{NL}(F_m,\ W_m + \bar{W})$$

7.2 Reduction to a Set of Algebraic Equations

If we represent the initial imperfections by

$$\bar{W} = t \sum \bar{W}_{io} \cos \frac{i\pi x}{L} + t \sum\sum \sin \frac{k\pi x}{L} \left(\bar{W}_{k\ell} \cos \frac{\ell y}{R} + \bar{W}'_{k\ell} \sin \frac{\ell y}{R}\right) \tag{68}$$

then Eqs. (66) and (67) admit separable solutions of the form

$$\begin{Bmatrix} W_m \\ \delta W_m \end{Bmatrix} = t \begin{Bmatrix} W_\nu \\ 0 \end{Bmatrix} + t \sum \begin{Bmatrix} W_{io} \\ \delta W_{io} \end{Bmatrix} \cos \frac{i\pi x}{L} + t \sum\sum \sin \frac{k\pi x}{L} \left(\begin{Bmatrix} W_{k\ell} \\ \delta W_{k\ell} \end{Bmatrix} \cos \frac{\ell y}{R} + \begin{Bmatrix} W'_{k\ell} \\ \delta W'_{k\ell} \end{Bmatrix} \sin \frac{\ell y}{R} \right) \tag{69}$$

$$\begin{Bmatrix} F_m \\ \delta F_m \end{Bmatrix} = \frac{ERt^2}{c} \begin{Bmatrix} -\lambda \frac{y^2}{2R^2} \\ 0 \end{Bmatrix} + \frac{ERt^2}{c} \sum \begin{Bmatrix} F_{io} \\ \delta F_{io} \end{Bmatrix} \cos \frac{i\pi x}{L} + \frac{ERt^2}{c} \sum\sum \sin \frac{k\pi x}{L} \left(\begin{Bmatrix} F_{k\ell} \\ \delta F_{k\ell} \end{Bmatrix} \cos \frac{\ell y}{R} + \begin{Bmatrix} F'_{k\ell} \\ \delta F'_{k\ell} \end{Bmatrix} \sin \frac{\ell y}{R} \right) \tag{70}$$

where

$$W_\nu = -\frac{\nu}{c} \frac{\bar{H}_{xx}}{1+\mu_1} \lambda$$

The unknown coefficients in these solutions are determined by Galerkin's procedure yielding a set linear algebraic equations in terms of the unknown correction terms, δW and δF.
In matrix notation

$$\begin{aligned} [\,A\,]\{\delta F\} + [\,B\,]\{\delta W\} &= -\{E^{(1)}\} \\ [\,C\,]\{\delta F\} + [\,D\,]\{\delta W\} &= -\{E^{(2)}\} \end{aligned} \tag{71}$$

For details the reader should consult Reference [16].
To obtain the buckling load for a given imperfect cylindrical shell one begins by making an initial guess for $\{W\}$ and $\{F\}$ at a small initial load level λ. Iteration is then carried out until the correction vectors are smaller than some preselected value. The converged solutions then are used as the initial guess at the next higher axial load level $\lambda + \Delta\lambda$.

The entire process is repeated for increasing values of the axial load parameter λ. The nonlinear analysis then will locate the limit point of the prebuckling state. By definition the value of the loading parameter λ_s corresponding to the limit point will be the theoretical buckling load (see Fig. 25).

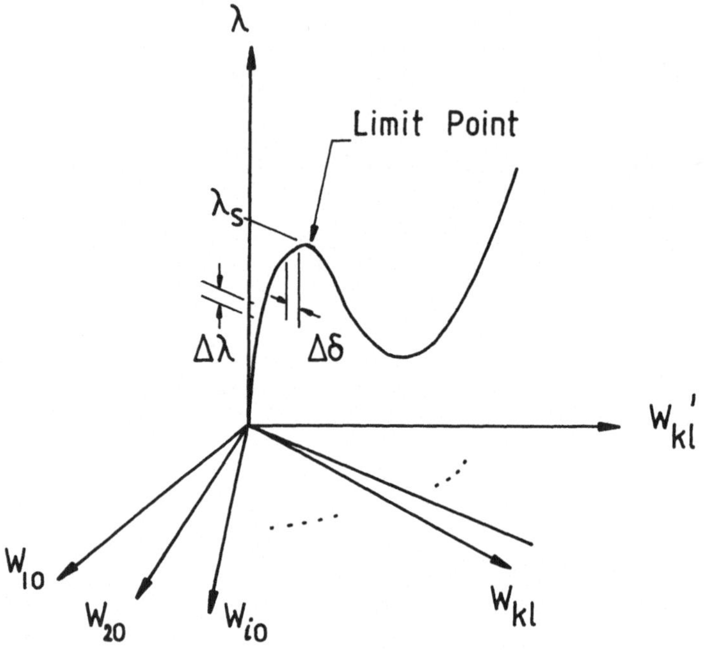

Fig. 25, Location of the limit point for a shell with general imperfections

7.3 Solution using Increments in End-shortening

Using load increments $\Delta\lambda$ the solution fails to converge beyond the limit point. A closer look at the solution curve presented in Fig. 25 reveals that one should be able to extend the response curve beyond the limit point by using increments in deformation instead of increments in loading. However, if in place of using increments in loading $\Delta\lambda$ one uses increments in (say) "end shortening" $\Delta\delta$, then one must increase the order of the equations by one (the axial load λ is now an unknown).

Following Ref. [17], let us define "unit end shortening" as

$$\varepsilon = -\frac{1}{2\pi RL}\int_0^{2\pi R}\int_0^{L} u_{,x}\, dxdy \tag{72}$$

where

$$u_{,x} = \varepsilon_x - \frac{1}{2} W_{,x}(W_{,x} + 2\bar{W}_{,x})$$

$$\varepsilon_x = \frac{1}{\alpha_2}\frac{(1-\nu^2)}{Et}(F,_{yy} - \frac{\nu}{1+\mu_2}F,_{xx}) + \frac{\chi_1}{\alpha_2}W,_{xx} - \frac{\nu}{\alpha_2}\frac{\chi_2}{1+\mu_2}W,_{yy}$$

and the notation used is defined in Appendix A.
Introducing these expressions into Eq. (72) and substituting for W, $\bar{W}$ and F from above yields, after carrying out the integrals involved

$$\delta = \frac{\varepsilon}{\varepsilon_{c\ell}} = \bar{H}_{yy}(\lambda+\Delta\lambda) + \delta_{NL} \tag{73}$$

where

$$\begin{aligned}\delta_{NL} &= c^2 \sum\alpha_i^2 (W_{io} + \bar{W}_{io})\, \delta W_{io} + \frac{c^2}{2}\sum\alpha_i^2 (W_{io} + 2\bar{W}_{io})\, W_{io} \\ &+ \frac{c^2}{2}\sum\sum\alpha_k^2 \left\{(W_{k\ell}+\bar{W}_{k\ell})\, \delta W_{k\ell} + (W'_{k\ell}+\bar{W}'_{k\ell})\, \delta W'_{k\ell}\right\} \\ &+ \frac{c^2}{4}\sum\sum\alpha_k^2 \left\{(W_{k\ell}+2\bar{W}_{k\ell})\, W_{k\ell} + (W'_{k\ell}+2\bar{W}'_{k\ell})\, W'_{k\ell}\right\}\end{aligned} \tag{74}$$

and

$$\varepsilon_{c\ell} = \frac{1}{c}\frac{t}{R}$$

This represents the one additional equation needed to increase the order of Eqs. (71) by one and be able to use increments in end-shortening $\Delta\delta$ in place of load increments $\Delta\lambda$. To obtain the buckling load for a given imperfect cylindrical shell, one begins by running first with load increments $\Delta\lambda$. If the solution fails to converge one switches to increments in end-shortening, $\Delta\delta_{NL}$. Now the initial guess for $\{W\}$ and $\{F\}$ are taken from the previous run, at one load level before that λ where the solution failed to converge. Iterations are then carried out using increments in end-shortening until the correction vectors are smaller than some preselected value. The converged solutions then are used as the initial guess at the next level of end-shortening $\delta_{NL} + \Delta\delta_{NL}$. The entire process is repeated for increasing values of δ_{NL} till the computed values of λ reach a maximum. The buckling load λ_s is taken as the maximum load level achieved in this run. Figures (26) and (27) show the behaviour of δ_{TOTAL} and δ_{NL} versus λ. Notice the relatively slowly changing curve of δ_{NL} vs. λ against the "sharp" corner observed in the curve of δ_{TOTAL} vs. λ.

7.4 Mode selection

The Multimode Method is available as a computer code called MIUTAM [18]. The number of modes of deformation, that can be included in the analysis, is thus limited by practical considerations, like the available core size and the time required for obtaining the solution. Since the shell buckling load is determined by obtaining an approximate solution of the governing equations using a particular set of modes, an attempt of

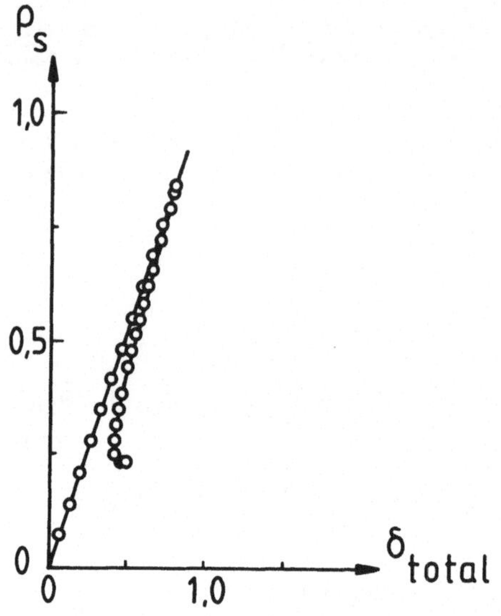

Fig. 26, Total end-shortening vs axial load (shell AS-2)

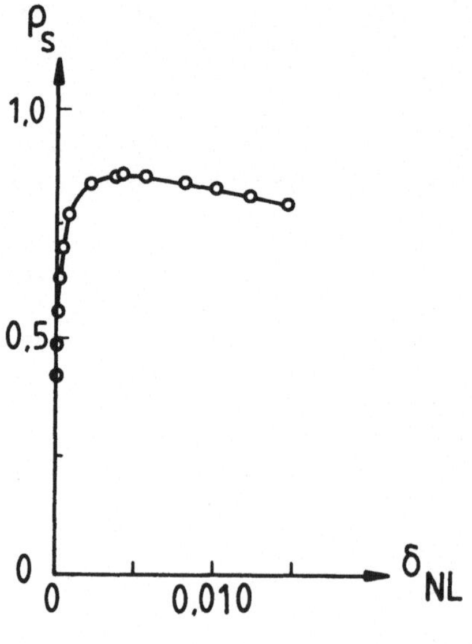

Fig. 27, Nonlinear part of end-shortening vs axial load

optimizing the selection of these modes must be made. That is, one must locate those modes which dominate the prebuckling and collapse behaviour of the shell in question.

Examples of attempts to locate "critical modes", defined as that combination of axisymmetric and asymmetric modes which yield the lowest buckling load, have been reported by Arbocz and Babcock[6,16,19], Imbert[20] and Singer et al[21,22]. These correlations studies have shown that in order to obtain a significant decrease from the buckling load of the perfect shell, at least one of the mode shapes used must have an eigenvalue close to the critical (lowest) eigenvalue of the perfect shell and a sizeable initial imperfection amplitude. Furthermore, it has been found that there are two general nonlinear coupling conditions which can lower the limit loads considerably, namely

1. coupling between one asymmetric mode (k, ℓ) and one axisymmetric mode $(i, 0)$; satisfied for any ℓ if $i = 2k$.
2. Coupling between three asymmetric modes (k_1, ℓ_1), (k_2, ℓ_2) and (k_3, ℓ_3); satisfied if $k_1 \pm k_2 \pm k_3$ = odd integer and $|\ell_1 \pm \ell_2| = \ell_3$

As pointed out by Koiter[5], for isotropic shells with a multiple lowest eigenvalue there is an additional strong coupling condition between two asymmetric modes (k_1, ℓ), (k_2, ℓ) and one axisymmetric mode $(i_{c\ell}, 0)$, if the wavenumbers are so chosen that the corresponding eigenvalues are all identical to $\lambda_c = 1.0$. This implies that for a given circumferential wave number ℓ the axial half-wave numbers k_1 and k_2 must be roots of Eq. (22)

$$\alpha_k^2 + \beta_\ell^2 - \alpha_k = 0$$

and that

$$i_{c\ell} = \frac{L}{\pi} \sqrt{\frac{2c}{Rt}}$$

These combinations of modes are called "Basic Triplets".

If the above coupling conditions are satisfied, then the resulting buckling load of the shell is generally lower than buckling load which one obtains with each of the modes considered separately. To account for the Poisson's effect like "breathing mode" of axially compressed cylinders some axisymmetric degrees of freedom coupled to the chosen asymmetric modes must always be included, otherwise the resulting mathematical model will be too stiff. Also since experimental evidence indicates that the prebuckling growth is dominated by long-wavelength asymmetric modes, one must always include at least that asymmetric mode with a single half-wave in the axial direction which has the lowest eigenvalue.

7.5 Correlation Studies

The results of the correlation studies reported in Reference[18] are summarized in Tables 2-4. In these tables the notation (2,0) denotes an axisymmetric mode with two half waves in the axial direction, whereas (1,9) stands for an asymmetric mode with a single half wave in the axial direction and nine full waves in the circumferential direction. The initial imperfections used are described in Appendix D.

As one sees in Table 2, for the isotropic shell A-8 the 15-modes imperfection model consists essentially of the full coupling of the two "Basic Triplets" (1,9) + (33,9) + (34,0) and (2,13) + (32,13) + (34,0). It can clearly be seen from Fig. 5 that the modes of the "Basic Triplets" all have eigenvalues equal to $\lambda_c = 1.0$. Notice also, that whereas the buckling load of the 4-modes imperfection model (one "Basic Triplet") is $\lambda_s = 0.846$, that of the 15-modes imperfection model is $\lambda_s = 0.693$, only slightly higher than the experimental value of $\lambda_{EXP} = 0.660$.

Table 2: Buckling loads calculated by the multimode analysis (shell A-8)

2-modes	ρ_s	15-modes	ρ_s
(2,0) + (1,9)	= 0.901	(1,11) (1,7) (33,11) (33,7)	
		(1,2) (33,2)	
4-modes		(2,0) + (1,9) + (33,9) + (34,0)	
(2,0)+(1,9)+(33,9)+(34,0)	= 0.846	(2,4) (32,4)	
		(4,0) + (2,13) + (32,13)	= 0.693

In table 3 the results of the buckling load calculations for the ring stiffened shell AR-1 are shown. The selection of the modes follows the same pattern as for the isotropic shell A-8. However, in this case the inclusion of additional modes results in only relatively small decreases in the predicted buckling loads. Considering the distribution of the eigenvalues for the perfect shell AR-1 shown in Fig. 6, one sees that all those modes, the associated eigenvalues of which are close to the lowest eigenvalue, have many half waves in the axial direction. However, as reported in Reference [23], the amplitudes of the corresponding initial imperfection Fourier coefficients are negligibly small. Conversely, the eigenvalues of the modes with sizeable initial amplitudes (modes with few half waves in the axial direction) are considerably higher than the lowest eigenvalue. Thus their contribution to the lowering of the calculated buckling loads is insignificant. Still the lowest calculated buckling load $\rho_s = \lambda_s/\lambda_{c_{17,0}} = 0.932$ compares reasonably well with the experimental buckling load of $\rho_{EXP} = 0.81$. It must be mentioned here that the calculated buckling load ρ_s has been normalized by $\lambda_{c_{17,0}} = 1.10$, whereas the experimental buckling load $\rho_{EXP} = N_{x_{EXP}}/N_{x_{C-4}}$ has been normalized by $N_{x_{C-4}} = -250.605$ N/cm. For a detailed discussion of the normalization procedure see Section 8 where the effect of boundary conditions is discussed.

The buckling load calculations for the imperfect stringer stiffened shell AS-2 are summarized in Table 4. Comparing the results of the 2-modes solution with that of the 4-modes solution it is evident that the inclusion of the additional short wavelength modes has only an insignificant effect. The reason for this becomes evident if one considers

Table 3: Buckling loads calculated by the multimode analysis (shell AR-1)

```
2-modes                    ρs      20-modes                              ρs

(16,0) + (8,16)       = 0.995          (1,13)    (15,13)
                                          |         |
                                        (1,3)     (15,3)
                                          |         |
4-modes                                 (1,10)    (15,10)
                                          |         |
(2,0)+(1,8)+(15,8)+(16,0)               (1,2)     (15,2)
                      = 0.985             |         |
                                 (2,0) + (1,8)  + (15,8) + (16,0) = 0.932
                                          |         |
                                        (1,15)    (15,15)
                                        /  |         |   \
                                 (1,9)  (1,7)     (15,7)   (15,9)
                                   |                          |
                                 (1,6)                     (15,6)
```

Table 4: Buckling loads calculated by the multimode analysis (shell AS-2)

```
2-modes                    ρs      14-modes                         ρs

(2,0) + (1,10)        = 0.904         (1,19)    (9,19)
                                         |         |
                                       (1,9)     (9,9)
                                         |         |
4-modes                         (2,0) + (1,10) + (9,10) + (10,0)
                                         |         |
(2,0)+(1,10)+(9,10)+(10,0)             (1,11)    (9,11)
                      = 0.903            |         |
                                       (1,21)    (9,21)
                                         |         |
                                       (1,2)     (9,2)          = 0.824
```

the distribution of eigenvalues for the perfect shell AS-2 displayed in Fig. 7. Only the eigenvalues of two asymmetric modes with long wavelength in the axial direction are close to the lowest eigenvalue, which in this case is the asymmetric mode (1,10). On the other hand, coupling of the three lowest asymmetric modes results in a significant decrease in the predicted buckling load. The lowest calculated buckling load $\rho_s = \lambda_s/\lambda_{c_{1,10}} = 0.824$ compares reasonably well with the experimental buckling load of $\rho_{EXP} = 0.715$. Also here different normalization factors have been used. Thus the calculated buckling load ρ_s has been normalized by

$\lambda_{c_{1,10}} = 1.44$, whereas the experimental buckling load $\rho_{EXP} = N_{x_{EXP}}/N_{x_{C-4}}$ has been normalized by $N_{x_{C-4}} = -320.831$ N/cm.

The MIUTAM program has been utilized extensively at the TECHNION [21,22] to carry out correlation studies on axially compressed stiffened shells, where not only the initial imperfections are measured but also Singer's vibration correlation technique[24,25,26] is employed to establish the appropriate (elastic) boundary conditions used during the experiments. As can be seen from Table 5[22], with a few exceptions the agreement between the predictions based on the measured initial imperfections and the experimental results is good. As pointed out by Singer[22], one deficiency of the imperfection measurement system used for the integrally stiffened shells at Technion is, that imperfections are measured before the shells are fixed in the testing machine in their final boundary conditions. To overcome this difficulty a new multipurpose scanning and measurement system has been built. In this system both the imperfection and the vibration measurements are carried out by the same closed-loop noncontact probe on the inside of the test specimen, once it has been fixed in its final position. It is expected that this new procedure will provide for an even better agreement between theoretical predictions and experimental results.

Table 5: Comparison between buckling loads predicted from measured initial imperfections and experimental results[22]

Shell	Predicted, ρ_{th}	Experimental, ρ_{EXP}	$\Delta\rho$
AB-5*	0.87	0.76	0.11
AB-6	0.72	0.75	-0.03
KR-1	0.84	0.68	0.16
SN-1	0.95	0.91	0.04
SN-4	0.88	0.82	0.06
SN-6*	0.73	0.63	0.10
DUD-2	0.86	0.63	0.23
DUD-3	0.70	0.58	0.12
DUD-4*	0.88	0.80	0.08
RS-36	0.91	0.76	0.15
DK-1*	0.89	0.83	0.06
DK-3*	0.78	0.71	0.07
DK-4*	0.89	0.73	0.16
DK-5*	0.80	0.84	-0.04

* significant load eccentricity

8. EFFECT OF BOUNDARY CONDITIONS

It has been shown by several investigators both analytically[24,27] and experimentally[28] that, in addition to initial geometric imperfections, the load-carrying capacity of stiffened shell structures can be influenced greatly by a variety of support conditions, by the nonlinear prebuckling deformations caused by the edge constraints and by the location of the load application points[29].

The simplified methods, used up to now to study the effects of initial imperfections, do approximately satisfy simply supported boundary conditions. In addition, it is assumed that the prebuckling behaviour can be adequately represented by a response pattern that is affine to the initial imperfection model used. To investigate the validity of this approach Arbocz and Sechler[13] in 1974 presented an analysis, where the axial dependence of the response functions is not specified à priori and where the experimental boundary conditions are rigorously enforced.

8.1 Reduction to an equivalent set of ordinary differential equations

In this EXTENDED 2-modes solution the initial imperfection is represented by

$$\bar{W} = t\, A_0(x) + t\, A_1(x) \cos \frac{ny}{R} \tag{75}$$

where $A_0(x)$ and $A_1(x)$ are given functions of x.

Any equilibrium state of the axially loaded cylinders is then represented by

$$W = tW_\nu + tw_0(x) + tw_1(x) \cos \frac{ny}{R} \tag{76}$$

$$F = \frac{ERt^2}{c} \left\{ -\frac{1}{2}\frac{y^2}{R^2}\lambda + f_0(x) + f_1(x)\cos\frac{ny}{R} + f_2(x)\cos\frac{2ny}{R} \right\} \tag{77}$$

where

$$W_\nu = -\frac{\nu}{c}\frac{\bar{H}_{xx}}{1+\mu_1}\lambda$$

By assuming the axial dependence of the response to contain unknown functions of x one can reduce the buckling problem to the solution of a set of nonlinear ordinary differential equations, which then allows the rigorous enforcing of boundary conditions.

Substituting the expressions for $\bar{W}$, W and F into the compatibility Eq. (1), using some trigonometric identities and finally equating coefficients of like terms results in a system of three nonlinear ordinary differential equations

$$\bar{H}_{xx} f_0^{iv} - \frac{1}{2}\frac{t}{R}\bar{Q}_{xx} w_0^{iv} + c\, w_0'' - \frac{c}{4}\frac{t}{R} n^2 \left[w_1''(w_1+2A_1) + 2w_1'(w_1'+2A_1') + w_1(w_1''+2A_1'') \right] = 0 \tag{78}$$

$$\bar{H}_{xx} f_1^{iv} - n^2\bar{H}_{xy}f_1'' + n^4\bar{H}_{yy}f_1 - \frac{1}{2}\frac{t}{R}(\bar{Q}_{xx}w_1^{iv} - n^2\bar{Q}_{xy}w_1'' + n^4\bar{Q}_{yy}w_1)$$

$$+ cw_1'' - \frac{c}{2}\frac{t}{R} n^2 \left[w_0''(w_1+2A_1) + w_1(w_0''+2A_0'')\right] = 0 \qquad (79)$$

$$\bar{H}_{xx}f_2^{iv} - 4n^2\bar{H}xyf_2'' + 16n^4\bar{H}_{yy}f_2$$

$$- \frac{c}{4}\frac{t}{R} n^2 \left[w_1''(w_1+2A_1) - 2w_1'(w_1'+2A_1') + w_1(w_1''+2A_1'')\right] = 0 \qquad (80)$$

where

$$(\ )' = \frac{d}{dx/R}$$

Substituting, in turn, the expressions assumed for $\bar{W}$, W and F into the equilibrium Eq. (2) and applying Galerkin's procedure yields the following system of two nonlinear ordinary differential equations

$$2\bar{Q}_{xx} f_0^{iv} + \frac{t}{R}\bar{D}_{xx} w_0^{iv} - 4c\frac{R}{t} f_0'' + 4c\lambda (w_0''+A_0'')$$

$$+ 2cn^2 \left[(w_1+A_1)f_1'' + 2(w_1'+f_1')f_1' + (w_1''+A_1'')f_1\right] = 0 \qquad (81)$$

$$2(\bar{Q}_{xx}f_1^{iv} - n^2\bar{Q}_{xy}f_1'' + n^4\bar{Q}_{yy}f_1) + \frac{t}{R}(\bar{D}_{xx}w_1^{iv} - n^2\bar{D}_{xy}w_1'' + n^4\bar{D}yyw_1)$$

$$- 4c\frac{R}{t} f_1'' + 4c\lambda (w_1''+A_1'') + 2cn^2\left[2(w_1+A_1)f_0''\right. \qquad (82)$$

$$\left. + 2(w_0''+A_0'')f_1 + (w_1+A_1)f_2'' + 4(w_1'+A_1')f_2' + 4(w_1''+A_1'')f_2\right]=0$$

Equation (78) can be integrated twice to yield

$$\bar{H}_{xx}f_0'' - \frac{1}{2}\frac{t}{R}\bar{Q}_{xx}w_0'' + cw_0 - \frac{c}{4}\frac{t}{R} n^2(w_1+2A_1)w_1 = 0 \qquad (83)$$

where the constants of integration are set equal to zero in order to satisfy periodicity[12].

With the help of equations (78) and (83) one can eliminate the terms f_0^{iv} and f_0'' from equations (81) and (82). Further, in order to be able to use the "shooting method"[30] it is necessary to eliminate the w_1^{iv} term from Eq. (79) and the f_1^{iv} term from Eq. (82). This can be accomplished with the help of Eqs. (82) and (79), repectively. Carrying out the indicated substitutions one can write after some further regrouping the following four nonlinear ordinary differential equations

$$f_1^{iv} = \Psi_1(f_1, f_1', f_1'', f_2, f_2', f_2'', w_0, w_0'', w_1, w_1', w_1'', \lambda)$$

$$f_2^{iv} = \Psi_2\ (f_2, f_2'', w_1, w_1', w_1'') \tag{84}$$

$$w_0^{iv} = \Psi_3\ (f_1, f_1', f_1'', w_0, w_0'', w_1, w_1', w_1'', \lambda)$$

$$w_1^{iv} = \Psi_4\ (f_1, f_1'', f_2, f_2', f_2'', w_0, w_0'', w_1, w_1', w_1'', \lambda)$$

The nonlinear functions $\Psi_1, \ldots, \Psi_4$ are listed in Reference [31]. The (say) C-3 boundary conditions ($W = W_{,x} = v = 0$, $N_x = -\sigma t$) expressed in terms of the assumed unknown functions become at $x = 0, L$

$$w_0 = -W_\nu$$

$$w_1 = w_0' = w_1' = f_1 = f_2 = f_1'' = f_2'' = 0 \tag{85}$$

A list of the reduced boundary conditions and details of the derivation can be found in Reference [12].

Introducing, as a unified variable, the 16-dimensional vector $\underset{\sim}{Y}$ defined as follows

$$Y_1 = f_1,\ Y_2 = f_2,\ Y_3 = w_0,\ Y_4 = w_1,\ Y_5 = f_1',\ \ldots,\ Y_{16} = w_1''' \tag{86}$$

then the system of equations (84) and the C-3 boundary conditions (Eq. 85) can be reduced to the following 2-point nonlinear boundary value problem

$$\frac{d}{d\bar{x}} \underset{\sim}{Y} = \underset{\sim}{f}\ (\bar{x}, \underset{\sim}{Y}; \lambda) \qquad \text{for } 0 \leq \bar{x} \leq \frac{L}{R} \tag{87}$$

$$Y_3 = -W_\nu \qquad \text{at } \bar{x} = 0, \frac{L}{R} \tag{88}$$

$$Y_1 = Y_2 = Y_4 = Y_7 = Y_8 = Y_9 = Y_{10} = 0$$

where

$$\bar{x} = \frac{x}{R}\ ; \ W_\nu = -\frac{\nu}{c} \frac{\bar{H}_{xx}}{1+\mu_1} \lambda \tag{89}$$

The solution of this highly nonlinear 2-point boundary value problem is obtained numerically by "parallel shooting over 8-intervals". For details of the numerical analysis the reader is referred to References [12], [13] and [31].

The limit point of the prebuckling states can then be obtained by plotting the maximum amplitude of w_1 (the asymmetric component of the prebuckling deformation) versus the axial load level λ. The axial load λ_s at the limit point is by definition the collapse load.

As mentioned earlier, using load increments the solution fails to converge close to and beyond the limit point. However, one is able to extend the response curve beyond the limit point by switching to increments in end-shortening[31].

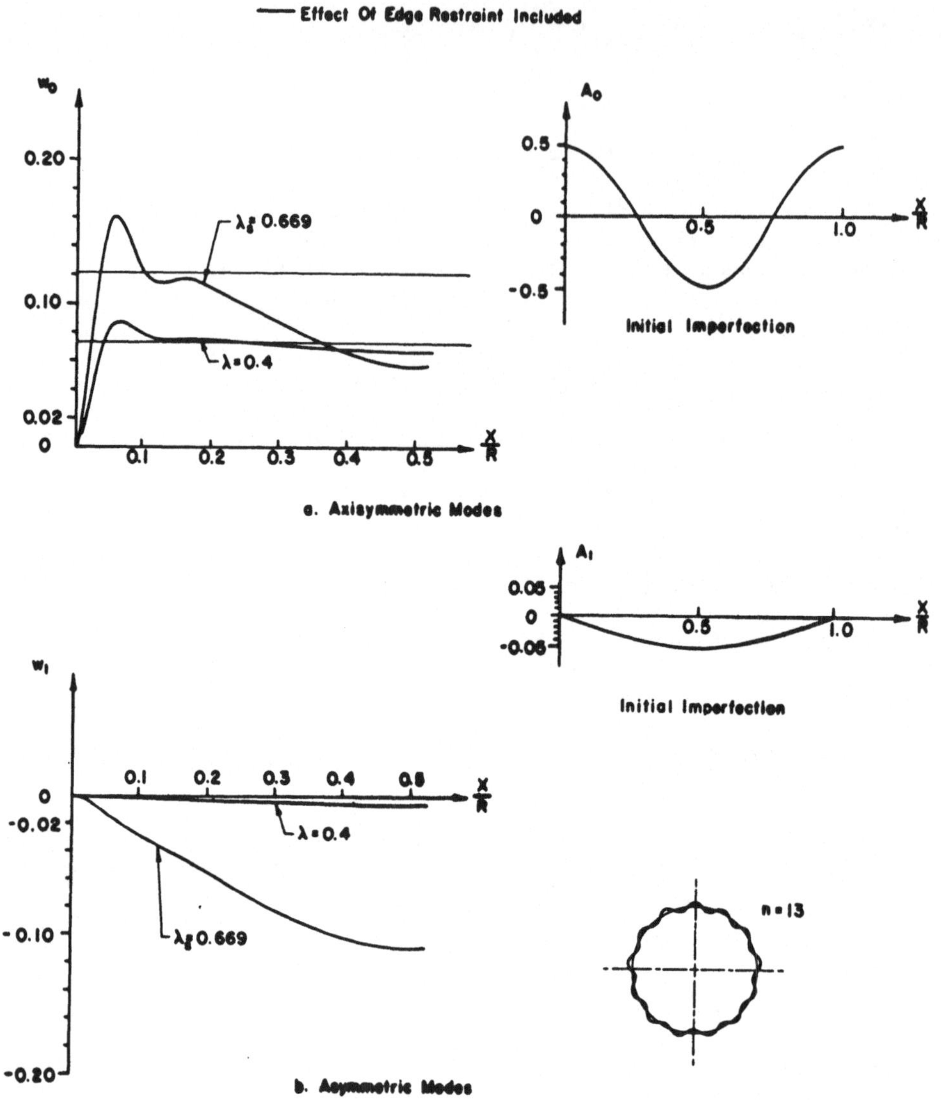

Fig. 28, Calculated radial displacements for C-3 boundary condition (isotropic shell, $\frac{L}{R} = 1.0$, $\frac{R}{t} = 1000$)

8.2 Effect of boundary conditions for isotropic shells

Using an isotropic shell with $\frac{L}{R} = 1.0$, $\frac{R}{t} = 1000$, C-3 boundary conditions and the following 2-modes initial imperfection model

$$\frac{\bar{W}}{t} = -0.5 \cos \frac{2\pi x}{L} + 0.05 \sin \frac{\pi x}{L} \cos 13 \frac{y}{R} \tag{90}$$

Arbocz and Sechler[13] obtained the results shown in Fig. 28. Because both modes are symmetric with respect to $\bar{x} = \frac{1}{2}\frac{L}{R}$ only half of the response curves is plotted. Notice that in this case the axisymmetric collapse mode at $\lambda_s = 0.669$ shown in Fig. 28a resembles very closely the assumed full wave cosine initial imperfection mode toward the center of the shell. The effects of the clamped C-3 boundary conditions and of the prebuckling deformations due to the edge constraint are restricted to a relatively narrow region next to the edge of the shell. The resemblance between the collapse mode and the assumed initial imperfection is even more striking for the asymmetric component shown in Fig. 28b. Here the effect of the clamped boundary condition is restricted to a very narrow region next to the edge of the shell and the collapse mode very closely resembles a half-wave sine mode.

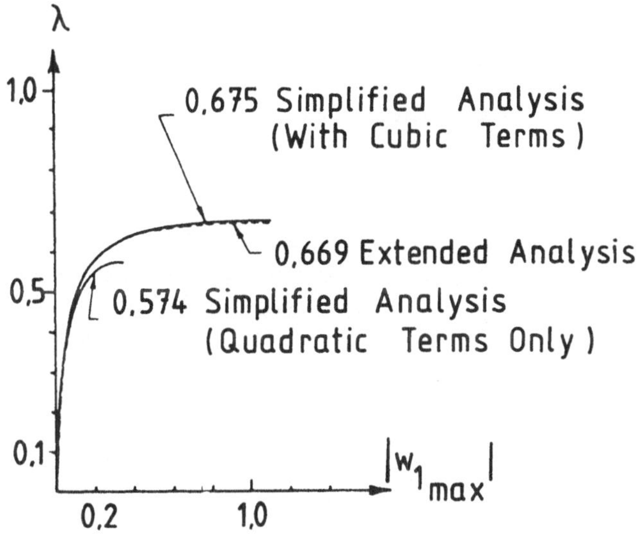

Fig. 29, Comparison between simplified and extended 2-modes solutions (isotropic shell, $\frac{L}{R} = 1.0$, $\frac{R}{t} = 1000$)

A comparison of the present results with those of the simplified 2-modes solution[19] reveal a very close agreement (see Fig. 29), if in the simplified analysis the half-wave cosine representation is used for the axisymmetric component and the half-wave sine representation for the asymmetric component. The comparison also indicates that the cubic terms in the simplified 2-modes solution cannot be neglected.

It is a well known experimental fact[32], that shells are more sensitive to inward than to outward axisymmetric imperfections. By inverting the sign of the axisymmetric imperfection in Eq. (90) Arbocz

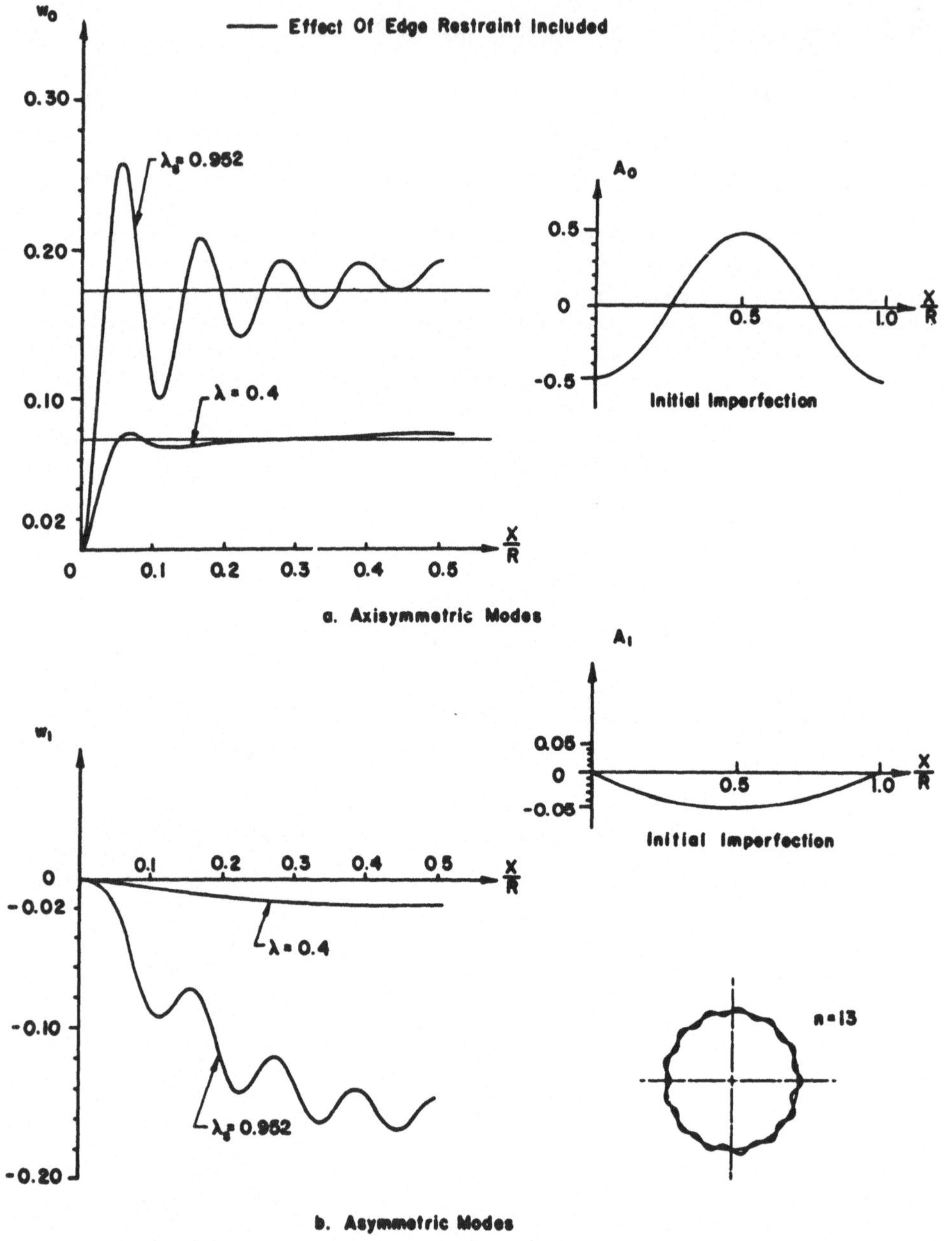

Fig. 30, Calculated radial displacement for C-3 boundary condition (isotropic shell, $\frac{L}{R} = 1.0$, $\frac{R}{t} = 1000$)

and Sechler[13] obtained the results shown in Fig. 30. Since now the full wave axisymmetric imperfection component is pointing outward at the center of the shell (at $\bar{x} = \frac{1}{2}\frac{L}{R}$), as expected, the limit point occurs at a much higher axial load level ($\lambda_s = 0.952$) than for the inward axisymmetric imperfection ($\lambda_s = 0.669$). The axisymmetric component of the collapse mode shown in Fig. 30a is dominated by the classic axisymmetric buckling mode consisting of 18 half waves. On the other hand, the asymmetric component of the collapse mode shown in Fig. 30b is clearly composed of 2 modes, the initial half-wave sine mode excited by the given initial imperfection and a short wavelength mode consisting of 16-17 half-waves in the axial direction. This second mode, which lies close to the Koiter circle (for $\ell = 13$ the two roots of Eq.(22)are $k_1 \simeq 1$ and $k_2 \simeq 17$), is excited because the limit load in this case is nearly equal to $\lambda_c = 1.0$

The results presented in Figures 28-30 confirm the practice of neglecting the effect of boundary conditions for not too short ($\frac{L}{R} > 1.0$, say) imperfect isotropic shells, if in the simplified analysis the cubic terms are not neglected in the resulting nonlinear algebraic equations (see Eqs. 51-52). Further, it was found that the usual assumption that the response pattern is affine to the initial imperfections (compare Eqs. 47 and 48) will only work satisfactorily if the axisymmetric initial imperfection points inward at the center (at $x = \frac{L}{2}$) of the shell.

8.3 Effect of boundary conditions for stringer stiffened shells

Singer and his coworkers have shown[24,25] that the different boundary conditions have a pronounced effect on the buckling load of stringer stiffened shells. To investigate these effects in the presence of initial imperfections Arbocz and Sechler[31] used the following 2-modes initial imperfection model

$$\frac{\bar{W}}{t} = - 0.01 \cos \frac{2\pi x}{L} + 0.50 \sin \frac{\pi x}{L} \cos n \frac{y}{R} \tag{91}$$

and obtained the results shown in Table 6 for the stringer stiffened shell AS-2. The numbers in parentheses are the number of circumferential waves n.

To be able to properly assess the effect of the initial imperfections, it is necessary first to calculate the buckling loads of the corresponding perfect shells. Here the computer code SRA[10] was used to calculate the axial load level at which bifurcation from a membrane or nonlinear prebuckling state into an asymmetric pattern with n waves in the circumferential direction will occur. Since the plot of bifurcation load vs. circumferential wave numbers can contain several local minima, it requires some engineering insight to find the lowest buckling load.

As can be seen from Table 6, for the stringer stiffened shell AS-2 the buckling loads with membrane prebuckling depend strongly on the boundary condition specified. Thus for the SS-3 boundary condition the buckling load is 229.80 N/cm. Stiffening this boundary condition raises the buckling load by 12% for the C-3, 31% for the SS-4 and by 40% for the C-4 boundary condition. Further the inclusion of the nonlinear prebuckling effect results in buckling loads which are a few percent lower than the corresponding buckling loads with membrane prebuckling. Notice

Table 6: Summary of the buckling load calculations for the stringer stiffened shell AS-2

	SS-3	SS-4	C-3	C-4
N_{x_M}	229.80(10)	300.52(14)	256.87(10)	320.81(14)
$N_{x_{NL}}$	223.99(10)	280.03(14)	256.38(10)	316.80(14)
$N_{s_{MM}}$	138.77(10)	...	...	...
$N_{s_{EXT}}$	138.58(10)	196.46(11)	157.00(10)	211.43(11)
ρ_s	0.60	0.65	0.61	0.66

also that the number of circumferential waves n at which bifurcation into an asymmetric buckling mode first occurs varies with the boundary conditions.

Hence, when analyzing the imperfect shells it becomes necessary to ascertain for each boundary condition that circumferential wave number n which for a given initial imperfection will yield the lowest buckling load. As can be seen from Table 6 the number of circumferential waves which yields the lowest buckling load can differ for perfect and for imperfect shells. Examining further the effect of the 2-modes initial imperfection model given by Eq. (91), it is seen that the simplified 2-modes solution[16] and the present "extended analysis" with SS-3 boundary condition give virtually the same buckling load. As can be seen from the results displayed in Table 6, the assumed initial imperfection results in a 40% decrease in buckling load when compared with the buckling load of a perfect shell using a membrane prebuckling state and the same SS-3 boundary condition. Further, using the same initial imperfection one obtains a 35% decrease in buckling load for the C-3 boundary condition, a 39% decrease for the SS-4 and a 34% decrease in buckling load for the C-4 boundary condition.

This similarity in the reduction of the buckling loads for different boundary conditions, when written as ratios of the buckling loads of the imperfect and the perfect shells using in both cases the same boundary conditions, has prompted Arbocz and Sechler[31] to propose the following normalization formula (here written for the C-4 boundary condition) in order to account for both the effect of initial imperfections and the appropriate boundary conditions

$$N_{s_{C-4}} = N_{s_{MM}} \left(\frac{N_{x_{C-4}}}{N_{x_{SS-3}}} \right) \tag{92}$$

where

$N_{s_{C-4}}$ = imperfect shell buckling load for C-4 boundary condition (N/cm)

$N_{s_{MM}}$ = imperfect shell buckling load by the multimode analysis[18] (N/cm)

$N_{x_{C-4}}$ = perfect shell buckling load using membrane prebuckling analysis and C-4 boundary condition[10,33] (N/cm)

$N_{x_{SS-3}}$ = perfect shell buckling load using membrane prebuckling analysis and SS-3 boundary condition[10,33] (N/cm)

Thus the effect of initial imperfections is taken into account by calculating the buckling load of the imperfect shell via the multimode analysis. This involves computer runs with the MIUTAM[18] computer code. To account for the effect of boundary conditions one executes two runs with the computer codes SRA[10] or BOSOR[33] for the perfect shell. In one run one uses the SS-3 boundary conditions, the same ones that are approximately satisfied by the multimode analysis. In the other run one uses the actual boundary conditions (in the above formula, for the purpose of illustration, the C-4 boundary conditions are used).

As an example of the successful application of the proposed normalization formula consider the following 7-modes imperfection model[34]:

$$\frac{\bar{W}}{t} = 0.0061 \cos\frac{2\pi x}{L} - \sin\frac{\pi x}{L}\left(0.5072 \cos\frac{2y}{R} + 0.0801 \cos\frac{9y}{R} + 0.0704 \cos\frac{10y}{R} + 0.0626 \cos\frac{11y}{R} + 0.0320 \cos\frac{19y}{R} + 0.0283 \cos\frac{21y}{R}\right) \tag{93}$$

Calculating the limit point of the imperfect shell AS-2 with the computer code MIUTAM yields the collapse load of $\rho_s = 0.825$. Recall that this value has been normalized by $N_{x_{SS-3}} = -229.8$ N/cm, the buckling load of the perfect shell AS-2 using membrane prebuckling and SS-3 ($N_x = v = w = M_x = 0$) boundary conditions.

To calculate the buckling load of an axially compressed cylinder, including the effect of general prescribed initial imperfections and enforcing the experimental boundary conditions rigorously, one can use the two-dimensional nonlinear shell analysis code STAGS[35]. In practice, the cost of performing such an analysis dictates that the size of the problem be cut down. To check the results of the 7-modes multimode solution the shell model shown in Fig. 31 was used. Based on an convergence study[34] the mesh selected consisted of 21 rows and 131 columns. Figure 32 shows the maximum displacement (here the radial displacement W at $x = \frac{L}{2}$ and $y = 0$) as a function of the axial load. The results of the nonlinear STAGS runs are indicated by circles. Above the last point the determinant of the stiffness matrix changes sign indicating the occurence of an instability. The limit load is $\rho_s = 0.829$, where the load parameter

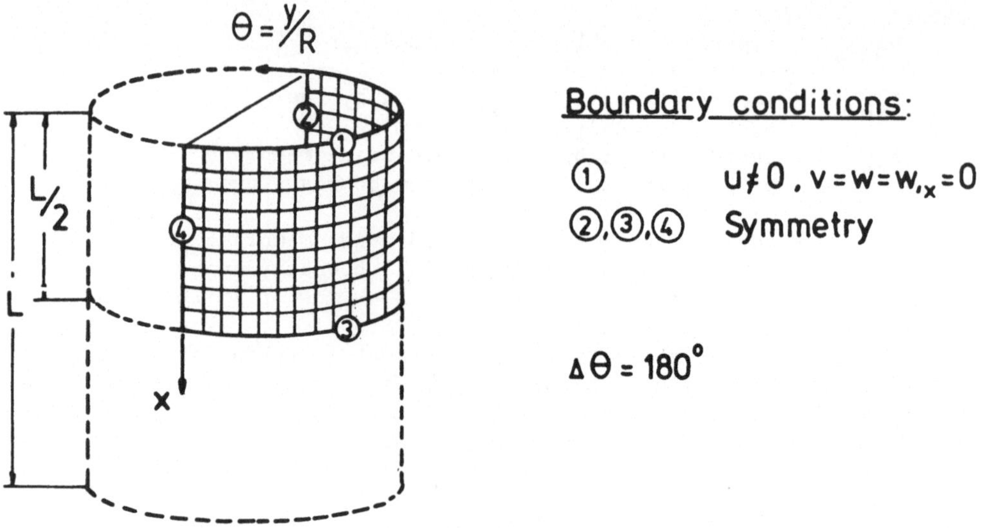

Fig. 31, Shell model used for collapse analysis with the 7-modes imperfection model (shell AS-2)

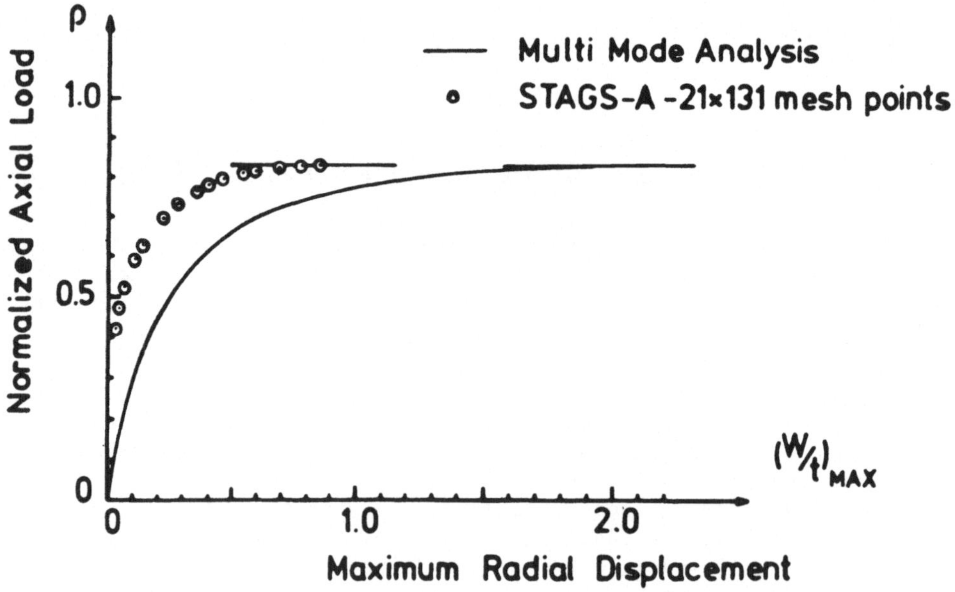

Fig. 32, Comparison of Multimode Analysis and STAGS (shell AS-2, 7-modes imperfection model)

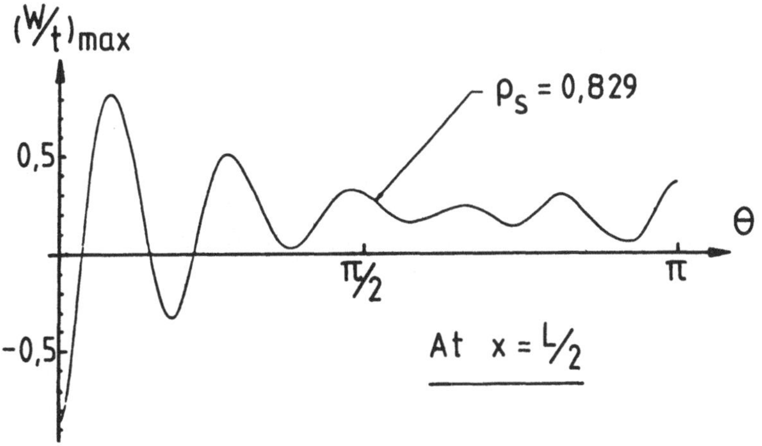

a By STAGS [35]

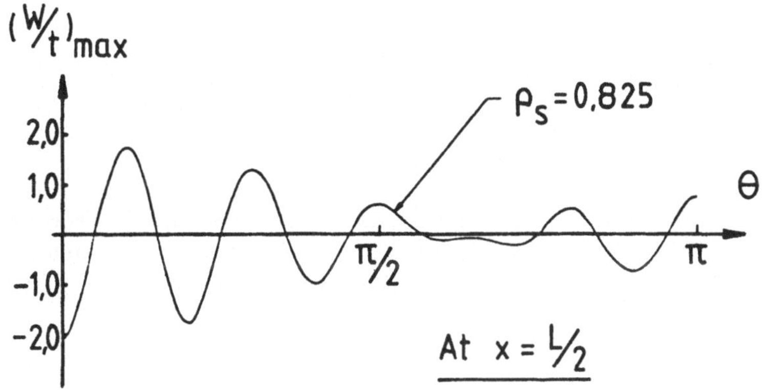

b By Multimode Analysis [18]

Fig. 33, Comparison of the calculated radial displacements at the limit point (shell AS-2)

ρ has been normalized by $N_{x_{C-4}}$ = - 320.8 N/cm, the buckling load of the perfect shell AS-2 using membrane prebuckling and C-4 ($u = v = w = w_{,x} = 0$) boundary conditions. The agreement between the two calculated limit points is remarkable. Also the shapes of the radial displacements at the center of the shell (at x = L/2) shown in Fig. 33 are very similar. Their amplitudes are, however, different by a factor of 2. This difference is thought to occur because of the different boundary conditions used, C-4 for the STAGS-run and SS-3 for the Multimode Analysis.

Despite this difference in the amplitudes of the displacements, the closeness of the predicted buckling loads tends to confirm the practice of initially using the simpler Multimode Analysis[16] to estimate the buckling loads of imperfect stringer stiffened shells[24,25,26]:

9. STOCHASTIC STABILITY ANALYSIS

Reading the proceeding chapters it must have become clear, that in most cases, for reliable buckling load predictions, one must know both the shape and the size of the initial imperfections, This brings up the critical question every shell designer must face, once it has been established that the buckling load of the structure under consideration is imperfection sensitive:

> "Is it cheaper to use a large "knockdown factor" and a large factor of safety to account for the uncertainties, or should one apply the Imperfection Sensitivity Theory in order to arrive at an optimal design?"

If the total weight of the structure and the cost of the material are of no major concern, then in many cases it may be easiest to rely on the current shell design manuals [36-38], which recommend the use of the following buckling formula

$$P_a \lesssim \frac{\gamma}{F.S.} P_{c\ell} \qquad (94)$$

where P_a is the allowable buckling load, $P_{c\ell}$ is the lowest buckling load of the perfect structure, γ is a "knockdown factor" and F.S. is a factor of safety. The empirical knockdown factor γ is so chosen that when it is multiplied with $P_{c\ell}$ a "lower bound" to all available experimental data is obtained. The design manuals contain charts from which the appropriate empirical knockdown factors γ can be determined and they make recommendations about the factors of safety to be used. This approach is called the "Lower Bound Design Philosophy" and it has been in use since the turn of the century.

If, however, the total weight of the structure is of critical importance, then a more sophisticated design approach is called for. That is, the designer must estimate how much the expected imperfections will decrease the buckling load of the chosen configuration. It is obvious, that the main difficulty in using the Imperfection Sensitivity Theory in practical design problems dealing with weight sensitive applications is related to the fact, that it requires some advanced knowledge of the geometric imperfections that will be present once the planned structure has been built, an information that is rarely available.

Since initial imperfections depend on the different fabrication processes used and are obviously random in nature, it has long been felt that for successful application the Imperfection Sensitivity Theory must be combined with some kind of probabilistic approach. The buckling of imperfection-sensitive structures with small random initial imperfections has been studied by several investigators [39-42]. In the absence of experimental evidence about the type of imperfections that occur in practice, and in order to reduce the mathematical complexities of the problem, all the early pioneering work has been done with some form of idealized imperfection distribution . However, it is not obvious how these methods can be extended to the general imperfection distributions observed in practice.

Thus, it was not until 1979 when Elishakoff[43] published his reliability study of the buckling of a finite column with random initial imperfections on nonlinear elastic foundation based on the Monte Carlo Method, that an approach has been presented which makes it possible to introduce the results of the initial imperfection surveys routinely into the Stochastic Stability Analysis.

When applying the Monte Carlo Method the measured initial imperfections are expanded in a double Fourier series, and then the Fourier coefficients are treated as random variables. Next by using a special numerical procedure[44] the Fourier coefficients of the desired large sample of random initial imperfection shapes are simulated. This is followed by a deterministic buckling load analysis of each of the simulated shells. Finally, the reliability function representing the probability (i.e. fraction of an ensemble) of the buckling load exceeding the specified load is calculated.

The relative ease with which one can derive the reliability function via the Monte Carlo Method, once a sufficiently large sample of initial imperfection measurements is available, will be demonstrated by considering the case of axially compressed cylindrical shells with random axisymmetric imperfections. Having N realizations of the measured initial imperfections

$$\bar{W}^{(m)}(x) = \frac{\bar{w}^{(m)}(x)}{t} = \sum_i A_i^{(m)} \phi_i(x) \qquad m = 1, 2, \ldots, N \tag{95}$$

where the ϕ_i's represent the complete set of orthogonal functions on [0,L], one calculates first, by taking "ensemble averages", the estimated mean of the Fourier coefficients $A_i^{(m)}$

$$\langle A_i \rangle^{(e)} = \frac{1}{N} \sum_{m=1}^{N} A_i^{(m)} \tag{96}$$

and then the estimated variance-covariance matrix

$$K^{(e)}_{A_j A_k} = \frac{1}{N-1} \sum_{m=1}^{N} \left[A_j^{(m)} - \langle A_j \rangle^{(e)} \right] \left[A_k^{(m)} - \langle A_k \rangle^{(e)} \right] \tag{97}$$

Since $K^{(e)}_{A_j A_k}$ is a positive-semidefinite matrix it can be decomposed in the form

$$K^{(e)}_{A_j A_k} = G G^T \tag{98}$$

where G is a lower triangular matrix found by the Cholesky decomposition algorithm. Next the vector $\underset{\sim}{A}$ of the simulated initial imperfections is obtained as follows:

$$\underset{\sim}{A} = G \, \underset{\sim}{\xi} + \langle \underset{\sim}{A} \rangle^{(e)} \tag{99}$$

where

$\langle \underset{\sim}{A} \rangle^{(e)}$ = estimated "mean" vector
$\underset{\sim}{\xi}$ = random vector

The components of the random vector $\underset{\sim}{\xi}$ are normally distributed random numbers with zero mean and unit variance generated by the computer. Taking, for example, 1000 different $\underset{\sim}{\xi}$'s one gets 1000 different $\underset{\sim}{A}$'s, that is, 1000 different simulated shells with $\underset{\sim}{A}$'s as the Fourier coefficients of the initial imperfections.

In the following, one carries out a deterministic buckling load calculation for each of the simulated shells generating the histogram of the buckling loads for the group of shells under consideration. Here reliability is defined as the probability that the random buckling load Λ will exceed the prescribed value λ. Thus

$$R(\lambda) = \text{Prob } (\Lambda \geq \lambda) \tag{100}$$

where λ is the normalized buckling load. Because of this definition one can calculate $R(\lambda)$ from the histogram of the buckling loads (see Fig. 34) by the frequency interpretation (that is, fraction of an ensemble) yielding the dots shown in Fig. 35. The accuracy of the Monte Carlo Method is demonstrated by the close coincidence of the dots with the solid curve which represents a closed form solution in terms of error functions[42], obtained for the infinitely long cylindrical shell with a single cosinusoidal axisymmetric imperfection using Koiter's equation (5.2) from Reference [7] as a nonlinear transfer function.

Knowing the reliability function makes it possible to find the allowable buckling load λ_a for the whole ensemble of shells produced by a given manufacturing process, defined as the buckling load for which the desired reliability (say 0.98) is attained. Notice that in Fig. 35 λ_a, the allowable buckling load ratio, is the same as the previously mentioned empirical knockdown factor γ in Eq. (94). However, contrary to the knockdown factors γ displayed in the charts of the current shell design manuals, which are arrived at as lower bounds to all available experimental data, λ_a is the allowable buckling load ratio (or the knockdown factor) only for a given imperfection distribution characteristic of a certain manufacturing process. Wether it will have a higher or lower value for another case depends on the characteristic initial imperfection distribution of the new manufacturing process under consideration. Thus by this procedure the lower bound depends clearly on the manufacturing process used to produce the particular shell. The feasibility of deriving reliability functions via the Monte Carlo Method has been demonstrated by Elishakoff and Arbocz both for axisymmetric[45] and for general asymmetric[46] imperfections.

Since, as mentioned above, the improvements in the currently recommended shell design procedures are primarily sought in a more selective approach when defining the knockdown factor γ, therefore innovative

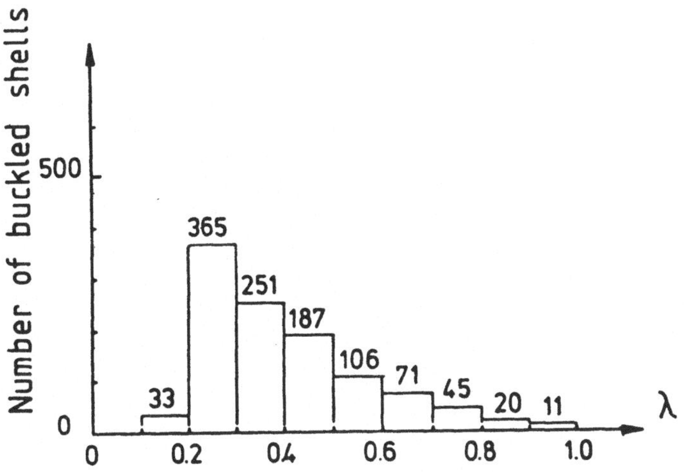

Fig. 34, Histogram of nondimensional buckling loads [45]

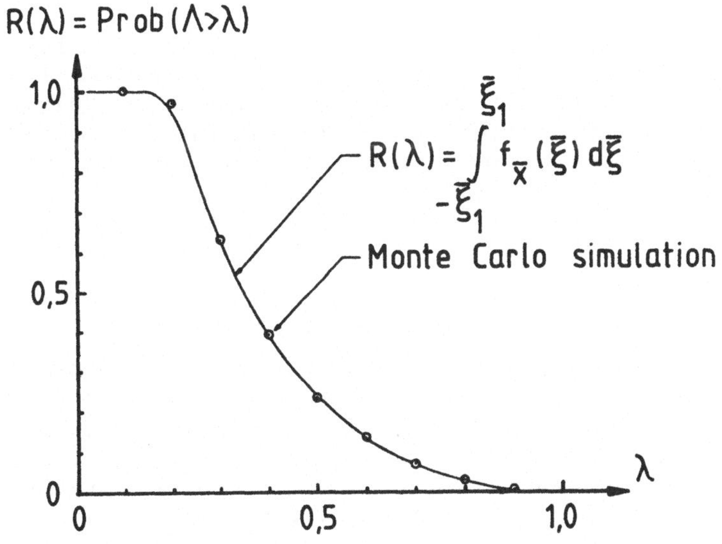

Fig. 35, Comparison of analytical reliability function with results of Monte Carlo Simulation[45]

design and selective use of fabrication processes are encouraged and rewarded. Thus, for instance, if a company takes great care in producing its shells very accurately and if it can show experimentally that the boundary conditions are defined in such a manner that no additional imperfections (especially at the shell edges) are introduced, then the use of an improved (higher) knockdown factor λ_a, derived by a stochastic approach, should be allowed. This new Improved Shell Design Procedure can be represented by the following formula:

$$P_a \leq \frac{\lambda_a}{F.S.} P_c \tag{101}$$

where

- P_a = allowable buckling load
- P_c = buckling load of the perfect structure
- λ_a = reliability based improved (higher) knockdown factor
- F.S. = factor of safety.

Here the buckling load of the perfect structure P_c is calculated with one of the advanced shell of revolution codes like SRA[10] or BOSOR[33]. These programs can handle very general wall constructions, combinations of different mechanical and thermal loadings and can model the actual boundary conditions accurately.

This new method is intended to replace, step by step, the old, in many cases overly conservative Lower Bound Design Philosophy. For its successful introduction one must have access to the following items:

(1) a large Initial Imperfection Data Bank, to make the derivation of characteristic imperfection distributions for the different fabrication processes possible;
(2) a numerical simulation procedure, to "create" the large number of shells with random imperfections needed for the Monte Carlo Method;
(3) an accurate and fast nonlinear code, to calculate the buckling loads of the "created" imperfect shells.

Since in this approach the probabilistic properties are not assumed, but the mean vector and the variance-covariance matrix of the Fourier coefficients are calculated from the actual measurements of the shell profiles, therefore for the successful introduction of this design procedure, a systematic expansion of the currently available imperfection data, classified according to the different fabrication processes, is a must.

Furthermore, the shape of the reliability curves is influenced strongly by the accuracy of the buckling load calculations, which, as has been pointed out by Arbocz and Babcock[47], depends heavily on the appropriate choice of the nonlinear model used, which in turn requires considerable knowledge by the user as to the physical behaviour of imperfect shell structures. This knowledge can be acquired by first using the series of imperfection sensitivity analysis of increasing complexity that have been discussed earlier in this paper.

10. CONCLUDING REMARKS

The readers may be surprised that in a paper dealing with numerical techniques the author has not mentioned finite element codes at all. This has been done intentionally. It is this author's believe that with the large scale introduction of computers in the practice and lately also at many technical schools, the teaching of and the approach to solving technical problems has been shifting in the wrong direction.

Twenty five years ago it was so, that numerical results were looked upon with a certain degree of distrust and they were only accepted if supported by some other facts. Now-a-days, as the older generation of engineers (the ones who have gotten their degrees before the advent of computers) is retiring and the younger ones with extensive training in the ever-so-popular finite element techniques take over, one begins to encounter in technical discussions a new mentality; the insight of how structures behave under loading of the older generation is being more and more replaced by the nearly religious faith of the younger ones in the predictions of their favorite computer codes.

A word of caution is in place here. A shell design specialist, who is aware of the latest theoretical developments and who is familiar with the theories upon which the nonlinear structural analysis codes he uses are based, can achieve very accurate modeling of the collapse behaviour of complex structures [48]. The danger lies in the use of the sophisticated computational tools by persons of inadequate theoretical background. Contrary to the beliefs of some, computer codes are no replacements for engineering know-how and engineering expertise.

Thus in this paper, the author tried to summarize the theoretical knowledge with computer enhancement he considers as the minimum background every structural analysist must have, before he begins to model the buckling behaviour of an axially compressed shell of complicated wall construction with one of the currently available general nonlinear finite element codes. In order to facilitate the use of the methods described in this paper the development of DISDECO, The Delft Interactive Shell Design Code [49], has been initiated recently at the Aeronautics Department of the Delft University of Technology. In this interactive code it will be attempted to combine in an optimal way the accumulated theoretical knowledge in the field of shell stability with the advanced interactive and computational facilities offered by todays high speed 32 bits workstations.

Once DISDECO becomes available the user can start the different programs described herein with nonlinear models of varying complexity from his workstation, preview and review the results, adjust the input as needed and execute the programs again. Finally, should the design requirements warrant it, the user can switch to running the refined numerical models of the final configuration with one of the general, thin-walled shell codes [35], which can handle complex geometries and both geometric and material nonlinearities.

ACKNOWLEDGEMENT

The author wishes to express his sincere thanks to Mrs. Marlene Lit for the skillful typing of the manuscript. The fine artwork by Mr. Jan de Vries is also very much appreciated.

REFERENCES

1. Hutchinson, J.W. and W.T. Koiter: Postbuckling theory. Appl. Mech. Rev. 23 (1970), 1353-1366.
2. Fung, Y.C. and E.E. Sechler (Editors): Thin-Shell Structures, Theory, Experiment and Design. Prentice Hall, Englewood Cliffs, N.J. 1974.
3. Singer, J.: The status of experimental buckling investigations, in: Buckling of Shells - A State-of-the-Art Colloquium (Ed. E. Ramm), Springer Verlag, Berlin 1982, 501-534.
4. Geier, B.: Das Beulverhalten Versteifter-Zylinderschalen. Teil 1, Differential-Gleichungen. Z. Flugwiss. 14 (1966), 306-323.
5. Koiter, W.T.: On the stability of elastic equilibrium. Ph. D. Thesis, in Dutch, TH Delft, Netherlands, H.T. Paris, Amsterdam 1945. English translation issued as NASA TT F-10 (1967), 833 p.
6. Arbocz, J.: The effect of initial imperfections on shell stability, in: Thin Shell Structures, Theory, Experiment and Design (Eds. Y.C. Fung and E.E. Sechler), Prentice Hall, Englewood Cliffs, N.J. (1974), 205-245.
7. Koiter, W.T.: The effect of axisymmetric imperfections on the buckling of cylindrical shells under axial compression. Koninkl. Ned. Akad. Wetenschap. Proc. B66 (1963), 265-279.
8. Hutchinson, J.W. and J.C. Amazigo: Imperfection sensitivity of eccentrically stiffened cylindrical shells. AIAA J. (1967), 392-401.
9. Budiansky, B. and J.W. Hutchinson: Dynamic buckling of imperfection-sensitive structures, in: Proc. XI Intern. Congr. Appl. Mech. (ed. H. Görtler), Springer Verlag, Berlin (1964), 636-651.
10. Cohen, G.A.: User document for computer programs for ring-stiffened shells of revolution. NASA CR-2086 (1973).
11. Hutchinson, J.W. and J.C. Frauenthal: Elastic postbuckling behaviour of stiffened and barreled cylindrical shells. J. Appl. Mech. 36 (1969), 784-790.
12. Arbocz, J. and E.E. Sechler: On the buckling of axially compressed ring and stringer stiffened imperfect cylindrical shells, GALCIT Report SM 73-10, California Institute of Technology, Pasadena (1973).
13. Arbocz, J. and E.E. Sechler: On the buckling of axially compressed imperfect cylindrical shells, J. Appl. Mech. 41 (1974), 737-743.
14. Hutchinson, J.W.: Axial buckling of pressurized imperfect cylindrical shells. AIAA J. 3 (1965), 1461-1466.
15. Thurston, G.A. and M.A. Freeland: Buckling of imperfect cylinders under axial compression. NASA CR-541 (1966).
16. Arbocz. J. and C.D. Babcock: A multimode analysis for calculating buckling loads of imperfect cylindrical shells. GALCIT Report SM 74-4, California Institute of Technology, Pasadena (1974).

17. Kempner, J.: Postbuckling behaviour of axially comppressed cylindrical shells. J. Aeron. Scien. 21 (1954), 329-335.
18. Arbocz, J. and C.D. Babcock: Prediction of buckling loads based on experimentally measured initial imperfections. in: Proc. IUTAM Symp. Buckling of Structures, Harvard University, Cambridge, Ma. 1974 (Ed. B. Budiansky), Springer Verlag, Berlin 1976 , 291-311.
19. Arbocz, J. and C.D. Babcock: The effect of general imperfections on the buckling of cylindrical shells. J. Appl. Mech. 36 (1969), 28-38.
20. Imbert, J.: The effect of imperfections on the buckling of cylindrical shells, Aeronautical Engineer Thesis, California Institute of Technology, Pasadena (1971).
21. Singer, J. Abramovich, H. and R. Yaffe: Initial imperfection measurements of stiffened shells and buckling predictions. In: Proc. 21st Israel Conference on Aviation and Astronautics, Israel J. of Techn. 17 (1979), 324-338.
22. Singer, J.: Vibrations and buckling of imperfect stiffened shells-recent developments. In: Proc. IUTAM Symp. Collapse: the buckling of structures in theory and practice, University College, London 1982 (Eds. J.M.T. Thompson and G.W. Hunt), Cambridge University Press, Cambridge 1983, 443-479.
23. Arbocz, J. and H. Abramovich: The initial imperfection data bank at the Delft University of Technology - Part I, Report LR-290, Delft University of Technology, 1979.
24. Singer, J. and A. Rosen: The influence of boundary conditions on the buckling of stiffened cylindrical shells. In: Proc. IUTAM Symp. Buckling of Structures, Harvard University, Cambridge, Ma., 1974 (Ed. B. Budiansky), Spinger Verlag, Berlin 1976, 227-250.
25. Singer, J. and H. Abramovich: Vibration techniques for definition of practical boundary conditions in stiffened shells. AIAA J. 17 (1979), 762-769.
26. Singer, J. and J. Prucz: Influence of imperfections on the vibrations of stiffened cylindrical shells. J. Sound and Vibration 80 (1982), 117-143.
27. Stein, M.: Some recent advances in the investigation of shell buckling. AIAA J. 6 (1968), 2339-2345.
28. Weller, T. Singer, J. and S.C. Batterman: Influence of eccentricity of loading on buckling of stringer-stiffened cylindrical shells. In: Thin Shell Structures, Theory, Experiment and Design (Eds. Y.C. Fung and E.E. Sechler), Prentice Hall, Englewood Cliffs, NJ. 1974, 305-324.
29. Block, D.L.: Influence of prebuckling deformations, ring stiffeners and load eccentricity on the buckling of stiffened cylinders. In: Proc. AIAA/ASME 8th Structures, Structural Dynamics and Materials Conference, 1967, 597-607.
30. Keller, H.: Numerical Methods for Two-Point Boundary Value Problems, Blaisdell Publishing Co., Waltham, Ma. 1968.
31. Arbocz, J. and E.E. Sechler: On the buckling of stiffened imperfect shells. AIAA J. 14 (1976), 1611-1617.
32. Babcock, C.D. and E.E. Sechler: The effect of initial imperfections on the buckling stress of cylindrical shells, NASA TN D-1510 (1962), 135-142.

33. Bushnell, D.: Stress, stability and vibration of complex branched shells of revolution: Analysis and user's manual for BOSOR-4. NASA CR-2116 (1972).
34. Arbocz, J. and C.D. Babcock: Utilization of STAGS to determine knockdown factors from measured initial imperfections, Report LR-275, Delft University of Technology, 1978.
35. Almroth, B.O., Brogan, F.A., Miller, E., Zele, F. and H.T. Peterson: Collapse analysis for shells of general shape, User's Manual for the STAGS-A computer code, Air Force Flight Dynamics Lab., Wright Patterson AFB, AFFDL-TR-71-8, 1973.
36. Anonymous: Buckling of Thin-Walled Cylinders. NASA Space Vehicle Design Criteria (Structures), NASA SP-8007, revised, 1968.
37. Anonymous: Rules for the Design, Construction and Inspection of Offshore Structures. Det Norkse Veritas, Oslo 1977.
38. Anonymous: Beulsicherheitsnachweise für Schalen. DASt Richtlinie 013, Deutscher Ausschuss für Stahlbau, 1980.
39. Bolotin, V.V.: Statistical Methods in the Nonlinear Theory of Elastic Shells. NASA TT F-85 (1962) Translation of a paper presented at a seminar in the Inst. of Mech. of the Acad. of Sciences USSR, 1957.
40. Makaroff, B.P.: Statistical analysis of the stability of imperfect cylindrical shells. In: Proc. 7th All-Union Conf. on the theory of Plates and Shells, Dnepropetrovsk 1969, 387-391 (in Russian).
41. Amazigo, J.C.: Buckling under axial compression of long cylindrical shells with random axisymmetric imperfections. Quart. Appl. Math. 26 (1969), 537-566.
42. Roorda, J. and J.S. Hansen: Random buckling behaviour in axially loaded cylindrical shells with axisymmetric imperfections. J. Spacecraft 9 (1972), 88-91.
43. Elishakoff, I.: Buckling of a stochastically imperfect finite column on a nonlinear elastic foundation - A reliability study. J. Appl. Mech. 46 (1979), 411-416.
44. Elishakoff, I.: Simulation of space-random fields for solution of stochastic boundary-value problems. J. Acoust. Soc. Amer. 65 (1979), 399-403.
45. Elishakoff, I. and J. Arbocz: Reliability of axially compressed cylindrical shells with random axisymmetric imperfections. Int. J. Solids and Structures, 18 (1982), 563-585.
46. Elishakoff, I. and J. Arbocz: Reliability of axially compressed cylindrical shells with general nonsymmetric imperfections. J. Appl. Mech. 52 (1985), 122-128.
47. Arbocz, J. and C.D. Babcock: The Buckling Analysis of Imperfection Sensitive Shell Structures. NASA CR-3310 (1980).
48. Bushnell, D.: Computerized Buckling Analysis of Shells. Kluwer Academic Publishers, Hingham, Ma. 1985.
49. Arbocz, J.: About the development of interactive shell design codes. In: Proc. Int. Conf. Spacecraft Structures, Toulouse, 1985 (In press).
50. Arbocz, J.: Shell stability analysis: theory and practice. In: proc. IUTAM Symp. Collapse: The buckling of structures in theory and practice, University College, London, 1982 (Eds. J.M.T. Thompson and G.W. Hunt), Cambridge University Press, Cambridge 1983, 43-74.

APPENDIX A: Definition of the Stiffener Parameters

Let

A_1, A_2 = cross-sectional area of stringer and ring, respectively

b,a = distance between stringers and rings, respectively

E = Young's modulus

e_1, e_2 = distance between centroid of stiffener cross-section and middle surface of shell

G = shear modulus

I_{11}, I_{22} = moment of inertia of stiffener cross-section about its centroidal axis

I_{01}, I_{02} = moment of inertia of stiffener cross-section about the middle surface of the shell

I_{t_1}, I_{t_2} = torsion constant of stiffener cross-section

t = thickness of shell

and introduce the following parameters:

$$\mu_1 = (1-\nu^2)\frac{A_1}{bt}, \quad \mu_2 = (1-\nu^2)\frac{A_2}{at}, \quad \chi_1 = (1-\nu^2)\frac{A_1}{bt}e_1,$$

$$\chi_2 = (1-\nu^2)\frac{A_2}{at}e_2, \quad \zeta_1 = \frac{EA_1e_1}{bD}, \quad \zeta_2 = \frac{EA_2e_2}{aD},$$

$$\eta_{01} = \frac{EI_{01}}{bD}, \quad \eta_{02} = \frac{EI_{02}}{aD} \quad \eta_{t_1} = \frac{GI_{t_1}}{bD},$$

$$\eta_{t_2} = \frac{GI_{t_2}}{aD}, \quad D = \frac{Et^3}{12(1-\nu^2)}, \quad \hat{\beta} = \frac{1}{(1+\mu_1)(1+\mu_2)-\nu^2}$$

Then the stiffener parameters are defined as follows:

$$D_{xx} = D.\bar{D}_{xx} = D\{1+\eta_{01} - \hat{\beta}(1+\mu_2)\,\zeta_1\chi_1\}$$

$$D_{xy} = D.\bar{D}_{xy} = D\{2+\eta_{t_1}+\eta_{t_2}+\hat{\beta}\nu(\zeta_1\chi_2+\zeta_2\chi_1)\}$$

$$D_{yy} = D.\bar{D}_{yy} = D\{1+\eta_{02} - \hat{\beta}(1+\mu_1)\zeta_2\chi_2\}$$

$$H_{xx} = \frac{1}{Et}\bar{H}_{xx} = \frac{1}{Et}\{(1-\nu^2)\hat{\beta}(1+\mu_1)\}$$

$$H_{xy} = \frac{1}{Et}\bar{H}_{xy} = \frac{1}{Et}\left\{2(1-\nu^2)\hat{\beta}\left[\frac{1+\nu}{\hat{\beta}(1-\nu^2)} - \nu\right]\right\}$$

$$H_{yy} = \frac{1}{Et}\bar{H}_{yy} = \frac{1}{Et}\{(1-\nu^2)\hat{\beta}(1+\mu_2)\}$$

$$Q_{xx} = \frac{t}{2c}\bar{Q}_{xx} = \frac{t}{2c}\left\{\nu\hat{\beta}\chi_1\frac{2c}{t}\right\}$$

$$Q_{xy} = \frac{t}{2c}\bar{Q}_{xy} = \frac{t}{2c}\left\{-\hat{\beta}\left[(1+\mu_2)\chi_1 + (1+\mu_1)\chi_2\right]\frac{2c}{t}\right\}$$

$$Q_{yy} = \frac{t}{2c}\bar{Q}_{yy} = \frac{t}{2c}\left\{\nu\hat{\beta}\chi_2\frac{2c}{t}\right\}$$

where $c = \sqrt{3(1-\nu^2)}$.

Further

$$\alpha_1 = \frac{1}{(1+\mu_1)\hat{\beta}} \quad ; \quad \alpha_2 = \frac{1}{(1+\mu_2)\hat{\beta}}$$

APPENDIX B: Definition of the coefficients in Eqs. (32), (51) and (52)

Recalling that

$$\alpha_i^2 = i^2 \frac{Rt}{2c}\left(\frac{\pi}{L}\right)^2, \qquad \alpha_{i-k}^2 = (i-k)^2 \frac{Rt}{2c}\left(\frac{\pi}{L}\right)^2$$

$$\alpha_k^2 = k^2 \frac{Rt}{2c}\left(\frac{\pi}{L}\right)^2, \qquad \alpha_{i+k}^2 = (i+k)^2 \frac{Rt}{2c}\left(\frac{\pi}{L}\right)^2$$

$$\beta_\ell^2 = \ell^2 \frac{Rt}{2c}\left(\frac{1}{R}\right)^2$$

we have

$$c_1 = \frac{c}{4}\frac{\beta_\ell^2}{\alpha_i^2}\frac{(1+4\alpha_k^2\bar{Q}_{xx})}{\bar{H}_{xx}}, \qquad c_2 = c\,\frac{\alpha_k^2\beta_\ell^2}{\alpha_i^2}\frac{(\bar{\gamma}_{Q,k,\ell}+\alpha_k^2)}{\bar{\gamma}_{H,k\,\ell}}$$

$$c_3 = \frac{c}{2}\frac{\alpha_i^2\beta_\ell^2}{\alpha_k^2}\frac{(\bar{\gamma}_{Q,k,\ell}+\alpha_k^2)}{\bar{\gamma}_{H,k,\ell}}, \qquad c_4 = \frac{c}{2}\frac{\beta_\ell^2}{\alpha_k^2}\frac{(1+\alpha_i^2\bar{Q}_{xx})}{\bar{H}_{xx}}$$

$$c_6 = \frac{c}{2}\frac{\alpha_i^2\beta_\ell^2}{\alpha_k^2}\frac{(\bar{\gamma}_{Q,i-k,\ell}+\alpha_{i-k}^2)}{\bar{\gamma}_{H,i-k,\ell}}, \qquad c_8 = \frac{c^2}{4}\alpha_i^2\beta_\ell^4\left\{\frac{1}{\bar{\gamma}_{H,i+k,\ell}}+\frac{1}{\bar{\gamma}_{H,i-k,\ell}}\right\}$$

$$c_9 = \frac{c^2}{4}\frac{\beta_\ell^4}{\alpha_k^2}\frac{1}{\bar{H}_{xx}}, \qquad c_{10} = \frac{c^2}{4}\alpha_k^2\frac{1}{\bar{H}_{yy}}$$

$$c_{11} = \frac{c^2}{2}\frac{\alpha_i^4\beta_\ell^4}{\alpha_k^2}\left\{\frac{1}{\bar{\gamma}_{H,i+k,\ell}}+\frac{1}{\bar{\gamma}_{H,i-k,\ell}}\right\}, \qquad c_{12} = \frac{c^2}{2}\alpha_i^2\beta_\ell^4\frac{1}{\bar{\gamma}_{H,i-k\,\ell}}$$

$$c_{13} = \frac{c^2}{2}\frac{\alpha_i^4\beta_\ell^4}{\alpha_k^2}\frac{1}{\bar{\gamma}_{H,i-k,\ell}}$$

where

$$\bar{\gamma}_{Q,k,\ell} = \bar{Q}_{xx}\,\alpha_k^4 + \bar{Q}_{xy}\,\alpha_k^2\beta_\ell^2 + \bar{Q}_{yy}\beta_\ell^4$$

$$\bar{\gamma}_{H,k,\ell} = \bar{H}_{xx}\,\alpha_k^4 + \bar{H}_{xy}\,\alpha_k^2\beta_\ell^2 + \bar{H}_{yy}\beta_\ell^4$$

$$\bar{\gamma}_{Q,i-k,\ell} = \bar{Q}_{xx}\alpha_{i-k}^4 + \bar{Q}_{xy}\,\alpha_{i-k}^2\beta_\ell^2 + \bar{Q}_{yy}\,\beta_\ell^4$$

$$\bar{\gamma}_{H,i-k,\ell} = \bar{H}_{xx}\alpha^4_{i-k} + \bar{H}_{xy}\alpha^2_{i-k}\beta^2_\ell + \bar{H}_{yy}\beta^4_\ell$$

$$\bar{\gamma}_{H,i+k,\ell} = H_{xx}\alpha^4_{i+k} + \bar{H}_{xy}\alpha_{i+k}\beta^2_\ell + \bar{H}_{yy}\beta^4_\ell$$

APPENDIX C: Coefficients for the computation of the b-factor in Eq. (45)

Using the previously defined stiffener and wave number parameters one gets

$$Aj = c\,\frac{\alpha_k^2\beta_\ell^2}{\alpha_j^2}\;\frac{(1 + \alpha_j^2\bar{Q}_{xx}) + 2\alpha_j^2\bar{H}_{xx}\bar{F}}{(1 + \alpha_j^2\bar{Q}_{xx})^2 + (\alpha_j^2\bar{D}_{xx} - 2\lambda_{c_{k\ell}})\alpha_j^2\bar{H}_{xx}}\left(\frac{4}{\pi}\,\frac{j}{4k^2 - j^2}\right)$$

$$B_j = c\alpha_k^2\beta_\ell^2\;\frac{(\bar{\gamma}_{Q,j,2\ell} + \alpha_j^2) + 2\bar{\gamma}_{H,j,2\ell}\bar{F}}{(\bar{\gamma}_{Q,j,2\ell} + \alpha_j^2)^2 + (\bar{\gamma}_{D,j,2\ell} - 2\alpha_j^2\lambda_{c_{k\ell}})\bar{\gamma}_{H,j,2\ell}}\left(\frac{4}{\pi}\,\frac{1}{j}\right)$$

$$C_j = c\,\frac{\alpha_k^2\beta_\ell^2}{\alpha_j^2}\;\frac{2(1+\alpha_j^2\bar{Q}_{xx})\bar{F} - (\alpha_j^2\bar{D}_{xx} - 2\lambda_{c_{k\ell}})}{(1 + \alpha_j^2\,\bar{Q}_{xx})^2 + (\alpha_j^2\bar{D}_{xx} - 2\lambda_{c_{k\ell}})\alpha_j^2\bar{H}_{xx}}\left(\frac{4}{\pi}\,\frac{j}{4k^2 - j^2}\right)$$

$$D_j = c\alpha_k^2\beta_\ell^2\;\frac{2(\bar{\gamma}_{Q,j,2\ell} + \alpha_j^2)\bar{F} - (\bar{\gamma}_{D,j,2\ell} - 2\alpha_j^2\lambda_{c_{k\ell}})}{(\bar{\gamma}_{Q,j,2\ell} + \alpha_j^2)^2 + (\bar{\gamma}_{D,j,2\ell} - 2\alpha_j^2\lambda_{c_{k\ell}})\bar{\gamma}_{H,j,2\ell}}\left(\frac{4}{\pi}\,\frac{1}{j}\right)$$

$$\bar{F} = \frac{\bar{\gamma}_{Q,k,\ell} + \alpha_k^2}{\bar{\gamma}_{H,k,\ell}}$$

where

$$\bar{\gamma}_{D,j,2\ell} = \bar{D}_{xx}\alpha_j^4 + \bar{D}_{xy}\alpha_j^2\beta_{2\ell}^2 + \bar{D}_{yy}\beta_{2\ell}^4$$

$$\bar{\gamma}_{H,j,2\ell} = \bar{H}_{xx}\alpha_j^4 + \bar{H}_{xy}\alpha_j^2\beta_{2\ell}^2 + \bar{H}_{yy}\beta_{2\ell}^4$$

$$\bar{\gamma}_{Q,j,2\ell} = \bar{Q}_{xx}\alpha_j^4 + \bar{Q}_{xy}\alpha_j^2\beta_{2\ell}^2 + \bar{Q}_{yy}\beta_{2\ell}^4$$

$$\alpha_j^2 = j^2\,\frac{Rt}{2c}\left(\frac{\pi}{L}\right)^2$$

$$\beta_{2\ell}^2 = (2\ell)^2\,\frac{Rt}{2c}\left(\frac{1}{R}\right)^2$$

APPENDIX D: The averaged imperfection model

For the correlation studies reported in Section 7.5 one needs the measured initial imperfections in the form of double Fourier series. The measurements were taken by equipment specifically designed for this purpose and the results are reported in the literature[19,21,23]. A careful analysis of the imperfection data accumulated sofar has shown that it is possible to define characteristic imperfection distributions for the different fabrication techniques used[50]. For the laboratory scale shells A-8, AR-1 and AS-2 the following Donnell-Imbert[20] imperfection models were found to be applicable

$$\bar{W}_{io} = \frac{\bar{X}_A}{i^q} \quad ; \quad \bar{W}_{k\ell} = \frac{\bar{X}}{k^r \ell^s}$$

where

$\bar{W}_{io}$ = amplitude of the ith axisymmetric Fourier harmonic

$\bar{W}_{k\ell}$ = amplitude of the k,ℓ-th asymmetric Fourier harmonic

$\bar{X}_A, \bar{X}, q, r, s,$
= coefficients determined by least-square fitting the experimental values

The numerical values of the imperfection models used in the study reported in Section 7.5 are

	Cosine representation Axisymmetric $\bar{X}_A$	q	Sine representation Asymmetric $\bar{X}$	r	s
A-8	0.1280	1.18	1.630	1.01	1.33
AR-1	0.0208	1.50	0.206	1.18	1.22
AS-2	0.0068	0.25	0.786	1.12	1.23

EFFECT OF PLASTICITY ON POST-BUCKLING BEHAVIOUR

Viggo Tvergaard

Department of Solid Mechanics
The Technical University of Denmark
Lyngby, Denmark

ABSTRACT

Plastic buckling theory is presented, starting with the general theory of uniqueness and bifurcation for elastic-plastic solids. Also asymptotic procedures for estimating the initial post-bifurcation behaviour and the imperfection-sensitivity are presented, based on the classical elastic-plastic solid. The effect of other elastic-plastic constitutive descriptions on bifurcation and post-bifurcation behaviour is discussed. Several numerical analyses of the plastic buckling of plate - or shell structures are reviewed, and special attention is given to the localization of buckling patterns that often occurs as a secondary bifurcation. Finally, the effect of material strain-rate sensitivity on the prediction of plastic instabilities is discussed in some detail.

1. INTRODUCTION

Plastic buckling of structures involves a complex interaction between geometric and material non-linearities. For elastic-plastic columns the first studies date back to the last century. However, it was not until 1947 that Shanley [1] explained the significance of the tangent modulus load as the critical bifurcation point, at which the straight column configuration looses its uniqueness, but not its stability.

A general theory of bifurcation and uniqueness in elastic-plastic solids was subsequently developed by Hill [2,3], and this theory now forms the basis of practically all investigations of structural buckling in the plastic range. An extensive listing of plastic bifurcation analyses, including the early discussion of the reduced modulus load vs. the tangent modulus load, has been given by Sewell [4].

Post-buckling analyses in the plastic range are considerably complicated by the necessity to account for elastic unloading regions that start at bifurcation and subsequently spread into the material. Therefore, a general theory of post-buckling behaviour and imperfection sensitivity, such as that developed by Koiter [5] for the elastic range, is not available in the plastic range.

An asymptotic initial post-bifurcation theory has been developed by Hutchinson [6,7], as an extension of Hill's [2,3] bifurcation theory. This asymptotic analysis gives an important basis for understanding the mechanisms active in plastic buckling. However, for structures with a more complex geometry, or for structures with initial imperfections of various types the current knowledge of load carrying capacities relies almost entirely on numerical solutions.

The present text is aimed at giving an impression of the methods used and the types of solutions obtained in the field of plastic buckling. Both bifurcation theory, post-bifurcation theory and numerical solutions will be presented, and the difference between predictions obtained by different elastic-plastic constitutive laws will be discussed. The types of imperfections considered include geometric imperfections as well as residual stresses.

No attempt is made here to give a complete review of the field of plastic buckling, since the emphasis is entirely on illustrating methods and phenomena. The readers interested in more complete surveys are referred to a number of papers that give detailed discussions of recent progress, e.g. Hutchinson [7] or Needleman and Tvergaard [8] for basic theory, Tvergaard [9] or Budiansky and Hutchinson [10] for considerations of both elastic and plastic buckling behaviour, and Bushnell [11] for an extensive list of papers with emphasis on numerical investigations.

2. BASIC EQUATIONS FOR ELASTIC-PLASTIC PLATES OR SHELLS

The equations governing the deformations of plate or shell structures are frequently formulated with specific reference to isotropic elastic material behaviour. In the case of elastic-plastic materials a number of additional complications arise from the necessity to account for arbitrary variations of the instantaneous stiffnesses through the material.

In the following a point on the shell middle surface is identified by the coordinates x^1 and x^2 , while x^3 is the coordinate along the normal that measures the distance from the middle surface. With the usual assumptions of a first order shell theory the in-plane components of the Lagrangian strain tensor are approximated by

$$\eta_{\alpha\beta} = \varepsilon_{\alpha\beta} - x^3 \kappa_{\alpha\beta} \tag{2.1}$$

where $\varepsilon_{\alpha\beta}$ and $\kappa_{\alpha\beta}$ denote the membrane strain tensor and the bending strain tensor, respectively, for the middle surface. Greek indices range from 1 to 2 , while Latin indices (to be employed subsequently) range from 1 to 3 , and the summation convention is adopted for repeated indices.

The three-dimensional incremental constitutive relations for the elastic-plastic material are taken to be of the form

$$\dot{\sigma}^{ij} = L^{ijk\ell} \dot{\eta}_{k\ell} \tag{2.2}$$

where σ^{ij} are the contravariant components of the stress tensor, $(\dot{\ })$ denotes an incremental quantity, and $L^{ijk\ell}$ are the instantaneous moduli that vary as a function of the stress history in each material point. Since the stress state in the shell is approximately plane, only in-plane stresses enter into (2.2). Thus, using $\eta_{\alpha 3} = 0$ and $\sigma^{33} = 0$, the constitutive relations can be written as

$$\dot{\sigma}^{\alpha\beta} = \hat{L}^{\alpha\beta\gamma\delta} \dot{\eta}_{\gamma\delta} \quad , \quad \hat{L}^{\alpha\beta\gamma\delta} = L^{\alpha\beta\gamma\delta} - \frac{L^{\alpha\beta 33} L^{33\gamma\delta}}{L^{3333}} \tag{2.3}$$

The membrane stress tensor $N^{\alpha\beta}$ and the moment tensor $M^{\alpha\beta}$ in the shell with thickness h are taken to be

$$N^{\alpha\beta} = \int_{-h/2}^{h/2} \sigma^{\alpha\beta} dx^3 \quad , \quad M^{\alpha\beta} = -\int_{-h/2}^{h/2} \sigma^{\alpha\beta} x^3 dx^3 \tag{2.4}$$

Then the requirement of equilibrium can be specified in terms of the principle of virtual work

$$\int_A \left\{ N^{\alpha\beta} \delta\varepsilon_{\alpha\beta} + M^{\alpha\beta} \delta\kappa_{\alpha\beta} \right\} dA = EVW \tag{2.5}$$

where A is the area of the middle surface, and EVW denotes the external virtual work.

The incremental constitutive relations written in terms of the stress and strain quantities of shell theory are found by substituting (2.3) and (2.1) into the incremental form of (2.4). This results in relations of the form

$$\dot{N}^{\alpha\beta} = H_{(1)}^{\alpha\beta\gamma\delta} \dot{\varepsilon}_{\gamma\delta} + H_{(2)}^{\alpha\beta\gamma\delta} \dot{\kappa}_{\gamma\delta} \quad , \quad \dot{M}^{\alpha\beta} = H_{(2)}^{\alpha\beta\gamma\delta} \dot{\varepsilon}_{\gamma\delta} + H_{(3)}^{\alpha\beta\gamma\delta} \dot{\kappa}_{\gamma\delta} \tag{2.6}$$

where the incremental moduli are given by

$$H_{(i)}^{\alpha\beta\gamma\delta} = \int_{-h/2}^{h/2} \hat{L}^{\alpha\beta\gamma\delta} (-x^3)^{i-1} dx^3 \quad , \quad i = 1,2,3 \tag{2.7}$$

For nonzero values of $H^{\alpha\beta\gamma\delta}_{(2)}$ there is a coupling between bending and stretching in the constitutive relations (2.6). This coupling disappears in cases where the plane stress moduli $\hat{L}^{\alpha\beta\gamma\delta}$ are constant through the thickness, e.g. in elastic shells.

The displacement components are denoted u^α on the surface base vectors and w in the normal direction. In buckling analyses a non-linear membrane strain tensor is required, and a linear bending strain tensor is sufficient in most cases. The first order shell theory expressions given by Koiter [12] are

$$\varepsilon_{\alpha\beta} = \frac{1}{2}(u_{\alpha,\beta} + u_{\beta,\alpha}) - d_{\alpha\beta}w + \frac{1}{2}a^{\gamma\delta}(u_{\gamma,\alpha} - d_{\gamma\alpha}w)(u_{\delta,\beta} - d_{\delta\beta}w) + \frac{1}{2}(w_{,\alpha} + d^\gamma_\alpha u_\gamma)(w_{,\beta} + d^\delta_\beta u_\delta) \tag{2.8}$$

$$\kappa_{\alpha\beta} = \frac{1}{2}\Big[(w_{,\alpha} + d^\gamma_\alpha u_\gamma)_{,\beta} + (w_{,\beta} + d^\gamma_\beta u_\gamma)_{,\alpha} - \frac{1}{2}d^\gamma_\alpha(u_{\beta,\gamma} - u_{\gamma,\beta}) - \frac{1}{2}d^\gamma_\beta(u_{\alpha,\gamma} - u_{\gamma,\alpha})\Big] \tag{2.9}$$

where $a_{\alpha\beta}$ and $d_{\alpha\beta}$ are the metric tensor and the curvature tensor, respectively, of the undeformed middle surface, and $(\;)_{,\alpha}$ denotes covariant differentiation. It is noted that strain measures proposed by Niordson [13] are identical with (2.8) and (2.9), except for small differences in the bending strain measure of the order of $d^\gamma_\alpha \varepsilon_{\gamma\beta}$.

With the strain measures (2.8) and (2.9) the incremental version of the principle of virtual work (2.5) takes the form

$$\int_A \Big\{\dot{N}^{\alpha\beta}\delta\varepsilon_{\alpha\beta} + \dot{M}^{\alpha\beta}\delta\kappa_{\alpha\beta} + N^{\alpha\beta}\Big[a^{\gamma\mu}(\dot{u}_{\gamma,\alpha} - d_{\gamma\alpha}\dot{w})(\delta u_{\mu,\beta} - d_{\mu\beta}\delta w) + (\dot{w}_{,\alpha} + d^\gamma_\alpha \dot{u}_\gamma)(\delta w_{,\beta} + d^\mu_\beta \delta u_\mu)\Big]\Big\} dA = (EVW)^{\bullet} \tag{2.10}$$

Approximate, incremental solutions of plastic buckling problems are often directly based on (2.10). The incremental solution can also be based on the Euler equations of (2.10); but it should be noted that these differential equations in terms of the displacement increments are quite complex in the general case, where nonuniform plastic straining gives nonzero values of all components of the moduli $H^{\alpha\beta\gamma\delta}_{(i)}$.

The simpler Donnell-Mushtari-Vlasov (DMV) shell equations give sufficiently accurate solutions in many cases, when the wavelength of the characteristic deformation pattern is small compared with the radii of curvature of the middle surface, and the nonlinearities in terms of the in-plane displacements are unimportant. In the context of DMV shell theory many of the terms in (2.8) and (2.9) are neglected so that the nonlinear membrane strain tensor and the linear bending strain tensor are given by

$$\varepsilon_{\alpha\beta} = \frac{1}{2}(u_{\alpha,\beta} + u_{\beta,\alpha}) - d_{\alpha\beta}w + \frac{1}{2}w_{,\alpha}w_{,\beta} \tag{2.11}$$

$$\kappa_{\alpha\beta} = w_{,\alpha\beta} \tag{2.12}$$

With these strain measures the incremental principle of virtual work takes the form

$$\int_A \left\{ \dot{N}^{\alpha\beta}\delta\varepsilon_{\alpha\beta} + \dot{M}^{\alpha\beta}\delta\kappa_{\alpha\beta} + N^{\alpha\beta}\dot{w}_{,\alpha}\delta w_{,\beta} \right\} dA = (EVW)^{\bullet} \tag{2.13}$$

The equations of plate theory appear from the shell theories, by substituting $d_{\alpha\beta} = 0$. In particular, the DMV shell theory reduces to the often used von Kármán plate equations.

3. BIFURCATION ANALYSIS

The general theory of uniqueness and bifurcation in elastic-plastic solids developed by Hill [2,3,14] applies to the small strain plasticity formulations considered here as well as to finite strain plasticity. In this section the bifurcation theory will be presented in the context of the relatively simple DMV shell theory (Hutchinson [7], Needleman and Tvergaard [8]); but it is noted that an extension of this formulation to the more accurate first order shell equations (2.8) - (2.10) is straightforward (e.g. Tvergaard [15]).

3.1 Solids with a smooth yield surface and normality

The bifurcation theory of Hill [2] was developed for materials governed by a general smooth yield surface constitutive relation, in which the instantaneous moduli are of the form

$$L^{ijk\ell} = \mathcal{L}^{ijk\ell} - \alpha g^{-1} m^{ij} m^{k\ell} \tag{3.1}$$

Here, $\mathcal{L}^{ijk\ell}$ are the linear elastic instantaneous moduli, m^{ij} is the outward unit normal to the yield surface in strain-rate space, and g depends on the deformation history and on the current stress state. The plastic part of the strain rate is parallel with the normal, and $\alpha = 1$ for $m^{ij}\dot{\eta}_{ij} \geq 0$, while $\alpha = 0$ for $m^{ij}\dot{\eta}_{ij} < 0$.

In particular for the simplest flow theory of plasticity, J_2 flow theory, the quantities m^{ij} and g are given by

$$m^{ij} = \sqrt{\frac{3}{2}}\,\frac{s^{ij}}{\sigma_e} \quad , \quad g^{-1} = \left(\frac{E}{1+\nu}\right)\frac{E/E_t - 1}{E/E_t - (1-2\nu)/3} \tag{3.2}$$

where E is Young's modulus, ν is Poisson's ratio, $s^{ij} = \sigma^{ij} - g^{ij}\sigma^k_k/3$ is the stress deviator, g^{ij} is the metric tensor, $\sigma_e = (\frac{3}{2}s^{ij}s_{ij})^{\frac{1}{2}}$ is the effective Mises stress, and the tangent modulus E_t is the slope of the uniaxial stress-strain curve at stress level σ_e .

Based on (3.1) the plane stress moduli can be written in the form (Hutchinson [7])

$$\hat{L}^{\alpha\beta\gamma\delta} = \hat{\mathcal{L}}^{\alpha\beta\gamma\delta} - \alpha\hat{g}^{-1}\hat{m}^{\alpha\beta}\hat{m}^{\gamma\delta} \tag{3.3}$$

where $\hat{\mathcal{L}}^{\alpha\beta\gamma\delta}$ are the in-plane elastic moduli and

$$\hat{m}^{\alpha\beta} = m^{\alpha\beta} - m^{33}\frac{\mathcal{L}^{\alpha\beta 33}}{\mathcal{L}^{3333}} \quad , \quad \hat{g}^{-1} = g^{-1}\frac{\mathcal{L}^{3333}}{L^{3333}} \tag{3.4}$$

In (3.3) $\alpha = 1$ for $\hat{m}^{\alpha\beta}\dot{\eta}_{\alpha\beta} \geq 0$ and $\alpha = 0$ for $\hat{m}^{\alpha\beta}\dot{\eta}_{\alpha\beta} < 0$.

Now, the prescribed dead loads or displacements are taken to be proportional to a single parameter λ . The equations governing bifurcation are formulated by assuming, at any point of the loading history, that there are at least two distinct solutions $\dot{u}_\alpha^a$, $\dot{w}^a$ and $\dot{u}_\alpha^b$, $\dot{w}^b$ corresponding to a given increment $\dot{\lambda}$ of the prescribed quantity. The difference between such two solutions is denoted by $(\tilde{\ }) = (\dot{\ })^a - (\dot{\ })^b$ and thus according to (2.11) and (2.12) the difference between the strain increments of DMV shell theory are

$$\tilde{\varepsilon}_{\alpha\beta} = \frac{1}{2}(\tilde{u}_{\alpha,\beta} + \tilde{u}_{\beta,\alpha}) - d_{\alpha\beta}\tilde{w} + \frac{1}{2}(w^0_{,\alpha}\tilde{w}_{,\beta} + \tilde{w}_{,\alpha}w^0_{,\beta}) \qquad (3.5)$$

$$\tilde{\kappa}_{\alpha\beta} = \tilde{w}_{,\alpha\beta} \qquad (3.6)$$

where $(\ )^0$ denotes the fundamental state of deformation. By the definition of λ the differences between load increments vanish where loads are prescribed, and $\tilde{u}_\alpha$, $\tilde{w}$ vanish where displacements are prescribed. Then according to (2.13) the following equation must be satisfied

$$H \equiv \int_A \left[\tilde{N}^{\alpha\beta}\tilde{\varepsilon}_{\alpha\beta} + \tilde{M}^{\alpha\beta}\tilde{\kappa}_{\alpha\beta} + N_0^{\alpha\beta}\tilde{w}_{,\alpha}\tilde{w}_{,\beta}\right] dA = 0 \qquad (3.7)$$

A *comparison solid*, with fixed instantaneous moduli $L_c^{ijk\ell}$ in (2.2), is defined by choosing these moduli equal to the plastic moduli wherever the stress state is on the yield surface (independent of the sign of $m^{ij}\dot{\eta}_{ij}$) and the elastic moduli elsewhere. The linear comparison moduli are used in (2.7) to obtain the DMV moduli $H_{c(i)}^{\alpha\beta\gamma\delta}$ for the comparison solid and the following quadratic functional is introduced

$$F = \int_A \left[H_{c(1)}^{\alpha\beta\gamma\delta}\tilde{\varepsilon}_{\alpha\beta}\tilde{\varepsilon}_{\gamma\delta} + 2H_{c(2)}^{\alpha\beta\gamma\delta}\tilde{\varepsilon}_{\alpha\beta}\tilde{\kappa}_{\gamma\delta} + H_{c(3)}^{\alpha\beta\gamma\delta}\tilde{\kappa}_{\alpha\beta}\tilde{\kappa}_{\gamma\delta} + N_0^{\alpha\beta}\tilde{w}_{,\alpha}\tilde{w}_{,\beta}\right] dA \qquad (3.8)$$

For positive g in (3.1) it can be proven that the integrand of H is nowhere smaller than that of F , and thus

$$H \geq F \qquad (3.9)$$

Clearly, the requirement $F > 0$ for any admissible nonvanishing displacement fields is a sufficient condition for uniqueness, as shown by Hill [2].

Bifurcation at the lowest eigenvalue λ_c of F occurs only if $H = 0$ simultaneously with $F = 0$, and this requires that both $m^{ij}\dot{\eta}_{ij}^a \geq 0$ and $m^{ij}\dot{\eta}_{ij}^b \geq 0$ wherever the current stress state is on the yield surface.

In many cases the fundamental solution involves plastic loading everywhere in the surrent plastic zone, so that $m^{ij}\dot{\eta}_{ij}^0 > 0$. Now, one of the distinct solutions, $(\dot{\ })^a$, is identified with the fundamental solution $(\dot{\ })^0$, and the eigenmode obtained from (3.8), normalized in some convenient fashion, is denoted by $(\ )^{(1)}$. Then the other distinct solution can be written as $\dot{u}_\alpha^b = \dot{u}_\alpha^0 + \dot{\xi}\, u_\alpha^{(1)}$, etc., where ξ is an amplitude (defined so that $\dot{\xi} \geq 0$) , and the variation of the load parameter λ immediately after bifurcation is

$$\lambda = \lambda_c + \lambda_1 \xi + \ldots \tag{3.10}$$

Here, λ_1 must be chosen sufficiently large so that the plastic loading condition $m^{ij}\dot{\eta}^b_{ij} \geq 0$ is satisfied everywhere. Generally, $\lambda_1 > 0$ is required, so that bifurcation takes place under increasing load in agreement with the result of Shanley [1] for columns.

3.2 Solids with non-normality

Normality of the plastic flow rule is generally accepted as a good approximation for the behaviour of structural metals, whereas elastic-plastic material models that represent a significant amount of non-normality are important in studies of ductile fracture by the growth of microscopic voids, or in studies of soils, rocks or concrete. Thus, non-normality is not very relevant to analyses of structural stability; but the implications for bifurcation will be briefly discussed here, to illustrate the significance of the assumptions in section 3.1.

Instead of (3.1) we now consider a material with the instantaneous moduli

$$L^{ijk\ell} = \mathcal{L}^{ijk\ell} - \alpha g^{-1} m_G^{ij} m_F^{k\ell} \tag{3.11}$$

where $\alpha = 1$ for $m_F^{ij}\dot{\eta}_{ij} \geq 0$; but $m_G^{ij} \neq m_F^{ij}$. In this case the relation (3.9) is not satisfied for the comparison solid defined in the usual manner, and thus the requirement $F > 0$ does not exclude bifurcation in the underlying elastic-plastic solid. However, a bifurcation found for this comparison solid is a possible solution for the elastic-plastic solid, provided that the two distinct solution increments, $(\dot{\ })^a$ and $(\dot{\ })^b$, can be chosen so that plastic loading takes place wherever the current stress state is on the yield surface. Thus, the usual comparison solid provides an *upper bound*.

Raniecki and Bruhns [16] have proposed an *alternative comparison solid*, in which the non-zero value of α is used for every material point currently on the yield surface, but both m_G^{ij} and m_F^{ij} are replaced by $\frac{1}{2}(m_G^{ij} + r m_F^{ij})/\sqrt{r}$, for $r > 0$. It can be proven [16] that the critical bifurcation point for this solid provides a *lower bound* to the first critical bifurcation point of the actual elastic-plastic solid.

The upper and lower bound bifurcation predictions have been applied by Bruhns and Raniecki [17] and Tvergaard [18] in cases of tensile instabilities at finite strain, and these studies have shown that the best lower bound (optimal choice of r) may differ significantly from the upper bound.

3.3 Solids that form a vertex on the yield surface

The elastic-plastic constitutive relations discussed so far are based on the assumption of a smooth yield surface; but there is the possibility that a vertex forms on subsequent yield surfaces. At such a vertex the direction of the plastic part of the strain rate is a function of the stress rate, which has a rather strong effect on bifurcation predictions.

The occurrence of a vertex is implied by physical models of polycrystalline metal plasticity, based on the concept of single crystal slip (Hill [19], Hutchinson [20]), at least when the yield surface is defined for small offset plastic strains. The experimental evidence of vertex formation on yield surfaces is, however, conflicting (Hecker [21]).

In the context of plastic buckling it was noted more than 30 years ago that the bifurcation predictions based on J_2 deformation theory gave much better agreement with experimentally obtained buckling loads than those based on J_2 flow theory (Bijlaard [22], Stowell [23], Gerard and Becker [24]). Batdorf [25] realized that the bifurcation predictions of deformation theory can be justified by appealing to a flow theory of plasticity, which develops a vertex on the yield surface. Since then the possibility of vertex formation has played a significant role in discussions of plastic buckling (see Hutchinson [7]).

Here, we shall focus on a phenomenological corner theory of plasticity, called J_2 corner theory, proposed by Christoffersen and Hutchinson [26]. In this theory the instantaneous moduli for nearly proportional loading are chosen equal to the J_2 deformation theory moduli, and for increasing deviation from proportional loading the moduli increase smoothly until they coincide with the elastic moduli for stress increments directed along or within the corner of the yield surface.

With $M^0_{ijk\ell}$ denoting the deformation theory compliances, so that $\dot{\eta}_{ij} = M^0_{ijk\ell}\dot{\sigma}^{k\ell}$, and $\mathcal{M}_{ijk\ell}$ denoting the linear elastic compliances, the plastic part of the compliances is $C_{ijk\ell} = M^0_{ijk\ell} - \mathcal{M}_{ijk\ell}$. The yield surface in the neighbourhood of the loading point is taken to be a cone in stress deviator space (Fig. 3.1) with the cone axis in the direction

$$\lambda^{ij} = s^{ij}(C_{mnpq}s^{mn}s^{pq})^{-\frac{1}{2}} \tag{3.12}$$

parallel with the stress deviator tensor s^{ij}. A positive angular measure θ of the stress-rate direction relative to the cone axis is defined by

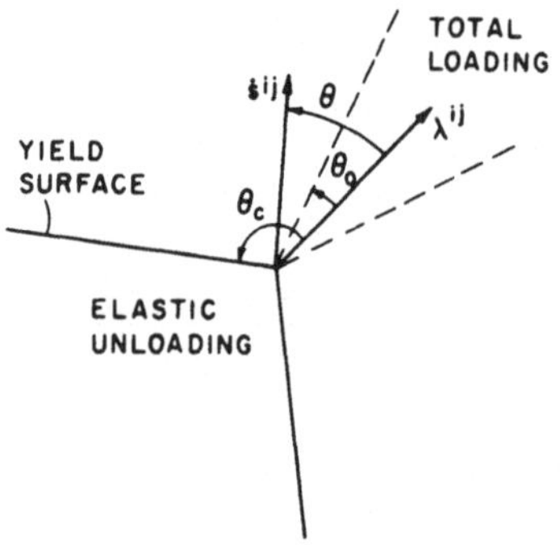

Fig. 3.1. Characterization of a vertex by the angular measure θ [8].

$$\cos\theta = C_{ijk\ell}\lambda^{ij}\overset{\bullet}{s}{}^{k\ell}(C_{mnpq}\overset{\bullet}{s}{}^{mn}\overset{\bullet}{s}{}^{pq})^{-\frac{1}{2}} \tag{3.13}$$

and a stress-rate potential at the vertex is formulated as

$$W = \frac{1}{2}M_{ijk\ell}\dot{\sigma}^{ij}\dot{\sigma}^{k\ell} + \frac{1}{2}f(\theta)C_{ijk\ell}\dot{\sigma}^{ij}\dot{\sigma}^{k\ell} \tag{3.14}$$

From this potential the strain-rate is obtained as

$$\dot{\eta}_{ij} = \frac{\partial^2 W}{\partial\dot{\sigma}^{ij}\partial\dot{\sigma}^{k\ell}}\dot{\sigma}^{k\ell} = M_{ijk\ell}(\theta)\dot{\sigma}^{k\ell} \tag{3.15}$$

Inverting (3.15) gives the θ - dependent moduli $L^{ijk\ell}(\theta)$ to be used in (2.2).

The angle of the yield surface cone is denoted θ_c , so that the transition function $f(\theta)$ in (3.14) is zero for $\theta_c \le \theta \le \pi$. In the total loading range, $0 \le \theta \le \theta_0$, $f(\theta)$ is unity, and in the transition region, $\theta_0 \le \theta \le \theta_c$, $f(\theta)$ is chosen to smoothly merge the deformation theory moduli with the elastic moduli in a way which ensures convexity of the incremental relation. The discussion in the following will refer to a function $f(\theta)$, which has been found in [26] to rather closely duplicate moduli obtained using a self-consistent model of a polycrystalline aggregate, and for which the moduli $L^{ijk\ell}(\theta)$ vary continuously with θ . The cone angle θ_c is often specified in terms of a different angular measure as $\beta = (\beta_c)_{max}$, where $\cos\beta = \dot{\sigma}_e(3\dot{s}_{ij}\dot{s}^{ij}/2)^{-\frac{1}{2}}$.

Bifurcation is again analysed by assuming that there are at least two incremental solutions $(\dot{\ })^a$ and $(\dot{\ })^b$ corresponding to a given increment of the prescribed quantity. The equation governing a non-zero difference $(\tilde{\ }) = (\dot{\ })^a - (\dot{\ })^b$ between such two solutions is still (3.7). Sewell [27] has shown for a pyramidal vertex that a *comparison solid* defined by the total loading moduli (all slip systems active) satisfies the inequality (3.9). In the context of J_2 corner theory the total loading moduli are those of deformation theory, valid for $0 \le \theta \le \theta_0$.

For many cases of interest the fundamental solution satisfies proportional loading or nearly proportional loading everywhere, so that $\theta < \theta_0$. Then, writing the variation of the load λ with the bifurcation mode amplitude ξ initially after bifurcation on the form

$$\lambda = \lambda_c + \lambda_1\xi + \ldots \tag{3.16}$$

the constant λ_1 can be chosen sufficiently large so that total loading is also satisfied on the initial part of the post-bifurcation path (3.16), as has been discussed by Needleman and Tvergaard [8]. In such cases the first critical bifurcation point for the comparison solid is identical with that of the elastic-plastic solid.

In the limit of a thoroughly nonlinear vertex description, $\theta_0 = 0$, the requirement of total loading at bifurcation gives rise to a smooth bifurcation $(\lambda_1 \to \infty$ in (3.16)), as has been found by Needleman and Tvergaard [8] for a cruciform column. It is noted that at this smooth bifurcation point the two incremental solutions $(\dot{\ })^a$ and $(\dot{\ })^b$ differ for the comparison solid, but not for the elastic-plastic solid.

If the fundamental solution has $\theta_0 < \theta < \theta_c$ in some material points, the total loading moduli may result in a rather poor lower bound, since bifurcation will be governed partly by the stiffer moduli of the transition range. Here, an *alternative comparison solid* may be defined by the instantaneous moduli associated with the current values of θ on the fundamental solution (see Tvergaard [15]). A bifurcation point found for this alternative comparison solid is also a bifurcation point of the underlying elastic-plastic solid, since the moduli $L^{ijk\ell}(\theta)$ vary continuously with θ, and since bifurcation stress-rate directions arbitrarily close to those of the fundamental solution can be enforced by superposing a sufficiently large fundamental solution increment on the eigenmode $(\tilde{\ })$. Thus, a bifurcation point predicted by this alternative comparison solid provides an *upper bound*. Determining the actual critical bifurcation point in cases where θ exceeds θ_0 is complex, since the corresponding critical value of θ must be determined simultaneously at every material point.

4. INITIAL POST-BIFURCATION BEHAVIOUR

A full understanding of bifurcation behaviour requires also insight in the post-bifurcation solution and the sensitivity to small imperfections, as is provided in the elastic range by Koiter's general theory of elastic stability [5]. For the plastic range an asymptotic theory of initial post-bifurcation behaviour has been developed by Hutchinson [6,7]. A considerable complication in this theory is the necessity to account for elastic unloading regions that start pointwise at bifurcation and subsequently spread into the material. The theory is limited to solids characterized by a smooth yield surface, with instantaneous moduli of the form (3.1).

4.1 Post-bifurcation theory

The post-bifurcation analysis is here presented rather briefly in the context of DMV shell theory (Hutchinson [7]), as a direct continuation of the bifurcation analysis given in section 3.1. The fundamental solution is taken to be associated with a monotonically increasing load parameter λ in the range of interest, and for simplicity consideration is restricted to cases in which the body has become fully plastic prior to bifurcation.

In the vicinity of the critical bifurcation load λ_c the asymptotically exact expression for the load parameter λ in terms of the bifurcation mode amplitude ξ is obtained of the form

$$\lambda = \lambda_c + \lambda_1\xi + \lambda_2\xi^{1+\beta} + \ldots \qquad (4.1)$$

for $\xi \geq 0$. For mode displacements in the opposite direction, the sign of ξ is changed, and then (4.1) still applies with different constants λ_1, λ_2, and β. In (4.1) λ_1 is positive, λ_2 is negative, and

$0 < \beta < 1$.

The expansion corresponding to (4.1) of the increments of the field quantities takes the form

$$\begin{Bmatrix} \dot{u}_{\gamma,\delta} \\ \dot{\eta}_{\gamma\delta} \\ \dot{\sigma}^{\gamma\delta} \\ \dot{N}^{\gamma\delta} \\ \vdots \end{Bmatrix} = \begin{Bmatrix} \dot{u}^{0}_{\gamma,\delta} \\ \dot{\eta}^{0}_{\gamma\delta} \\ \dot{\sigma}^{\gamma\delta}_{0} \\ \dot{N}^{\gamma\delta}_{0} \\ \vdots \end{Bmatrix} + \begin{Bmatrix} \overset{(1)}{u}_{\gamma,\delta} \\ \overset{(1)}{\eta}_{\gamma\delta} \\ \overset{(1)}{\sigma}{}^{\gamma\delta} \\ \overset{(1)}{N}{}^{\gamma\delta} \\ \vdots \end{Bmatrix} + \xi^{\beta} \begin{Bmatrix} \overset{*}{u}_{\gamma,\delta} \\ \overset{*}{\eta}_{\gamma\delta} \\ \overset{*}{\sigma}{}^{\gamma\delta} \\ \overset{*}{N}{}^{\gamma\delta} \\ \vdots \end{Bmatrix} + \begin{Bmatrix} \overset{d}{u}_{\gamma,\delta}(\xi\ ,x^{\alpha}) \\ \overset{d}{\eta}_{\gamma\delta}(\xi\ ,x^{i}) \\ \overset{d}{\sigma}{}^{\gamma\delta}(\xi\ ,x^{i}) \\ \overset{d}{N}{}^{\gamma\delta}(\xi\ ,x^{\alpha}) \\ \vdots \end{Bmatrix} \tag{4.2}$$

Here, $(\dot{\ })$ is taken to denote $d(\ )/d\xi$ and, since the fundamental solution is a function of λ , $(\dot{\ })^{0} \equiv (d(\ )^{0}/d\lambda)(d\lambda/d\xi)$. The quantities denoted by $(^{(1)})$ in (4.2) are the eigenmodal quantities associated with the lowest eigenvalue λ_c of $F = 0$. This mode also satisfies the variational equation

$$\int_A \left\{ \overset{(1)}{N}{}^{\alpha\beta}\delta^{c}\varepsilon_{\alpha\beta} + \overset{(1)}{M}{}^{\alpha\beta}\delta\kappa_{\alpha\beta} + N^{\alpha\beta}_{0c}\overset{(1)}{w}_{,\alpha}\delta w_{,\beta} \right\} dA = 0 \tag{4.3}$$

where

$$\delta^{c}\varepsilon_{\alpha\beta} \equiv \frac{1}{2}(\delta u_{\alpha,\beta} + \delta u_{\beta,\alpha}) - d_{\alpha\beta}\delta w + \frac{1}{2}(\overset{0c}{w}_{,\alpha}\delta w_{,\beta} + \overset{0c}{w}_{,\beta}\delta w_{,\alpha}) \tag{4.4}$$

The quantities denoted by $(\overset{*}{\ })$ in (4.2) result from boundary layer terms due to elastic unloading, and $(^{d})$ denotes higher order terms that satisfy $\xi^{-\beta}(^{d}) \to 0$ for $\xi \to 0$.

As mentioned in connection with (3.10), the initial slope λ_1 must be chosen large enough so that no elastic unloading occurs at bifurcation in any material point

$$\hat{m}^{\alpha\beta}_{c}\left(\lambda_1 \left.\frac{d\overset{0}{\eta}_{\alpha\beta}}{d\lambda}\right|_{c} + \overset{(1)}{\eta}_{\alpha\beta} \right) \geq 0 \tag{4.5}$$

If the possibility of elastic unloading is neglected, so that the material is considered as hypoelastic (see section 5), the expansion (4.1) is replaced by $\lambda = \lambda_c + \lambda_1^{he}\xi + O(\xi^2)$. Now, if the minimum value of λ_1 that is needed to satisfy the equality in (4.5) is less than λ_1^{he} , then in the immediate vicinity of the bifurcation point no elastic unloading will occur and, in this vicinity, the elastic-plastic solid behaves as the hypoelastic solid. The more usual case is that the minimum value of λ_1 that satisfies the equality in (4.5) is greater than λ_1^{he} . Clearly, choosing a smaller value of λ_1 than this minimum violates (4.5). Now suppose a value of λ_1 greater than this minimum were chosen. Then, no elastic unloading would occur in some vicinity of the bifurcation point. Hence, in this vicinity the elastic-plastic solid would behave as the hypoelastic solid for which $\lambda_1 = \lambda_1^{he}$, but this choice of λ_1 would violate (4.5), leading to a contradiction. Therefore, λ_1 must be taken as the minimum value that satisfies (4.5). In the following discussion attention is confined to the usual case where $\lambda_1 > \lambda_1^{he}$.

Neutral loading at bifurcation and subsequent elastic unloading may start at one isolated point, at several points simultaneously, or along a curve. In a case where unloading starts at one isolated point x_c^i on a smooth shell surface (Fig. 4.1a) a set of local Cartesian coordinates z_i are introduced, centered at x_c^i, such that z_3 is directed along the outward normal to the surface. In the analysis a Taylor expansion is needed for the function on the left-hand side in (4.5)

$$\hat{m}_c^{\alpha\beta}\left(\lambda_1 \frac{d\overset{0}{\eta}_{\alpha\beta}}{d\lambda}\bigg|_c + \overset{(1)}{\eta}_{\alpha\beta}\right) = C_3 z_3 + C_{\alpha\beta} z_\alpha z_\beta + \ldots \tag{4.6}$$

and for the shell surface

$$D_3 z_3 + D_{\alpha\beta} z_\alpha z_\beta + \ldots = 0 \tag{4.7}$$

In the following $C_3 < 0$ is assumed, and it is noted that there are cases in which some of the linear terms with coefficients C_1, C_2, D_1, and D_2 in (4.6) and (4.7) do not vanish.

Now, stretched boundary layer coordinates $\overset{*}{z}_i$ are chosen (Hutchinson [7]) such that the surfaces enclosing the elastically unloaded region are independent of ξ to lowest order when written in terms of $\overset{*}{z}_i$. Anticipating $\lambda_2 < 0$, we choose

$$\overset{*}{z}_i = z_i \left[\frac{\xi^{-\beta}}{-\lambda_2(1+\beta)}\right]^{a_i}, \tag{4.8}$$

where $a_i = 1$ if the lowest order z_i-term in (4.6) or (4.7) is linear, $a_i = \frac{1}{2}$ if the lowest order z_i-term is quadratic, and $a_i = 0$ if the two surfaces (4.6) and (4.7) are parallel in the z_i-direction.

Expanding the expression $\hat{m}^{\gamma\delta}\dot{\eta}_{\gamma\delta} = 0$ for the instantaneous neutral-loading surface and introducing the stretched coordinates the following lowest order expression appears

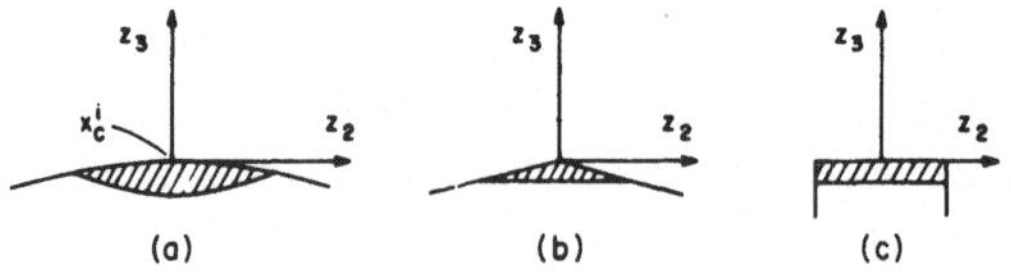

Fig. 4.1. Cross-sections of elastic unloading regions (hatched areas). (a) Unloading starts at a single point of smooth surface. (b) Unloading starts at the top of a sharp edge on the surface. (c) Unloading starts simultaneously along the z_2-axis [8].

$$0 = \hat{m}^{\gamma\delta}\dot{\eta}_{\gamma\delta} = \xi^{\beta}[(1+\beta)\lambda_2 f(\overset{*}{z}_i) + \hat{m}_c^{\gamma\delta}\overset{*}{\eta}_{\gamma\delta}] + \dots , \tag{4.9}$$

where, for fixed $\overset{*}{z}_i$, the terms in square brackets have values independent of ξ . In particular, for (4.6)

$$f(\overset{*}{z}_i) = \left[\hat{m}_c^{\gamma\delta}\frac{d\overset{0}{\eta}_{\gamma\delta}}{d\lambda}\bigg|_c\right]_{x_c^i} - C_3\overset{*}{z}_3 - C_{\alpha\beta}\overset{*}{z}_\alpha\overset{*}{z}_\beta . \tag{4.10}$$

The strains are constrained by (2.1) to vary linearly through the thickness, which means that a nonzero $\overset{*}{\eta}_{\gamma\delta}$, resulting from nonzero $\overset{*}{\varepsilon}_{\gamma\delta}$ and $\overset{*}{\kappa}_{\gamma\delta}$, would vary linearly through the entire thickness, and also outside the region of elastic unloading. Therefore, we must require $\overset{*}{\eta}_{\gamma\delta} = \overset{*}{\varepsilon}_{\gamma\delta} = \overset{*}{\kappa}_{\gamma\delta} = 0$, and consequently the lowest order expression (4.9) for the neutral loading surface reduces to $f(\overset{*}{z}_i) = 0$ in stretched coordinates.

Hutchinson's analysis of the lowest order changes of the stress rate due to elastic unloading leads to the expression

$$\overset{*}{\sigma}{}^{\gamma\delta} = (1+\beta)\lambda_2[\hat{g}_c^{-1}\,\hat{m}_c^{\gamma\delta}]_{x_c^i}\, f(\overset{*}{z}_i) \tag{4.11}$$

inside the unloading region and $\overset{*}{\sigma}{}^{\gamma\delta} = 0$ outside. The contribution of these stresses to the increment of the membrane stress tensor is

$$\overset{*}{N}{}^{\gamma\delta} = \int_{-h/2}^{h/2}\overset{*}{\sigma}{}^{\gamma\delta}dx^3 = -(1+\beta)\lambda_2\xi^{\beta}\int\overset{*}{\sigma}{}^{\gamma\delta}d\overset{*}{z}_3 + \dots$$

$$= -(1+\beta)^2\lambda_2^2\,\xi^{\beta}[\hat{g}_c^{-1}\,\hat{m}_c^{\gamma\delta}]_{x_c^i}\int f(\overset{*}{z}_i)d\overset{*}{z}_3 + \dots \tag{4.12}$$

and similarly for $\overset{*}{M}{}^{\gamma\delta}$, where the last integrals extend through the unloading region.

The incremental principle of virtual work (2.13) is satisfied by the fundamental solution as well as by the bifurcated solution. When the fundamental solution is used to eliminate the external virtual work from (2.13), and the expansion (4.2) is substituted into the resulting equation, using (4.1), (4.3), and (4.4), we obtain

$$\int_A\left\{\overset{d}{N}{}^{\gamma\delta}\delta^c\varepsilon_{\gamma\delta} + \overset{d}{M}{}^{\gamma\delta}\delta\kappa_{\gamma\delta} + N_{0c}^{\gamma\delta}\overset{d}{w}_{,\gamma}\delta w_{,\delta}dA\right\} +$$

$$\xi^{\beta}\int_A\left\{\overset{*}{N}{}^{\gamma\delta}\delta^c\varepsilon_{\gamma\delta} + \overset{*}{M}{}^{\gamma\delta}\delta\kappa_{\gamma\delta}\right\}dA + \xi\int_A\left\{2\lambda_1\overset{(1)}{N}{}^{\gamma\delta}\frac{d\overset{0}{w}_{,\gamma}}{d\lambda}\bigg|_c\delta w_{,\delta} + \right.$$

$$\left.2\left(\lambda_1\frac{dN_0^{\gamma\delta}}{d\lambda}\bigg|_c + \overset{(1)}{N}{}^{\gamma\delta}\right)\overset{(1)}{w}_{,\gamma}\delta w_{,\delta}\right\}dA + \dots = 0 . \tag{4.13}$$

This variational equation for the $\overset{d}{(\;)}$ - quantities is used to identify the value of β and to determine λ_2 .

The second term in (4.13) can be rewritten using (4.8) and (4.12) as

$$\xi^{\beta}\int_A\left\{\overset{*}{N}{}^{\gamma\delta}\,\delta^c\varepsilon_{\gamma\delta}+\overset{*}{M}{}^{\gamma\delta}\,\delta\kappa_{\gamma\delta}\right\}dA =$$

$$-\,[\xi^{\beta}(1+\beta)(-\lambda_2)]^{(2+a_1+a_2)}\,[\hat{g}_c^{-1}\,\hat{m}_c^{\gamma\delta}\,\delta\eta_{\gamma\delta}]_{x_c^i}\int_{\overset{*}{V}} f(\overset{*}{z}_i)\,d\overset{*}{V}+\ldots \tag{4.14}$$

Here, $d\overset{*}{V} = d\overset{*}{z}_1 d\overset{*}{z}_2 d\overset{*}{z}_3$, and $\overset{*}{V}$ is the elastic unloading region enclosed by (4.10) and the stretched coordinate version of (4.7). Now substituting (4.14) into (4.13) it can be argued that all three terms must be of order ξ , with $\beta = (2+a_1+a_2)^{-1}$. Suppose $\beta > (2+a_1+a_2)^{-1}$, so that (4.14) is of order higher than ξ . Then the first term of (4.13) must be of order ξ and balance the last term; but this is only possible if $\lambda_1 = \lambda_1^{he}$, i.e. without elastic unloading. If $\beta < (2+a_1+a_2)^{-1}$, the first two terms in (4.13) must balance each other and the $(^d)$ - terms be of order $\xi^{\beta(2+a_1+a_2)}$; but then the bifurcation mode would be the only solution for $(^d)$ and $\lambda_2 = 0$ would be required (implying that the boundary terms vanish).

Thus, $\beta = (2+a_1+a_2)^{-1}$ must be chosen, and (4.13) leads to the following equation from which λ_2 is determined

$$\{-\lambda_2(1+\beta)\}^{1/\beta}[\hat{g}_c^{-1}\,\hat{m}_c^{\gamma\delta}\,\overset{(1)}{\eta}_{\gamma\delta}]_{x_c^i}\int_{V^*} f(\overset{*}{z}_i)\,d\overset{*}{V} = \mathcal{A} + \lambda_1\mathcal{B} \tag{4.15}$$

where $\mathcal{A}$ and $\mathcal{B}$, to be specified in section 5, are determined from a hypoelastic analysis.

In the most common case, in which unloading starts at an isolated point on a smooth surface, as anticipated in (4.6), (4.7), and Fig. 4.1a, $a_1 = a_2 = \frac{1}{2}$, which leads to $\beta = \frac{1}{3}$. If unloading starts simultaneously along every point of the z_2 - axis, such as on a flat surface of a column (Fig. 4.1c), or in axisymmetric bifurcation of a circular cylindrical shell, $a_1 = \frac{1}{2}$, $a_2 = 0$ results in $\beta = \frac{2}{5}$ (Hutchinson [7]). A wedgeshaped unloading region (Fig. 4.1b), due to linear z_2 - terms in the expression corresponding to either (4.6) or (4.7), requires $a_1 = \frac{1}{2}$, $a_2 = 1$, and thus $\beta = \frac{2}{7}$ (Tvergaard and Needleman [28,29,30]).

Since $\lambda_2 < 0$, the first three terms of (4.1) can be employed to obtain the following estimates of the maximum support load λ_{max} and the corresponding bifurcation mode amplitude ξ_{max}

$$\xi_{max} = \left(\frac{\lambda_1}{-\lambda_2(1+\beta)}\right)^{1/\beta} \quad , \quad \lambda_{max} = \lambda(\xi_{max}) \tag{4.16}$$

It must be emphasized, however, that although (4.1) is an asymptotically exact expression, the values given by (4.16) are not asymptotic in any sense, since the maximum occurs at a finite (perhaps small) value of ξ . In most cases numerical computations of post-bifurcation behaviour show that both ξ_{max} and the difference $\lambda_{max} - \lambda_c$ are somewhat larger than predicted by (4.16).

4.2 Results for integrally stiffened panels

The asymptotic post-bifurcation theory will be illustrated here by results obtained by Tvergaard and Needleman [28,29] for integrally stiffened panels under axial compression. The bifurcation analysis loading to the results will be briefly described as well.

The stiffened panel is assumed to be infinitely wide in the x_2-direction, with a constant spacing b between the stiffeners, and is under axial compression in the x_1-direction (Fig. 4.2). The plate thickness is h, and the eccentricity e of the stiffeners is positive in the x_3-direction. The panel is either finite in the x_1-direction with the distance a between the simply supported edges (Fig. 4.2a) or it is continuous over several transverse supports at distance a (Fig. 4.2b).

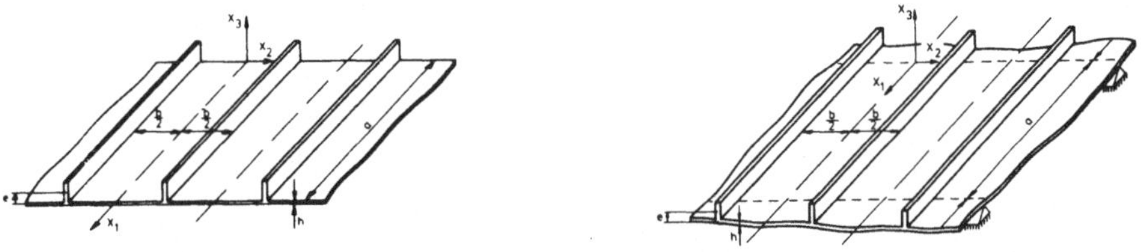

(a) (b)

Fig. 4.2. Part of integrally stiffened panel. (a) On two simple supports at distance a . (b) Continuous over several supports at distance a [28,29].

The stress state in a perfect panel prior to bifurcation is a pure membrane state with the only nonvanishing stress component being a constant axial stress $\sigma_{11} = \lambda\sigma_{11}^0$ at every point of the plate and the stiffeners. Thus, in general the relationship between the in-plane stress rates and strain rates stops being isotropic as soon as the absolute value of $\lambda\sigma_{11}^0$ exceeds the yield stress σ_y. Then the lowest bifurcation load can be determined as that of an elastic stiffened orthotropic plate with moduli equal to the instantaneous plastic moduli.

In the following we use the expressions $E_1 = \hat{L}_{1111}$, $E_2 = \hat{L}_{2222}$, $E_{12} = \hat{L}_{1122}$, $E_G = \hat{L}_{1212}$ and $E_s = \hat{L}_s$ for the plastic branch of the nonvanishing components of instantaneous moduli in the prebuckling state at the bifurcation point. We also use the area A_s and the area moment of inertia I_s of the stiffener cross-section, and the expression $D_1 = E_1h^3/12$, $D_2 = E_2h^3/12$, $D_{12} = E_{12}h^3/12$ and $D_G = E_Gh^3/6$. The bifurcation mode, consisting of $\overset{(1)}{u}_1$, $\overset{(1)}{u}_2$, $\overset{(1)}{w}$, is found directly from the Euler equations of the variational equation $\delta F = 0$, where F is given by (3.8). These differential equations are of the form

$$E_1 \overset{(1)}{u}_{1,11} + E_G \overset{(1)}{u}_{1,22} + (E_{12} + E_G) \overset{(1)}{u}_{2,12} = 0 \tag{4.17}$$

$$(E_{12} + E_G) \overset{(1)}{u}_{1,12} + E_G \overset{(1)}{u}_{2,11} + E_2 \overset{(1)}{u}_{2,22} = 0 \tag{4.18}$$

$$D_1 \overset{(1)}{w}_{,1111} + 2(D_{12}+D_G)\overset{(1)}{w}_{,1122} + D_2 \overset{(1)}{w}_{,2222} - \lambda_c N_{11}^0 \overset{(1)}{w}_{,11} = 0 \qquad (4.19)$$

and the corresponding linear discontinuity conditions at a stiffener are

$$-E_G h\left(\overset{(1)+}{u}_{1,2} - \overset{(1)-}{u}_{1,2} + \overset{(1)+}{u}_{2,1} - \overset{(1)-}{u}_{2,1}\right) - E_s A_s\left(\overset{(1)-}{u}_{1,11} - e\overset{(1)-}{w}_{,111}\right) = 0 \qquad (4.20)$$

$$-E_2\left(\overset{(1)+}{u}_{2,2} - \overset{(1)-}{u}_{2,2}\right) - E_{12}\left(\overset{(1)+}{u}_{1,1} - \overset{(1)-}{u}_{1,1}\right) = 0 \qquad (4.21)$$

$$D_2\left(\overset{(1)+}{w}_{,222} - \overset{(1)-}{w}_{,222}\right) + E_s I_s \overset{(1)-}{w}_{,1111} - eE_s A_s\left(\overset{(1)-}{u}_{1,111} - e\overset{(1)-}{w}_{,1111}\right) -$$
$$\lambda_c N_s^0 \overset{(1)-}{w}_{,11} = 0 \qquad (4.22)$$

$$-D_2\left(\overset{(1)+}{w}_{,22} - \overset{(1)-}{w}_{,22}\right) - G_s K_s \overset{(1)-}{w}_{,112} - \lambda_c e^2 N_s^0 \overset{(1)-}{w}_{,112} = 0 \qquad (4.23)$$

$$\overset{(1)+}{u}_1 = \overset{(1)-}{u}_1 , \quad \overset{(1)+}{u}_2 = \overset{(1)-}{u}_2 , \quad \overset{(1)+}{w} = \overset{(1)-}{w} , \quad \overset{(1)+}{w}_{,2} = \overset{(1)-}{w}_{,2} \qquad (4.24)$$

Here superscript + refers to the side of the stiffener in the positive x_2 - direction and superscript - refers to the other side. The pre-buckling unit membrane stress and stiffener force are $N_{11}^0 = \sigma_{11}^0 h$ and $N_s^0 = \sigma_{11}^0 A_s$, respectively.

For the panel of finite length in the x_1 - direction the simple support boundary conditions at the edges $x_1 = 0 , a$ are taken to be

$$\overset{(1)}{u}_{1,1} = \overset{(1)}{u}_2 = \overset{(1)}{w} = \overset{(1)}{w}_{,11} = 0 \qquad (4.25)$$

For the continuous panel over several transverse supports at distance a (Fig. 4.2b) the bifurcation load and the bifurcation mode between two supports are identical with the solution of equations (4.17) - (4.25).

In the infinitely wide periodic structure, the buckling mode displacements $\overset{(1)}{u}_1$ and $\overset{(1)}{w}$ are symmetric about the centre line $x_2 = 0$ between two stiffeners, and $\overset{(1)}{u}_2$ is antisymmetric about this line. After rather lengthy calculations the solution of equations (4.17) - (4.25) is found to be of the form

$$\overset{(1)}{w} = \left\{c_1 \cosh(r_1 x_2) + c_2 \cos(r_2 x_2)\right\}\sin\frac{k\pi x_1}{a} \qquad (4.26)$$

$$\overset{(1)}{u}_1 = \left\{c_3 \cosh(r_3 x_2)\cos(r_4 x_2) + c_4 \sinh(r_3 x_2)\sin(r_4 x_2)\right\}\cos\frac{k\pi x_1}{a} \qquad (4.27)$$

$$\overset{(1)}{u}_2 = \Big\{(c_3 b_1 + c_4 b_2)\sinh(r_3 x_2)\cos(r_4 x_2)$$
$$+ (-c_3 b_2 + c_4 b_1)\cosh(r_3 x_2)\sin(r_4 x_2)\Big\}\sin\frac{k\pi x_1}{a} \qquad (4.28)$$

in which k is a positive integer, and the constants r_1, r_2, r_3, r_4, b_1 and b_2 depend on k, a, λ_c and the instantaneous moduli. The expressions for these parameters are lengthy and will not be given here. The same mode expressions (4.26) - (4.28) apply in the local coordinate system of a neighbouring plate section between two stiffeners, but with different constants $c_5 - c_8$ instead of $c_1 - c_4$. Substituting these buckling mode expressions in the discontinuity conditions (4.20) - (4.24), we obtain eight linear, homogeneous equations, from which the constants $c_1 - c_8$ and the smallest critical bifurcation load λ_c are determined.

Buckling as a wide Euler column, with $k = 1$ and identical modes for all plate sections between two stiffeners, is critical as long as the stiffeners are relatively weak. When the bending stiffness of the stiffeners is sufficiently high, local buckling of the plate between the stiffeners occurs first, usually with a value of k somewhat above a/b. The local buckling displacement fields in two neighbouring plate sections are usually identical with opposite sign.

The results to be shown are obtained for panels specified by $a/b = 4$, $e/b = 0.1$, $(A_s + hb)/b^2 = 0.0256$, with integral stiffeners of rectangular cross-section. The ratio between the amount of material in the plate and that in the whole panel is specified in terms of the parameter $\alpha = (1 + A_s/bh)^{-1}$. The uniaxial stress-strain curve is represented as a power hardening law with a well defined yield stress σ_y and continuous tangent modulus

$$\varepsilon = \begin{cases} \dfrac{\sigma}{E} & , \quad \text{for} \quad \sigma \leq \sigma_y \\ \dfrac{\sigma_y}{E}\left(\dfrac{1}{n}\left(\dfrac{\sigma}{\sigma_y}\right)^n - \dfrac{1}{n} + 1\right) & , \quad \text{for} \quad \sigma > \sigma_y \end{cases} \tag{4.29}$$

For the panel simply supported at the two edges $x_1 = 0, a$, Fig. 4.3 indicates the shape of the elastic unloading zones that propagate into the material for the case of local buckling and also for the case of wide column buckling either in the positive x_3-direction or in the negative x_3-direction. Fig. 4.3 also gives the corresponding values of β. For the continuous panel on several transverse supports at distance

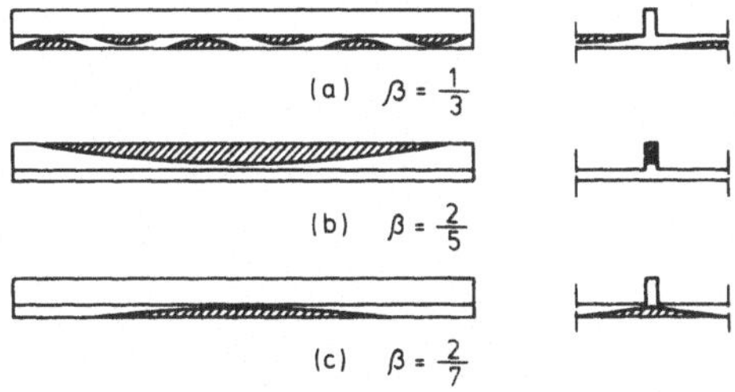

Fig. 4.3. Shape of elastic unloading zones in panel simply supported at two edges (plate thickness exaggerated). (a) Local buckling with 6 half sine waves in x_1-direction. (b) Wide column buckling in positive x_3-direction. (c) Wide column buckling in negative x_3-direction [29].

a , the shape of the elastic unloading zones at local buckling are identical with Fig. 4.3a. When the continuous panel bifurcates in the wide column buckling mode the unloading zones shown in Fig. 4.3b appear in half of the bays that buckle in positive x_3 - direction whereas no elastic unloading occurs at bifurcation in the other half of the bays that buckle in negative x_3 - direction.

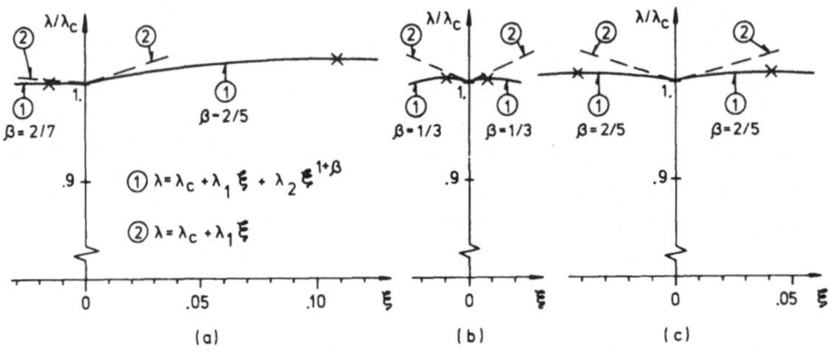

Fig. 4.4. Load-mode displacement relationship. (a) Wide column buckling in panel simply supported at the two edges. Buckling in positive x_3 - direction for positive ξ ($\alpha = 0.65$, $\sigma_y/E = 0.001$, $n = 10$). (b) Local buckling ($\alpha = 0.525$, $\sigma_y/E = 0.001$, $n = 10$) . (c) Wide column buckling for continuous panel on several transverse supports ($\alpha = 0.65$, $\sigma_y/E = 0.001$, $n = 10$) [29].

Fig. 4.4 shows the asymptotic relationship between load and mode displacement amplitude for wide column mode bifurcation in a panel simply supported at the two edges, for wide column mode bifurcation in a continuous panel on several transverse supports and for local mode bifurcation in a panel with either of these two sets of boundary conditions. The buckling mode is normalized so that for $\xi = 1$ the maximum deflection w is equal to the plate thickness. For bifurcation in the local buckling mode the plastic post-bifurcation behaviour is symmetric, and the same is true for wide column mode bifurcation of the continuous panel. At wide column mode bifurcation for a panel simply supported at the two edges the post-bifurcation behaviour is asymmetric as shown in Fig. 4.4a.

Comparison with numerically computed post-bifurcation solutions shows that the general shape of the curves is well represented by the asymptotic estimates [28,29]. However, as also mentioned in connection with (4.16), the actual values of $\lambda_{max} - \lambda_c$ and ξ_{max} are significantly underestimated by the asymptotic solutions.

One of the most important outcomes of the asymptotic post-bifurcation analysis is the detailed description it gives of the influence of elastic unloading just after bifurcation. This forms the main basis for understanding the difference between elastic and elastic-plastic post-bifurcation behaviour.

5. HYPOELASTIC ESTIMATES OF IMPERFECTION-SENSITIVITY

The effect of a small initial imperfection on a structure compressed into the plastic range has not been described by a simple asymptotic formula such as those obtained by Koiter [5] for the elastic range. The main difficulty is that an asymptotic expansion of the initial part of the equilibrium solution is only valid up to the point at which elastic unloading starts, while a second expansion that accounts for the growing elastic unloading region is required to represent the remaining part of the equilibrium path (Hutchinson [7]). It is also a complication that the maximum support load of a perfect elastic-plastic structure is attained at a limit point after finite buckling mode deflections and not at the bifurcation point as in the elastic range.

The use of a hypoelastic asymptotic analysis to estimate the imperfection-sensitivity of an elastic-plastic structure was first introduced by Hutchinson and Budiansky [31], for the case of a cruciform column. This structure is exceptional in that often no strain rate reversal occurs before the maximum load, so that neglecting elastic unloading, and thus replacing the elastic-plastic material model by a corresponding hypoelastic material model, is completely justified. For most other structures compressed into the plastic range elastic unloading does occur before the load maximum.

The possibility of also using the hypoelastic asymptotic analysis in cases where neglecting elastic unloading is not rigorously justified has been investigated by Needleman and Tvergaard [32]. It was found, for an axially compressed square plate that the hypoelastic asymptotic estimates can reveal the 'overall' behaviour when small initial imperfections are present.

Another approximate approach that accounts for elastic unloading by using an asymptotic estimate of the locus of maxima around the reduced modulus load has been suggested by van der Heijden [33]. This method gives a useful representation of the imperfection-sensitivity in the case of a model problem, but has not yet been generalised to more realistic structures.

The asymptotic analysis, based on neglecting elastic unloading will be described here rather briefly in the context of DMV-shell theory (Tvergaard [34], Needleman and Tvergaard [8]). The incremental constitutive relations are still of the form (2.3); but the plastic loading branch $\hat{L}^{\alpha\beta\gamma\delta}$ is taken to be active everywhere in the current plastic zone regardless of the direction of $\dot{\eta}_{\gamma\delta}$ in strain-rate space.

The perfect shell made of hypoelastic material has the same bifurcation stress as the corresponding elastic-plastic shell, but instead of (4.1) the initial post-bifurcation expansion in the case of a unique buckling mode takes the form

$$\lambda = \lambda_c + \lambda_1^{he}\xi + \lambda_2^{he}\xi^2 + \ldots \tag{5.1}$$

$$\left.\begin{aligned} u_\alpha &= u_\alpha^0 + \overset{(1)}{u}_\alpha \xi + \overset{(2)}{u}_\alpha \xi^2 + \ldots \\ w &= w^0 + \overset{(1)}{w}\xi + \overset{(2)}{w}\xi^2 + \ldots \end{aligned}\right\} \tag{5.2}$$

where the higher order contributions are taken to be orthogonal to the bifurcation mode

$$\int_A N_c^{\alpha\beta} \overset{(1)}{w}_{,\alpha} \overset{(i)}{w}_{,\beta}\, dA = 0 \quad , \quad \text{for} \quad i > 1 . \tag{5.3}$$

The expansion of the strains, analogous to (5.2), is obtained by substituting (5.2) into (2.1), (2.11) and (2.12), and the corresponding expansion of the stresses is obtained from (2.3), using the following Taylor expansion of the tensor of moduli about the bifurcation point λ_c

$$\begin{aligned} \hat{L}^{\alpha\beta\gamma\delta} = \hat{L}_c^{\alpha\beta\gamma\delta} &+ \left.\frac{\partial \hat{L}^{\alpha\beta\gamma\delta}}{\partial \sigma^{\mu\nu}}\right|_c (\sigma^{\mu\nu} - \sigma_c^{\mu\nu}) \\ &+ \frac{1}{2} \left.\frac{\partial^2 \hat{L}^{\alpha\beta\gamma\delta}}{\partial \sigma^{\mu\nu}\partial \sigma^{\rho\omega}}\right|_c (\sigma^{\mu\nu} - \sigma_c^{\mu\nu})(\sigma^{\rho\omega} - \sigma_c^{\rho\omega}) + \ldots \end{aligned} \tag{5.4}$$

Expressions for the parameters λ_1^{he} , λ_2^{he} etc. are obtained by substituting the expansions into the incremental principle of virtual work (2.13). Hutchinson [7] has given the expression for λ_1^{he} , which is

$$\lambda_1^{he} = - \mathcal{A}/\mathcal{B} \tag{5.5}$$

$$\mathcal{A} = \int_V \left\{ 3 \overset{(1)}{\sigma}{}^{\alpha\beta} \overset{(1)}{w}_{,\alpha} \overset{(1)}{w}_{,\beta} + \overset{(1)}{\eta}_{\alpha\beta} \overset{(1)}{\eta}_{\gamma\delta} \left.\frac{\partial \hat{L}^{\alpha\beta\gamma\delta}}{\partial \sigma^{\mu\nu}}\right|_c \overset{(1)}{\sigma}{}^{\mu\nu} \right\} dV \tag{5.6}$$

$$\begin{aligned} \mathcal{B} = \int_V \Bigg\{ & 4 \overset{(1)}{\sigma}{}^{\alpha\beta} \left.\frac{\partial \overset{0}{w}_{,\alpha}}{\partial \lambda}\right|_c \overset{(1)}{w}_{,\beta} + 2 \left.\frac{\partial \overset{0}{\sigma}{}^{\alpha\beta}}{\partial \lambda}\right|_c \overset{(1)}{w}_{,\alpha} \overset{(1)}{w}_{,\beta} + \overset{(1)}{\eta}_{\alpha\beta} \overset{(1)}{\eta}_{\gamma\delta} \left.\frac{\partial \hat{L}^{\alpha\beta\gamma\delta}}{\partial \sigma^{\mu\nu}}\right|_c \left.\frac{\partial \overset{0}{\sigma}{}^{\mu\nu}}{\partial \lambda}\right|_c \\ & + \overset{(1)}{\eta}_{\alpha\beta} \left.\frac{\partial \overset{0}{\eta}_{\gamma\delta}}{\partial \lambda}\right|_c \left.\frac{\partial \hat{L}^{\alpha\beta\gamma\delta}}{\partial \sigma^{\mu\nu}}\right|_c \overset{(1)}{\sigma}{}^{\mu\nu} \Bigg\} dV \end{aligned} \tag{5.7}$$

It is noted that $\mathcal{A}$ and $\mathcal{B}$ are the two constants appearing in (4.15). The expression for λ_2^{he} is considerably simplified if $\lambda_1^{he} = 0$, in which case we find

$$\lambda_2^{he} = - \mathcal{C}/\mathcal{D} \tag{5.8}$$

$$\begin{aligned} \mathcal{C} = & \int_A \left\{ 3 \overset{(2)}{N}{}^{\alpha\beta} \overset{(1)}{w}_{,\alpha} \overset{(1)}{w}_{,\beta} + 6 \overset{(1)}{N}{}^{\alpha\beta} \overset{(1)}{w}_{,\alpha} \overset{(2)}{w}_{,\beta} \right\} dA + \int_V \Bigg\{ \overset{(2)}{\sigma}{}^{\mu\nu} \left.\frac{\partial \hat{L}^{\alpha\beta\gamma\delta}}{\partial \sigma^{\mu\nu}}\right|_c \overset{(1)}{\eta}_{\alpha\beta} \overset{(1)}{\eta}_{\gamma\delta} \\ & + 2 \overset{(1)}{\sigma}{}^{\mu\nu} \left.\frac{\partial \hat{L}^{\alpha\beta\gamma\delta}}{\partial \sigma^{\mu\nu}}\right|_c \overset{(1)}{\eta}_{\alpha\beta} \overset{(2)}{\eta}_{\gamma\delta} + \frac{1}{2} \overset{(1)}{\sigma}{}^{\mu\nu} \overset{(1)}{\sigma}{}^{\rho\omega} \left.\frac{\partial^2 \hat{L}^{\alpha\beta\gamma\delta}}{\partial \sigma^{\mu\nu}\partial \sigma^{\rho\omega}}\right|_c \overset{(1)}{\eta}_{\alpha\beta} \overset{(1)}{\eta}_{\gamma\delta} \Bigg\} dV \end{aligned} \tag{5.9}$$

$$\mathcal{D} = \int_A \left\{ 6 \overset{(1)}{N}{}^{\alpha\beta} \left.\frac{\partial \overset{0}{w}_{,\alpha}}{\partial\lambda}\right|_c \overset{(1)}{w}_{,\beta} + 3 \left.\frac{\partial N_0^{\alpha\beta}}{\partial\lambda}\right|_c \overset{(1)}{w}_{,\alpha} \overset{(1)}{w}_{,\beta} \right\} dA$$
$$+ \int_V \left\{ \left.\frac{\partial \sigma_0^{\mu\nu}}{\partial\lambda}\right|_c \left.\frac{\partial \hat{L}^{\alpha\beta\gamma\delta}}{\partial\sigma^{\mu\nu}}\right|_c \overset{(1)}{\eta}_{\alpha\beta} \overset{(1)}{\eta}_{\gamma\delta} + 2 \overset{(1)}{\sigma}{}^{\mu\nu} \left.\frac{\partial \hat{L}^{\alpha\beta\gamma\delta}}{\partial\sigma^{\mu\nu}}\right|_c \left.\frac{\partial \overset{0}{\eta}_{\alpha\beta}}{\partial\lambda}\right|_c \overset{(1)}{\eta}_{\gamma\delta} \right\} dV \quad (5.10)$$

In (5.6) - (5.10) V is the volume of shell material corresponding to the middle surface area A .

The expressions (5.9) and (5.10) depend on the second order contributions to the asymptotic expansions, which are found by solution of a variational equation obtained from the principle of virtual work. For $\lambda_1^{he} = 0$ the variational equation takes the form

$$\int_A \left\{ \overset{(2)}{M}{}^{\alpha\beta} \delta\kappa_{\alpha\beta} + \overset{(2)}{N}{}^{\alpha\beta} \delta \overset{c}{\varepsilon}_{\alpha\beta} + N_c^{\alpha\beta} \overset{(2)}{w}_{,\alpha} \delta w_{,\beta} + \overset{(1)}{N}{}^{\alpha\beta} \overset{(1)}{w}_{,\alpha} \delta w_{,\beta} \right\} dA = 0 \quad (5.11)$$

$$\overset{(2)}{\sigma}{}^{\alpha\beta} = \hat{L}_c^{\alpha\beta\gamma\delta} \overset{(2)}{\eta}_{\gamma\delta} + \frac{1}{2} \overset{(1)}{\sigma}{}^{\mu\nu} \left.\frac{\partial \hat{L}^{\alpha\beta\gamma\delta}}{\partial\sigma^{\mu\nu}}\right|_c \overset{(1)}{\eta}_{\gamma\delta} \quad (5.12)$$

$$\overset{(2)}{\eta}_{\gamma\delta} = \frac{1}{2}\left(\overset{(2)}{u}_{\gamma,\delta} + \overset{(2)}{u}_{\delta,\gamma} \right) - d_{\gamma\delta} \overset{(2)}{w} + \frac{1}{2} \overset{(1)}{w}_{,\gamma} \overset{(1)}{w}_{,\delta} - x^3 \overset{(2)}{w}_{,\gamma\delta} \quad (5.13)$$

where $\overset{(2)}{N}{}^{\alpha\beta}$ and $\overset{(2)}{M}{}^{\alpha\beta}$ are obtained from (2.6) using (5.12) and (5.13).

The imperfections considered in this analysis are taken in the shape of the critical bifurcation mode with amplitude $\bar{\xi}$. The asymptotic analysis of the influence of such imperfections is divided into a singular perturbation analysis valid for λ near λ_c , a regular perturbation analysis valid for $\lambda < \lambda_c$, and finally the matching of these two results [31,8].

In the regular perturbation analysis the lowest order effect of a small imperfection is determined for $\lambda < \lambda_c$. The displacements, strains and stresses are written as

$$u_\alpha = \overset{0}{u}_\alpha + \tilde{u}_\alpha \quad , \quad w = \overset{0}{w} + \tilde{w} \quad ,$$
$$\eta_{\alpha\beta} = \overset{0}{e}_{\alpha\beta} + \tilde{\eta}_{\alpha\beta} \quad , \quad \sigma^{\alpha\beta} = \sigma_0^{\alpha\beta} + \tilde{\sigma}^{\alpha\beta} \quad (5.14)$$

where the solution for the perfect shell at the current load level λ is denoted by $(\ )^0$, while $(\tilde{\ })$ denotes small perturbation quantities that vanish at $\lambda = 0$. All equations are linearized with respect to the small perturbation- and imperfection quantities. Thus, the constitutive equation (2.3) yields the linearized relation

$$\frac{d\tilde{\sigma}^{\alpha\beta}}{d\lambda} = \hat{L}_0^{\alpha\beta\gamma\delta} \frac{d\tilde{\eta}_{\gamma\delta}}{d\lambda} + \left.\frac{\partial \hat{L}^{\alpha\beta\gamma\delta}}{\partial\sigma^{\mu\nu}}\right|_0 \tilde{\sigma}^{\mu\nu} \frac{d\overset{0}{e}_{\gamma\delta}}{d\lambda} \quad (5.15)$$

and the linearized principle of virtual work takes the form

$$\int_A \left\{ \tilde{N}^{\alpha\beta} \delta e_{\alpha\beta} + \tilde{M}^{\alpha\beta} \delta\kappa_{\alpha\beta} + N_0^{\alpha\beta} \left(\tilde{w}_{,\alpha} + \bar{\xi}\, \overset{(1)}{w}_{,\alpha} \right) \delta w_{,\beta} \right\} dA = 0 \qquad (5.16)$$

where $e_{\alpha\beta}$ denotes the linear part of $\varepsilon_{\alpha\beta}$.

Now, we write $\tilde{u}_\alpha = \xi \overset{(1)}{u}_\alpha + \hat{\xi}\hat{u}_\alpha$, $\tilde{w} = \xi \overset{(1)}{w} + \hat{\xi}\hat{w}$, etc., where $\hat{u}_\alpha$ and $\hat{w}$ are orthogonal to the eigenmode, and we express the variation of the amplitude ξ of the eigenmode as

$$\xi(\lambda) = \bar{\xi}\left(1 - \frac{\lambda}{\lambda_c}\right)^{-\psi} p(\lambda) \qquad (5.17)$$

The exponent ψ is determined such that the function $p(\lambda)$ is regular at λ_c (the detailed expression for ψ is given in [8,34]). For an elastic (linear or non-linear) solid, $\psi = 1$, while for a path dependent hypoelastic solid $\psi > 1$. The value $p(\lambda_c)$ is determined from a numerical incremental solution of (5.16), using (5.14) and (5.17).

A singular perturbation analysis gives an asymptotic expression for λ as a function of ξ , in the vicinity of λ_c . Some parameter values in this expression are determined by a matching with (5.17) in the vicinity of $\lambda = \lambda_c$. Then, for imperfection-sensitive structures, this expansion may be used to estimate the value of the snap buckling load λ_s (smaller than λ_c) as a function of the imperfection amplitude $\bar{\xi}$.

For a structure with asymmetric bifurcation $(\lambda_1^{he} \neq 0)$ the following asymptotic estimate of λ_s is found, corresponding to $\lambda_1^{he}\bar{\xi} < 0$

$$\left.\begin{aligned} \frac{\lambda_s}{\lambda_c} &= 1 - \mu|\bar{\xi}|^{1/(\psi+1)} \\ \mu &= (\psi+1)\psi^{-\psi/(\psi+1)} \left[\frac{|\lambda_1^{he}|}{\lambda_c}\right]^{1/(\psi+1)} [p(\lambda_c)]^{1/(\psi+1)} \end{aligned}\right\} \qquad (5.18)$$

whereas no λ_s is predicted, to lowest order, for $\lambda_1^{he}\bar{\xi} > 0$. If the post-bifurcation behaviour is symmetric $(\lambda_1^{he} = 0)$ no imperfection-sensitivity is predicted for $\lambda_2^{he} \geq 0$, but the following estimate is obtained for $\lambda_2^{he} < 0$ and $\lambda_1^{he} = 0$

$$\left.\begin{aligned} \frac{\lambda_s}{\lambda_c} &= 1 - \mu\bar{\xi}^{2/(2\psi+1)} \\ \mu &= (2\psi+1)(2\psi)^{-2\psi/(2\psi+1)} \left[\frac{-\lambda_2^{he}}{\lambda_c}\right]^{1/(2\psi+1)} [p(\lambda_c)]^{2/(2\psi+1)} \end{aligned}\right\} \qquad (5.19)$$

In the special case of an elastic solid the expressions (5.18) and (5.19) reduce to well-known results of Koiter's [5] elastic post-buckling theory.

Even though elastic unloading is neglected in the analysis in the present section, it has been found by Needleman and Tvergaard [32] that the hypoelastic asymptotic estimate (5.19) gives a good indication of the imperfection-sensitivity of elastic-plastic square plates under axial compression. These plates are simply supported, with the edges free to move in the in-plane directions, or with the edges constrained to remain straight. It is found that the hypoelastic asymptotic analyses distinguish between the cases where there is imperfection-sensitivity and the cases with no imperfection-sensitivity.

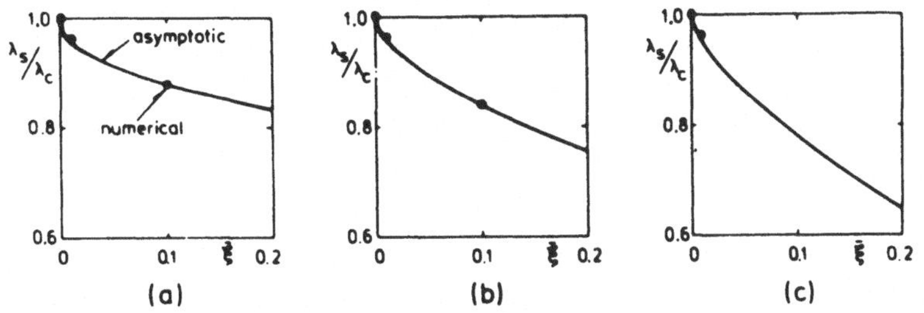

Fig. 5.1. Comparison of numerical results and asymptotic hypoelastic predictions for the imperfection-sensitivity of elastic-plastic cylindrical panels. (a) Panel with $\theta = 0.5$, $\sigma_y/E = 0.002$, $n = 10$. (b) Panel with $\theta = 0.75$, $\sigma_y/E = 0.0028$, $n = 10$. (c) Panel with $\theta = 0.75$, $\sigma_y/E = 0.0028$, $n = 3$ [34].

Another example is provided by Tvergaard's [34] investigation of the postbuckling behaviour of an axially compressed elastic-plastic cylindrical panel. This investigation concentrates on local buckling between stiffeners in a longitudinally stiffened cylindrical shell. The comparison of numerical results and asymptotic hypoelastic predictions in Fig. 5.1 shows that, at least for small imperfections, there is a reasonable agreement. Here, $\theta = 0.289b/\sqrt{Rh}$ is a measure of the curvature of the panel, where h , b and R are the thickness, width and radius of the panel, respectively.

The hypoelastic analysis has been used by Tvergaard and Needleman [35] to analyse the imperfection-sensitivity of elastic-plastic columns. In this case, with a uniaxial stress state, the hypoelastic solid is actually a nonlinear elastic solid, and thus $\psi = 1$ in (5.17). Columns with an asymmetric cross-section have an asymmetric post-bifurcation behaviour, so that here the imperfection-sensitivity estimates are based on (5.18). Fig. 5.2 shows examples of such behaviour, where the tendency of the numerical results is well represented by the asymptotic estimates. Here, $\bar{\xi}$ is the initial imperfection amplitude, normalized by the radius of gyration of the cross-section $r = \sqrt{I/A}$. For the high hardening material, $n = 4$, there is rather good agreement, whereas for $n = 10$ the range of validity of the asymptotic results is not so large.

For a column with a rectangular cross-section the post-bifurcation behaviour is symmetric, $\lambda_1^{he} = 0$. Here, Tvergaard and Needleman [35] find the surprising result that $\lambda_2^{he} > 0$ for any $n > 1$, so that no imperfection-sensitivity is predicted according to (5.19), even though

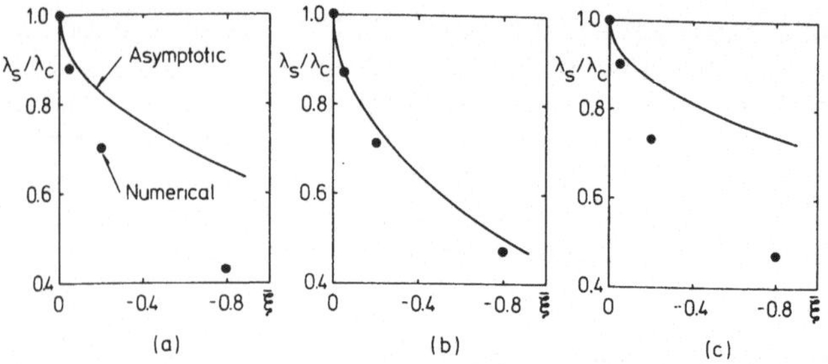

Fig. 5.2. Comparison of numerical results and asymptotic hypoelastic predictions for the imperfection-sensitivity of elastic-plastic columns. (a) U - shaped cross-section, $n = 10$ and $-\sigma_c/\sigma_y = 1.07$. (b) U - shaped cross-section, $n = 4$ and $-\sigma_c/\sigma_y = 1.19$. (c) Triangular cross-section, $n = 10$ and $-\sigma_c/\sigma_y = 1.07$ [35].

numerical analyses show that imperfections reduce the load carrying capacity. The positive value of λ_2^{he} has been confirmed by a numerical analysis; but the range of validity of the asymptotic expansion is very small in this case [35]. The numerical investigation also shows that the inability of the asymptotic estimate to predict the imperfection-sensitivity in this case is not a result of neglecting elastic unloading.

6. NUMERICAL ANALYSES OF PLASTIC BUCKLING

The determination of the load carrying capacity of an elastic-plastic structure requires the solution of a complex nonlinear problem. Asymptotic solutions as those discussed in sections 4 and 5 give a valuable understanding of the phenomena; but in general the current knowledge of plastic postbuckling behaviour and imperfection-sensitivity relies heavily on numerical solutions.

Such solutions are often obtained by a linear incremental procedure based on the variational equation (2.10) (or (2.13) if DMV theory is used). In most cases the variational equation corresponding to a small load increment is solved approximately by the finite element method. The plastic loading history is followed in several integration points within each element, so that the values of the instantaneous moduli $\hat{L}^{\alpha\beta\gamma\delta}$ are known in these points. The integrals through the thickness in (2.7) are evaluated numerically based on the integration points, and so are the integrals over the middle surface in (2.10). A number of numerical results will be shown in the following to illustrate plastic buckling behaviour.

Fig. 6.1 shows solutions for the eccentrically stiffened panels illustrated in Fig. 4.2, in a case where the wide column buckling mode is first critical (Tvergaard and Needleman [28,29]). The numerically determined post-bifurcation behaviour is seen to agree qualitatively with the asymptotic results in Fig. 4.4a and b, respectively. Mode interaction leads to imperfection-sensitivity in a single bay panel with

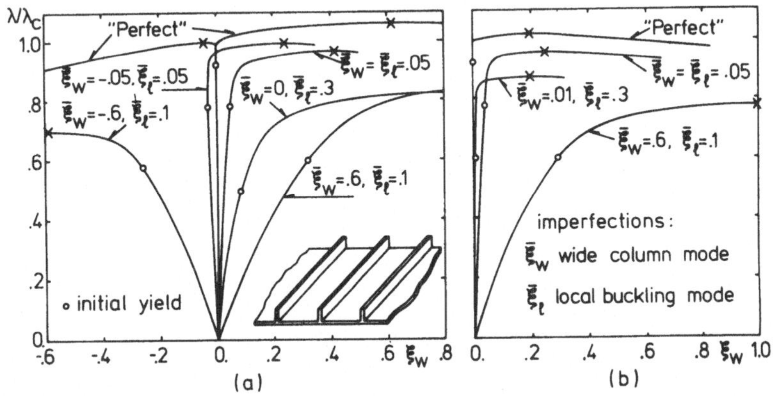

Fig. 6.1. Load versus wide column mode deflection in eccentrically stiffened elastic-plastic panels with $\sigma_E/\sigma_\ell = 0.85$, $\alpha = 0.65$, $n = 10$. (a) Panel over one bay. (b) Multi-bay panel [29,9].

positive column mode deflections (Fig. 6.1a), where the skin is being further compressed by bending; but for negative column mode deflections the skin is being stretched, so that the considerable imperfection-sensitivity is entirely due to material non-linearity. In the multi-bay panel (Fig. 6.1b) both these mechanisms are active, in neighbouring bays. Fig. 6.2 shows solutions for a single-bay panel with simultaneous plastic bifurcation in the wide column and local buckling modes. Structures with simultaneous buckling modes have attracted much interest

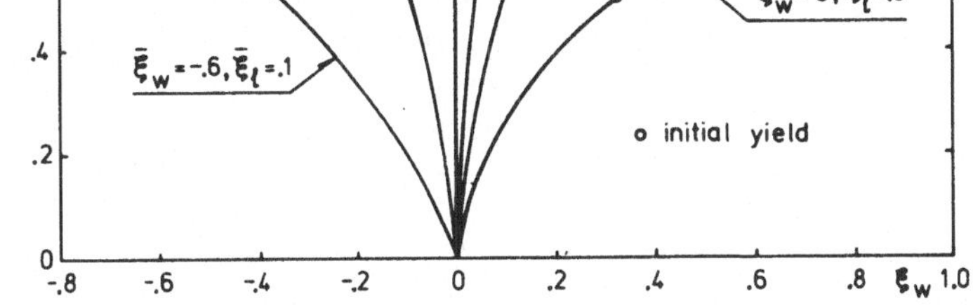

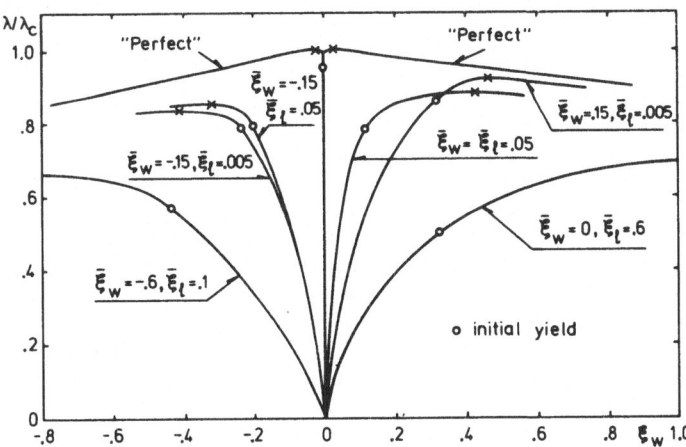

Fig. 6.2. Load versus wide column mode displacement for panel with simultaneous plastic bifurcation in wide column and local buckling modes ($\sigma_y/E = 0.0015$, $n = 10$, $\alpha = 0.653$, $a/b = 4$, $e/b = 0.1$, $(A_s + hb)/b^2 = 0.0256$) [28].

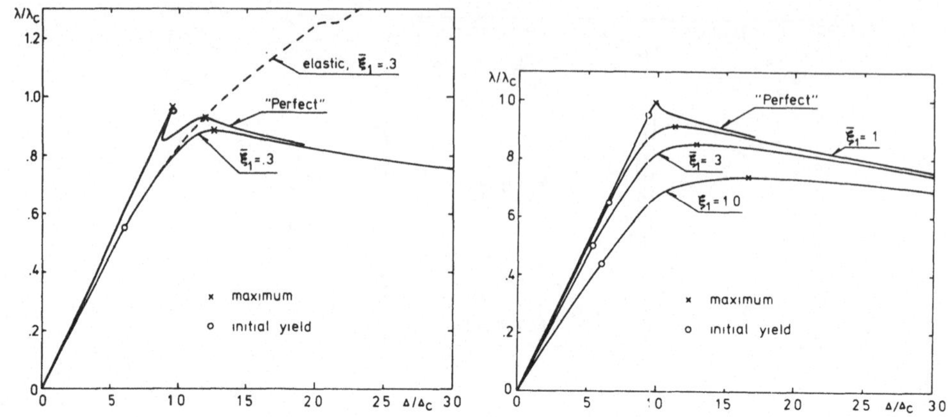

(a) (b)

Fig. 6.3. Load vs end-shortening for elliptic cylindrical shell with $b/a = 0.5$, $\sigma_y/E = 0.0025$, $n = 10$ and $\nu = 0.3$. (a) Shell that bifurcates in the elastic range, $\sigma_E/\sigma_y = 0.92$. (b) Shell that bifurcates in the plastic range, $\sigma_E/\sigma_y = 1.058$ [36].

in the elastic range [5,9], due to the strong imperfection-sensitivity that often results from non-linear mode interaction. However, in the plastic range asymptotic studies have so far been limited to cases with only a single critical buckling mode.

Fig. 6.3 shows examples of load vs. end-shortening curves for oval cylindrical shells under axial compression, with the ratio $b/a = 0.5$ of the axes of the elliptic cross-section (Tvergaard [36]). For elastic shells the initial postbuckling behaviour is strongly unstable; but subsequently, due to load carrying capacity of the strongly curved parts of the shell, support loads well above the primary bifurcation load are found (Fig. 6.3a). As shown by the solid curves in Figs. 6.3a and b this high post-buckling strength is not available in the elastic-plastic shells, where plastic yielding eliminates the extra load carrying capacity of the strongly curved parts of the shell.

The cruciform column is perhaps the most extreme example of the discrepancy between flow theory and deformation theory bifurcation predictions that has been mentioned in section 3.3 (Stowell [23], Hutchinson [7], Hutchinson and Budiansky [31]). Here, the bifurcation load (for the twisting mode) predicted by J_2 flow theory is always identical to the elastic bifurcation load P_E . For this case the effect of a vertex on the yield surface has been analysed by Needleman and Tvergaard [8], based on J_2 corner theory (see section 3.3). The bifurcation load predicted by J_2 corner theory is identical to that of deformation theory (since there is proportional loading), and Fig. 6.4 shows results for a case where this bifurcation load is $0.5P_E$. Here, the amplitude ξ is a measure of the twisting angle.

Fig. 6.4a shows the numerically determined post-bifurcation behaviour corresponding to two different vertex descriptions. For the material

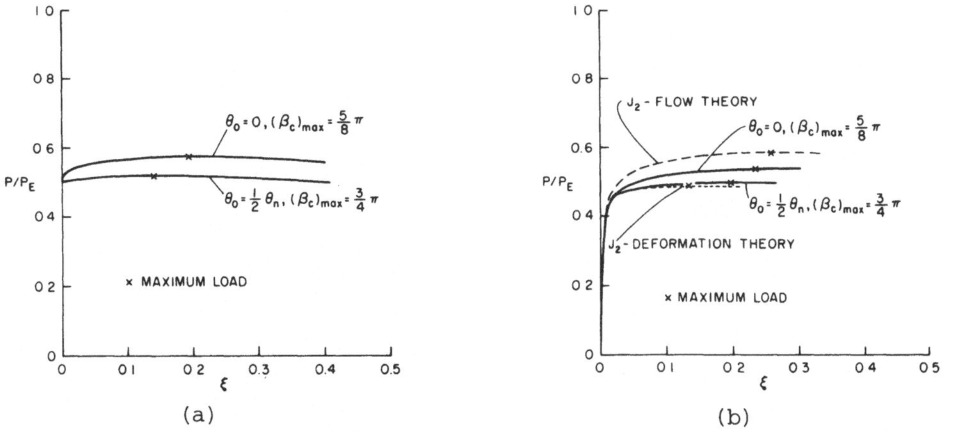

Fig. 6.4. Behaviour of a cruciform column for various constitutive descriptions, when $\sigma_y/E = 0.002$, $\nu = 0.3$, $n = 4$ and $P_D/P_E = 0.5$. (a) Postbifurcation behaviour for perfect column according to two vertex descriptions. (b) Influence of an initial imperfection [8].

with a finite total loading range limited by $\theta_0 = \frac{1}{2}\theta_n$, where θ_n specifies the cone of normals, a positive value of the post-bifurcation slope λ_1 in (3.16) is found. However, for $\theta_0 = 0$ a smooth bifurcation is found, as has been discussed in section 3.3. For a small initial imperfection the predictions of the different material models are shown in Fig. 6.4b. Here, the load carrying capacity predicted by J_2 flow theory is far below the bifurcation load P_E , whereas only a weak imperfection-sensitivity is associated with the other material models. It is noted that the various vertex predictions have the J_2 flow theory predictions and the deformation theory predictions as upper and lower limits, respectively.

J_2 corner theory has also been used to analyse the plastic buckling of axially compressed circular cylindrical shells (Tvergaard [15]). For shells with initial axisymmetric imperfections in the shape of the critical buckling mode bifurcation into a non-axisymmetric shape is analysed. Fig. 6.5a shows the bifurcation load vs. imperfection amplitude for various radius to thickness ratios, including the elastic result of Koiter [37]. The material considered is taken to have $\theta_0 = 0$, and thus, due to the non-proportional loading in the present fundamental solution, the bifurcation points shown are actually the upper bound solutions discussed in section 3.3.

Fig. 6.5b shows results for shells with both axisymmetric and non-axisymmetric imperfections, where $\bar{\xi}_m$ denotes the amplitude of an imperfection with circumferential wave number m . Here, the calculation for $\bar{\xi}_0 = 0.1$ and $\bar{\xi}_9 = 10^{-5}$ gives a good indication of the post-bifurcation behaviour for the axisymmetric shell with $\bar{\xi}_0 = 0.1$, and the calculations for larger non-axisymmetric imperfections illustrate the imperfection-sensitivity associated with the bifurcation loads shown

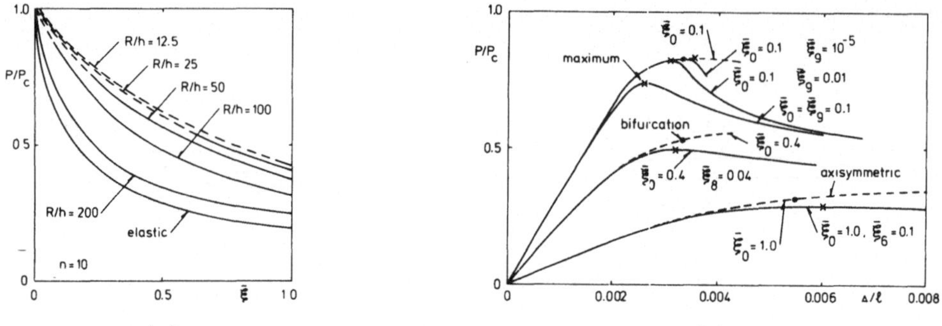

(a) (b)

Fig. 6.5. Buckling of cylindrical shell that forms a vertex on the yield surface ($\sigma_y/E = 0.0025$, $\nu = 0.3$, $n = 10$, $\theta_0 = 0$, $(\beta_c)_{max} = 100^0$) . (a) Bifurcation load or maximum load (dashed curves) vs. axisymmetric imperfection amplitude. (b) Axial load vs. end shortening for $R/h = 100$ and non-axisymmetric imperfections [15].

in Fig. 6.5a.

The influence of residual stresses on plastic buckling is an important topic that has received some attention over the years, mainly in connection with weld-induced residual stresses in plates (Dwight and Moxham [38], Graves Smith [39], Little [40]). Bushnell [11] has discussed various sources of residual stresses, and has also shown axisymmetric results for a pressure vessel with ring-stiffeners welded to it. A relatively simple three-spring model of a column has recently been used by Caotim and Roorda [41] to study some effects of residual stresses on buckling, both experimentally and theoretically.

A detailed study of the interaction between weld-induced residual

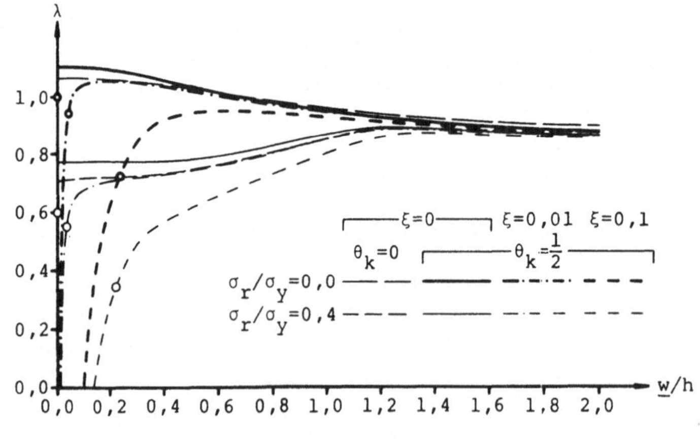

Fig. 6.6. Buckling behaviour of flat plates, $\theta_k = 0$, or cylindrical panels, $\theta_k = ½$, with $n = 10$, geometric imperfection $\bar{\xi}$, and residual compressive stress σ_r [42].

stresses and geometrical imperfections in plate and shell structures has been carried out by Ravn-Jensen [42]. Fig. 6.6 shows some of his results for local buckling between axial stiffeners welded to a circular cylindrical shell under axial compression. Here, $\theta_k = 0.5$ denotes the same curvature as that of the cylindrical panel considered in Fig. 5.1a, $\bar{\xi}$ is the imperfection amplitude, and σ_r is the value of the residual compressive stress in the central part of the panel between the stiffeners. The load parameter λ is here normalized by the load corresponding to initial yielding (for $\sigma_r = 0$) . In addition to the results for the curved panel, post-bifurcation solutions are also shown for a flat plate $(\theta_k = 0)$, with or without residual stresses. It is seen that both types of imperfections reduce the load carrying capacity relative to that of the perfect panel. However, the figure also indicates that for geometrical imperfections slightly larger than $\bar{\xi} = 0.1$ the load carrying capacity is not further reduced by weld-stresses, which is important. The results in [42] show the general trend that reductions of the load carrying capacity due to residual stresses are most significant for structures made of a low hardening material.

7. LOCALIZATION OF BUCKLING PATTERNS

A common feature of many structures subject to compressive loading is that the critical buckling mode is periodic in the direction of compression. However, the final buckled configuration in such structures is frequently a localized one, involving only one or a few buckles. For example, Moxham [43] carried out over 100 tests on axially compressed mild steel plate strips and invariably observed a localized buckling pattern rather than the periodic one associated with the critical mode.

In fact, quite general considerations (Tvergaard and Needleman [44]) indicate that a wide variety of structures are susceptible to this type of localization. These structures have the common property that the applied load versus shortening curve achieves a maximum. The mechanism of localization involves a bifurcation, at or subsequent to the attainment of a maximum load, where the initial periodic buckling pattern loses uniqueness. After bifurcation, the periodic deflection pattern is replaced, often rather abruptly, by a localized one.

The localization of buckling patterns is in various ways analogous to the necking of tensile bars and first a simple one-dimensional model (Tvergaard and Needleman [44]) is used to illustrate the basic mechanism involved.

In Fig. 7.1, the response of a buckled structure is represented as that of a bar constrained to remain straight, but free to slide in the

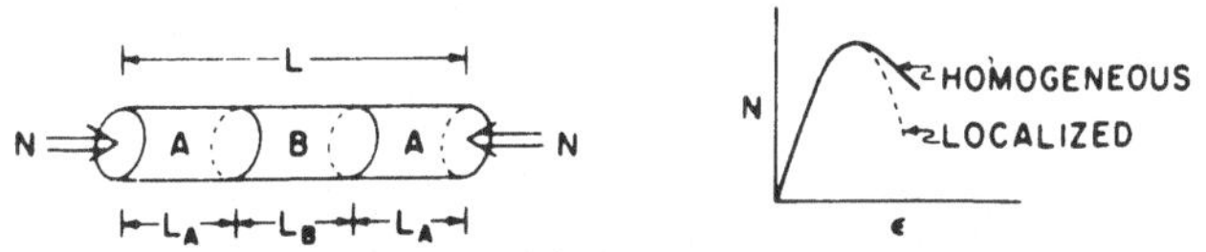

Fig. 7.1. The axially compressed bar model [44].

axial direction. The axial stress-strain response of the bar is taken to be nonlinear as expressed by the incremental relation

$$\dot{N} = C\dot{\varepsilon} \ , \tag{7.1}$$

where N is the axial force, ε is the axial strain, $(\dot{\ })$ denotes incremental quantities and $C(\varepsilon)$ is the instantaneous modulus. For example, in an axially compressed plate strip with a periodic buckling-deflection pattern, N can be identified with the axial load and ε with the node-to-node shortening divided by the distance between nodes. In this manner a periodic buckling pattern in the actual structure corresponds to the bar (Fig. 7.1) being uniformly compressed, and the nonlinear axial stiffness of the bar model accounts for the softening due to buckling as well as that resulting from any plastic yielding.

We now investigate whether a local region (B in Fig. 7.1) can undergo incremental straining different from that in its surroundings (A in Fig. 7.1), for prescribed end-displacements. Incremental equilibrium requires

$$\dot{N}_A = \dot{N}_B \ . \tag{7.2}$$

Thus, using the same instantaneous modulus, $C(\varepsilon)$, in each region (since prior to bifurcation $\varepsilon_A = \varepsilon_B$), (7.2) in conjunction with (7.1) implies

$$C(\dot{\varepsilon}_A - \dot{\varepsilon}_B) = 0 \ . \tag{7.3}$$

Bifurcation $(\dot{\varepsilon}_A \neq \dot{\varepsilon}_B)$ is possible when $C = 0$, that is, at the maximum load point. If no maximum load is attained, localization does not occur.

The same conclusion concerning the onset of localization can be reached by a variational formulation. With $\dot{\varepsilon}$ in (7.1) identified as the axial displacement-gradient increment, $\dot{u}_{,x}$, and with displacements taken to be prescribed at the ends of the bar, bifurcation away from the homogeneously deformed state is governed by the variational equation (analogous to (3.8))

$$\delta I = 0 \quad , \quad I = \int_0^L C\tilde{\varepsilon}\tilde{\varepsilon}\, dx \ , \tag{7.4}$$

where $L = 2L_A + L_B$ in Fig. 7.1, and $(\tilde{\ })$ denotes the difference between two solutions corresponding to the same prescribed end-displacements. Bifurcation modes of the form

$$\tilde{u} = \sin\frac{n\pi x}{L} \quad , \quad n = 1,2,3,\ldots \ , \tag{7.5}$$

are all available when $C = 0$, that is, at the maximum load point.

A similar simple one-dimensional model of a bent tube can also be constructed (Tvergaard and Needleman [44]). In tube bending a maximum moment is reached, mainly due to ovalization of the cross-section (the Brazier effect). The one-dimensional tube model predicts localization, for prescribed end-rotations, when the maximum moment is attained, thus initiating the well-known final collapse mode, in which the tube forms a sharp kink.

These simple models can only illustrate the basic mechanism of buckling localization. As an example of an actual structural model Tvergaard and Needleman [44] have considered an elastic column on a softening elastic foundation, which is taken to be bilinear with stiffness K_1 for deflections smaller than w_0 and stiffness K_2 for larger deflections. Then, for a perfect column, simply-supported at its ends, the critical compressive force and corresponding bifurcation mode are given by

$$P_c = 2\sqrt{K_1 EI} \quad , \quad w_c = \sin\frac{m\pi x}{\ell} \; , \tag{7.6}$$

where EI is the bending stiffness, and ℓ , the length of the column, is chosen so that $m = (K_1/EI)^{1/4}\ell/\pi$ is an integer. Analytical solutions to the governing differential equation have been used to determine the development of the periodic buckling pattern for columns with an initial imperfection in the shape of the critical mode (7.6) and to analyse bifurcation away from this periodic pattern.

Fig. 7.2 displays curves of load versus periodic buckling-mode amplitude for various initial imperfection amplitudes for a case where $K_2/K_1 = 0.1$. Bifurcation occurs somewhat beyond the load maximum, with the delay being larger for short columns (small m) that for long

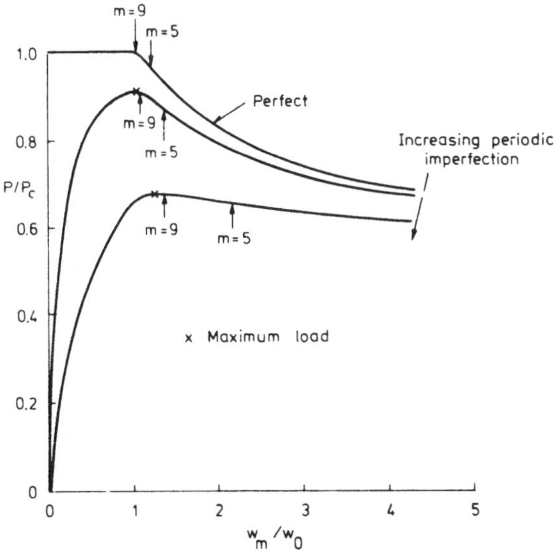

Fig. 7.2. Load versus lateral deflection for columns on a softening foundation with $K_2/K_1 = 0.1$. The curves depict the behaviour for periodic deflections and the arrows mark the point of bifurcation into the localized mode [44].

columns (large m) . The delay increases also with increasing periodic imperfection magnitude. The mode associated with bifurcation from the periodic deflection pattern has a number of half-waves different from m . Numerical solutions for columns supported on softening foundations with slightly nonperiodic imperfections have shown that localization actually does occur rather than, say, mode snapping into a mode with a number of half-waves different from m (Needleman and Tvergaard [8]).

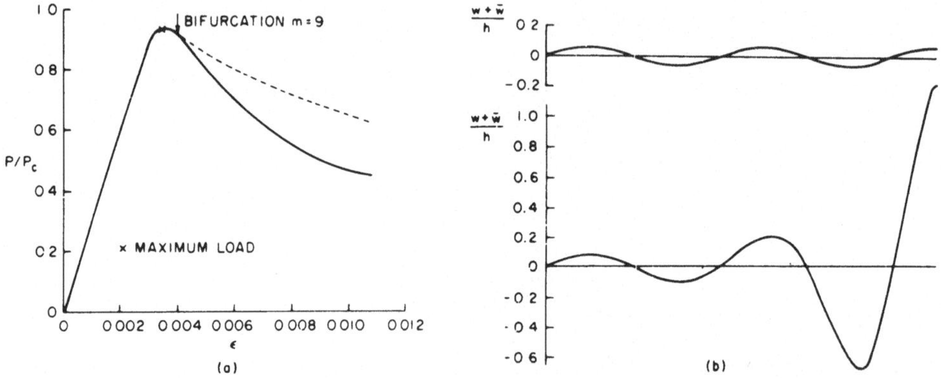

Fig. 7.3. (a) Load versus average axial strain for a multiply-supported elastic-plastic column. The dashed curve shows the periodic solution and the solid curve shows the response when localization develops in a column with nine bays. (b) The deflection pattern at the maximum load and after localization has developed [8].

The development of buckling-pattern localization in another column structure, a multi-bay elastic-plastic column, has also been studied in [8]. Numerical results depicting the development of localization in a column spanning nine bays are shown in Fig. 7.3. The column material is characterized by a power-law stress-strain relation with a hardening exponent representative of a lightly hardening material. When the column has an initially periodic imperfection, the bifurcation that leads to localization occurs a little beyond the maximum load point. A small additional nonperiodic imperfection has little influence on the column behaviour until near the bifurcation point where the deformations begin to localize. Fig. 7.3b shows this nearly periodic deflection pattern at the maximum load point and the localized pattern that develops subsequently. Here, the deflections are normalized by the height h of the column cross-section.

Buckling localization in columns under dynamic loading has also been studied (Tvergaard and Needleman [45]), in connection with shock-absorbing devices. Furthermore, the thermal buckling of railroad tracks (Tvergaard and Needleman [46]) has been studied by modelling the rails and cross-ties as a beam, which rests on a softening foundation, representing the underlying crushed stone layer (the ballast). This study shows that also the localized buckling mode observed for railroad tracks is explained as a bifurcation from a periodic deflection mode.

The structure in which Moxham [43] observed localized buckling is a long plate simply-supported at the edges and subject to axial compression. Tvergaard and Needleman [44] have carried out full finite element analyses of elastic-plastic plate strips. For a lightly hardening material, $n = 10$, a load maximum is achieved for the periodic solution. In this case, a bifurcation point is found somewhat beyond the maximum load point and localization occurs. For a high hardening material, $n = 2.5$, no maximum load point is reached, and localization does not occur, as anticipated from the simple bar model. The plates analysed, with aspect ratio $a/b = 3$ have three half-waves along the length, and Fig. 7.4 shows the ratio of the maximum deflections in the central buckle and the first buckle, respectively. Initially this ratio is 1.1, due to a small localized imperfection superposed on the periodic imperfection; but it is interesting to note that prior to the load maximum this ratio approaches unity rather than growing, as one might expect in the case where localization is going to occur.

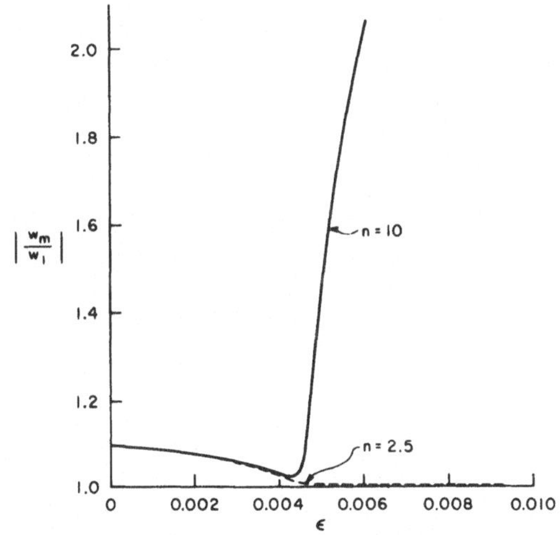

Fig. 7.4. Ratio between lateral deflection amplitudes at the central buckle and at the first buckle versus average axial strain for an elastic-plastic plate with $a/b = 3$, $\bar{\xi}_1 = 0.01$, $\bar{\xi}_2 = 0.001$, $\sigma_y/E = 0.00337$, $h/b = 0.035$, $\nu = 0.3$ [44].

Both Fig. 7.4 and Fig. 7.3 illustrate that small localized imperfections superposed on larger periodic imperfections have little effect prior to localization, and that they do not significantly affect the structure's load-carrying capacity. Many post-buckling analyses have been based on the assumption of periodicity, and therefore it is important to know that these predictions are not invalidated by a small deviation from periodicity, even when localization is going to occur after the load maximum.

Several finite element analyses of axially compressed elastic-plastic plate strips, with a number of differently shaped imperfections, have been presented recently by Dow and Smith [47]. In particular, they have analysed a plate with a local dimple imperfection and have found that the reduction of the plate strength is only slightly smaller than that corresponding to a periodic imperfection with the same amplitude and wavelength. This finding could be seen as a contradiction of the conclusion in the preceding paragraph. However, these two situations are different, and the conclusion in the preceding paragraph relates to small deviations from periodic imperfections and thus only to imperfection shapes that are still predominantly periodic. The result that a local dimple imperfection is nearly as detrimental as the corresponding modal imperfection has been found previously for a number of elastic and elastic-plastic imperfection-sensitive structures (e.g. discussion by Tvergaard [9]).

For axially compressed circular cylindrical shells with a periodic axisymmetric imperfection, as those considered in Fig. 6.5a, localization also occurs, when a maximum load is reached due to plastic yielding of the material. Analyses of the bifurcation from a periodic axisymmetric pattern into an asymmetric mode (Gellin [48], Tvergaard [15]) have shown that bifurcation occurs for all radius to thickness ratios considered, whereas experimental investigations show that sufficiently thick-walled shells collapse in a purely axisymmetric mode. Fig. 6.5a shows that these bifurcations occur prior to the load maximum for relatively thin-walled shells. In the cases where the load maximum is reached first, the bifurcation analyses in [15,48] are not valid, since the assumed fundamental solution is changed by localization.

The extent to which localization of the axisymmetric prebifurcation deformations affects the onset of bifurcation into a non-axisymmetric mode has been investigated by Tvergaard [49]. Fig. 7.5 shows the axial

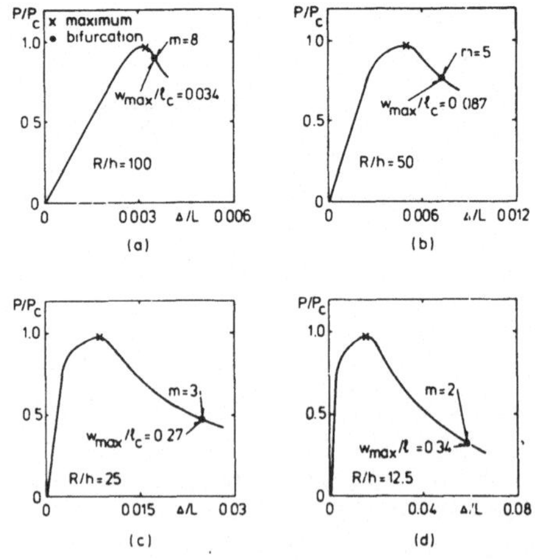

Fig. 7.5. Bifurcation into a diamond pattern collapse mode indicated on curves of axial load vs shortening for cylindrical shells with $n = 10$ and $\sigma_Y/E = 0.0025$ [49].

load vs. shortening for four different shells, with small (partly localized) initial imperfections, $\bar{\xi} = 0.02$, and with the length $5\ell_c$, where ℓ_c is the half-wave length of the critical axisymmetric bifurcation mode. Bifurcation occurs after the maximum in all four cases (in agreement with Fig. 6.5a). However, for the most thick-walled shells the maximum axisymmetric deflection w_{max} grows so large prior to bifurcation (around $0.3\ell_c$) that here the already well developed axisymmetric buckle is expected to collapse completely in a shape that remains essentially axisymmetric. Thus, Fig. 7.5 indicates that a transition from a diamond mode to an axisymmetric mode of collapse will occur as the radius to thickness ratio is increased to around 25 , in the case considered. The influence of the hardening exponent n , the yield strain σ_y/E , and the boundary conditions, on the localization-induced transition in collapse mode has been studied in [49].

8. EFFECT OF MATERIAL RATE-SENSITIVITY

The elastic-plastic buckling analyses discussed in the previous sections are all based on time-independent elastic-plastic material models. For most purposes such models give a sufficiently accurate description of the inelastic behaviour of metals. However, experimental measurements tend to show a dependence of the flow stress on the rate of strain, even at low strain-rates, and it has been emphasized by a number of authors that microstructural slip in metals and thus macroscopic inelastic deformations are inherently time-dependent (Rice [50], Clifton [51], Frost and Ashby [52]). Here, the effect of material rate-sensitivity on plastic buckling predictions will be discussed rather briefly.

When strain-rate sensitivity is taken into account, the material behaviour is represented in terms of a viscoplastic constitutive description, where the plastic part of the strain-rate $\dot{\eta}^P_{ij}$ is a function of the current stresses and strains (but not of the stress-rate). The relationship may be of the form

$$\dot{\eta}^P_{ij} = F(\sigma_e , \varepsilon_e) \frac{\partial \Phi}{\partial \sigma^{ij}} \tag{8.1}$$

where Φ is a plastic potential function, σ_e and ε_e are the effective stress and strain, respectively, and $(\dot{\ })$ denotes the time derivative. Often the function F is taken to be proportional with $[\sigma_e/g(\varepsilon_e)]^{1/m}$, where the function $g(\varepsilon_e)$ incorporates the strain hardening (at some reference strain-rate $g(\varepsilon_e)$ is equal to σ_e) . The rate-hardening exponent m is a measure of the degree of strain-rate sensitivity. Thus, for a rate-sensitive version of the classical J_2 flow theory an expression for the inelastic part of the strain-rate may be assumed of the form

$$\dot{\eta}^P_{ij} = \dot{\varepsilon}_o \left[\frac{\sigma_e}{g(\varepsilon_e)}\right]^{1/m} \frac{3}{2} \frac{s_{ij}}{\sigma_e} \tag{8.2}$$

where $\dot{\varepsilon}_o$ denotes the reference value of the effective strain-rate. The limiting case, $m = 0$, corresponds to time-independent plasticity; but for any positive value of m an inelastic deformation increment according to (8.1) requires an increment of time.

Bifurcation in a rate-sensitive elastic-plastic solid is entirely governed by the time-independent part of the deformations. Assuming the total strain-rate given by the sum of the elastic and plastic parts, $\dot{\eta}_{ij} = \dot{\eta}^E_{ij} + \dot{\eta}^P_{ij}$, with the viscous expression (8.1) for $\dot{\eta}^P_{ij}$, the first critical bifurcation point is that predicted by linear elasticity, even for a very small degree of strain-rate sensitivity, and this elastic bifurcation point is hardly ever reached in the range of problems considered here. Thus, it turns out that the central role played by the critical bifurcation point and the post-bifurcation behaviour in the case of time-independent plasticity, is here replaced by a very strong sensitivity to small initial imperfections.

Some insight in the influence of material rate-sensitivity on plastic buckling has been obtained by a column mode analysis for an eccentrically stiffened panel, as that shown in Fig. 4.2a (Tvergaard [53]). For simplicity only the column mode was considered in [53], so that in fact this is an investigation for a column with an asymmetric cross-section, analogous to [30]. This structure has been analysed for different rates of compression and for different values of the rate-hardening exponent m in (8.2), assuming that the rates are so low that dynamic (inertia) effects can be neglected. Both numerical analyses and a perturbation procedure have been used to study the problem.

Corresponding to a fixed speed of average axial shortening, $\dot{\varepsilon}_a = -\dot{\varepsilon}_o$, Fig. 8.1 gives computed load maxima versus imperfection amplitude $\bar{\xi}$. The load is normalized by the bifurcation load λ_c corresponding to time-independent plasticity and $\bar{\xi}$ denotes the imperfection amplitude relative to the flange thickness. Naturally, the maximum load attained is quite sensitive to the speed of axial compression, particularly for the more rate-sensitive material, $m = 0.05$,

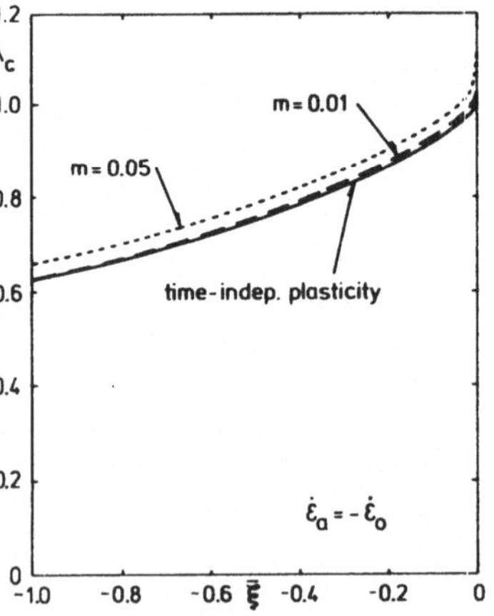

Fig. 8.1. Maximum load versus imperfection amplitude for wide column with $\lambda_E/\lambda_c = 2.18$ [53].

where the maximum load is typically increased by about 10% if the strain-rate $\dot{\varepsilon}_a$ is increased from $-\dot{\varepsilon}_o$ to $-10\dot{\varepsilon}_o$. Therefore it should be noted that the uniaxial stress-strain curve for the time-independent plasticity model in Fig. 8.1 is chosen such that the three materials compared in the figure have identical response when strained uniformly at the strain-rate $-\dot{\varepsilon}_o$.

For the rate-sensitive materials in Fig. 8.1 the first critical bifurcation point corresponds to the elastic value λ_E , which is quite high relative to the bifurcation load λ_c predicted by time-independent plasticity, $\lambda_E = 2.18\lambda_c$. However, the results in Fig. 8.1 show that although the role of bifurcation is strongly changed by accounting for strain-rate sensitivity, the classical elastic-plastic buckling predictions remain good approximations for imperfection amplitudes that are not extremely small, as long as the rate hardening exponent does not exceed values of the order of 0.01 to 0.05 .

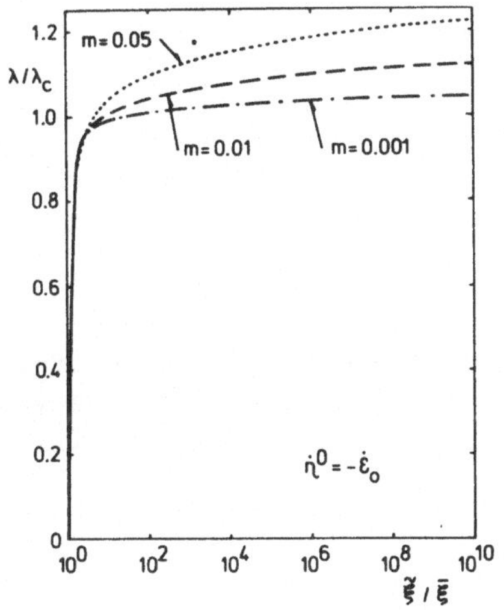

Fig. 8.2. Load versus amplitude growth according to perturbation solution for wide column with $\lambda_E/\lambda_c = 2.18$ [53].

The results of the perturbation procedure are based on assuming small values of the mode amplitude $\tilde{\xi}$. Fig. 8.2 shows such perturbation solutions normalized by the imperfection amplitude $\bar{\xi}$. For small imperfections the load maxima are typically reached at amplitudes $\tilde{\xi}$ of the order of 0.1 (although this is not found by the perturbation solution); but this means that the value of λ for $\tilde{\xi}/\bar{\xi} = 10^{10}$ in Fig. 8.2 is representative of the load maximum corresponding to $\bar{\xi} \simeq 10^{-11}$. This extremely small imperfection amplitude is less than 10^{-8} times any realistic value, and thus Fig. 8.2 shows that even for unrealistically small imperfections the load carrying capacity is far below the critical bifurcation load λ_E . In fact the bifurcation load λ_c for time-independent plasticity remains a more relevant estimate of the load carrying capacity than the bifurcation load λ_E corresponding to the viscoplastic material model.

It is finally noted that the effect of strain-rate sensitivity has also been studied for tensile instabilities, e.g. by Hutchinson and Neale [54]. The behaviour found is analogous to that found for buckling; but the localized mode of instability found in tensile loading tends to give a stronger influence of rate-sensitivity.

REFERENCES

1. Shanley, F.R.: Inelastic column theory, J. Aeronaut. Sci., 14 (1947), 261-267.
2. Hill, R.: A general theory of uniqueness and stability in elastic-plastic solids, J. Mech. Phys. Solids, 6 (1958), 236-249.
3. Hill, R.: Bifurcation and uniqueness in nonlinear mechanics of continua, Problems of Continuum Mechanics, Soc. Ind. Appl. Math., Philadelphia, Pennsylvania, (1961), 155-164.
4. Sewell, M.J.: A survey of plastic buckling, Stability (ed. H. Leipholz), Univ. of Waterloo Press, Ontario, (1972), 85-197.
5. Koiter, W.T.: Over de stabiliteit van het elastisch evenwicht, Thesis, Delft, H.J. Paris, Amsterdam (1945). English translations (a) NASA TT-F10, 833 (1967), (b) AFFDL-TR-70-25 (1970).
6. Hutchinson, J.W.: Postbifurcation behavior in the plastic range, J. Mech. Phys. Solids, 21 (1973), 163-190.
7. Hutchinson, J.W.: Plastic buckling, Advan. Appl. Mech., 14 (1974), 67-144.
8. Needleman, A. and V. Tvergaard: Aspects of plastic post-buckling behaviour, Mechanics of Solids, The Rodney Hill 60th Anniversary Volume (eds. H.G. Hopkins and M.J. Sewell), Pergamon Press, Oxford, (1982), 453-498.
9. Tvergaard, V.: Buckling behaviour of plate and shell structures, Proc. 14th IUTAM Congress (ed. W.T. Koiter), North-Holland, (1976), 233-247.
10. Budiansky, B. and J.W. Hutchinson: Buckling: progress and challenge, Trends in Solid Mechanics, Dedicated to the 65th Birthday of W.T. Koiter (eds. J.F. Besseling and A.M.A. van der Heijden), Delft University Press, (1979), 93-116.
11. Bushnell, D.: Plastic buckling, Pressure Vessels and Piping: Design Technology (eds. S.Y. Zamrik and D. Dietrich), ASME, New York, (1982), 47-117.
12. Koiter, W.T.: On the nonlinear theory of thin elastic shells, Proc. Kon. Ned. Akad. Wetensch., Ser. B69 (1966), 1-54.
13. Niordson, F.I.: Shell theory, North-Holland, Amsterdam 1985.
14. Hill, R.: Aspects of invariance in solid mechanics, Advan. Appl. Mech., 18 (1978), 1-75.
15. Tvergaard, V.: Plastic buckling of axially compressed circular cylindrical shells, Int. J. Thin-Walled Struct., 1 (1983), 139-163.
16. Raniecki, B. and O.T. Bruhns: Bounds to bifurcation stresses in solids with non-associated plastic flow law at finite strain, J. Mech. Phys. Solids, 29 (1981), 153-172.

17. Bruhns, O. and B. Raniecki: Ein schrankenverfahren bei verzweigungsproblemen inelastischer formänderungen, ZAMM, 62 (1982), T111-T113.
18. Tvergaard, V.: Influence of void nucleation on ductile shear fracture at a free surface, J. Mech. Phys. Solids, 30 (1982), 399-425.
19. Hill, R.: Generalized constitutive relations for incremental deformation of metal crystals by multislip, J. Mech. Phys. Solids, 14 (1966), 95-102.
20. Hutchinson, J.W.: Elastic-plastic behavior of polycrystalline metals and composites, Proc. Roy. Soc. London, A318 (1970), 247-272.
21. Hecker, S.S.: Experimental studies of yield phenomena in biaxially loaded metals, Constitutive Equations in Visco-plasticity, ADM-Vol. 20, ASME, New York, (1976), 1-33.
22. Bijlaard, P.P.: Theory and tests on the plastic stability of plates and shells, J. Aeronaut. Sci., 16 (1949), 529-541.
23. Stowell, E.Z.: A unified theory of plastic buckling of columns and plates, Nat. Adv. Comm. Aeronaut. Rep. 898 (1948).
24. Gerard, G. and H. Becker: Handbook of structural stability: Part I - buckling of flat plates, Nat. Adv. Comm. Aeronaut. Tech. Note 3781 (1957).
25. Batdorf, S.B.: Theories of plastic buckling, J. Aeronaut. Sci., 16 (1949), 405-408.
26. Christoffersen, J. and J.W. Hutchinson: A class of phenomenological corner theories of plasticity, J. Mech. Phys. Solids, 27 (1979), 465-487.
27. Sewell, M.J.: A plastic flow rule at a yield vertex, J. Mech. Phys. Solids, 22 (1974), 469-490.
28. Tvergaard, V. and A. Needleman: Mode interaction in an eccentrically stiffened elastic-plastic panel under compression, Buckling of Structures (ed. B. Budiansky), Springer-Verlag, Berlin, (1976), 160-171.
29. Tvergaard, V. and A. Needleman: Buckling of eccentrically stiffened elastic-plastic panels on two simple supports or multiply supported, Int. J. Solids Struct., 11 (1975), 647-663.
30. Tvergaard, V. and A. Needleman: On the buckling of elastic-plastic columns with asymmetric cross-sections, Int. J. Mech. Sci., 17 (1975), 419-424.
31. Hutchinson, J.W. and B. Budiansky: Analytical and numerical study of the effects of initial imperfections on the inelastic buckling of a cruciform column, Buckling of Structures (ed. B. Budiansky), Springer-Verlag, Berlin, (1976), 98-105.
32. Needleman, A. and V. Tvergaard: An analysis of the imperfection-sensitivity of square elastic-plastic plates under axial compression, Int. J. Solids Struct., 12 (1976), 185-201.
33. van der Heijden, A.M.A.: A study of Hutchinson's plastic buckling model, J. Mech. Phys. Solids, 27 (1979), 441-464.
34. Tvergaard, V.: Buckling of elastic-plastic cylindrical panel under axial compression, Int. J. Solids Struct., 13 (1977), 957-970.

35. Tvergaard, V. and A. Needleman: On the foundations of plastic buckling, Developments in Thin-Walled Structures - 1 (eds. J. Rhodes and A.C. Walker), Appl. Science Publishers, (1982), 205-233.
36. Tvergaard, V.: Buckling of elastic-plastic oval cylindrical shells under axial compression, Int. J. Solids Struct., 12 (1976), 683-691; Errata, ibid. 14 (1978), 329.
37. Koiter, W.T.: The effect of axisymmetric imperfections on the buckling of cylindrical shells under axial compression, Proc. Kon. Ned. Ak. Wet., 66B (1963), 265-279.
38. Dwight, J.B. and K.E. Moxham: Welded steel plates in compression, The Structural Engineer, 47 (1969), 49-66.
39. Graves Smith, T.R.: The effect of initial imperfections on the strength of thin-walled box columns, Int. J. Mech. Sci., 13 (1971), 911-925.
40. Little, G.H.: Rapid analysis of plate collapse by liveenergy minimization, Int. J. Mech. Sci., 19 (1977), 725-744.
41. Caotim, D. and J. Roorda: On the effect of residual stresses in the plastic buckling of columns - A model study, Solid Mechanics, Univ. of Waterloo, Paper No. 193 (1984).
42. Ravn-Jensen, K.: Influence of residual stresses on the stability of plate- and shell structures (in danish), Thesis, Department of Solid Mechanics, Technical Univ. of Denmark (1984).
43. Moxham, K.E.: Buckling tests on individual welded steel plates in compression, Cambridge University Engineering Department Report CUED/C-Struct/Tr.3 (1971).
44. Tvergaard, V. and A. Needleman: On the localization of buckling patterns, J. Appl. Mech., 47 (1980), 613-619.
45. Tvergaard, V. and A. Needleman: On the development of localized buckling patterns, Collapse (eds. J.M.T. Thompson and G.W. Hunt), Cambridge University Press, (1983), 1-17.
46. Tvergaard, V. and A. Needleman: On localized thermal track buckling, Int. J. Mech. Sci., 23 (1981), 577-587.
47. Dow, R.S. and C.S. Smith: Effects of localized imperfections on compressive strength of long rectangular plates, J. Construct. Steel Research, 4 (1984), 51-76.
48. Gellin, S.: Effect of an axisymmetric imperfection on the plastic buckling of an axially compressed cylindrical shell, J. Appl. Mech., 46 (1979), 125-131.
49. Tvergaard, V.: On the transition from a diamond mode to an axisymmetric mode of collapse in cylindrical shells, Int. J. Solids Structures, 19 (1983), 845-856.
50. Rice, J.R.: On the structure of stress-strain relations for time-dependent plastic deformation in metals, J. Appl. Mech., 37 (1970), 728-737.
51. Clifton, R.J.: Comments on microscopic mechanisms of plastic flow, Plasticity of Metals at Finite Strain (eds. E.H. Lee and R.L. Mallett), Division of Applied Mechanics, Stanford University, (1982), 623-628.

52. Frost, H.J. and M.F. Ashby: Deformation mechanism maps - The plasticity and creep of metals and ceramics, Pergamon Press, Oxford (1982).
53. Tvergaard, V.: Rate-sensitivity in elastic-plastic panel buckling, Aspects of the Analysis of Plate Structures, A volume in honour of W.H. Wittrick (eds. D.J. Dawe *et al.*), Clarendon Press, Oxford (1985), 293-308.
54. Hutchinson, J.W. and K.W. Neale: Influence of strain-rate sensitivity on necking under uniaxial tension, Acta Metallurgica, 25 (1977), 839-846.

EXPERIMENTAL TECHNIQUES AND COMPARISON WITH THEORETICAL RESULTS

Josef Singer
L. Shirley Tark Professor of Aircraft Structures
Department of Aeronautical Engineering
Technion - Israel Institute of Technology
Haifa 32000, Israel

ABSTRACT

The role of experiments in buckling and postbuckling studies of structures is examined. The essential elements of the experimental approach are discussed for a simple buckling test - a column under axial compression - and typical modern techniques are enumerated. The theory of modeling and its applications are summarized. The problems of buckling and postbuckling experiments for plates and shells are then discussed in detail, with emphasis on comparison with theoretical results and recent developments like initial imperfection measurements, definition of boundary condition and nondestructive methods.

1. INTRODUCTION

The importance of the study of the postbuckling behavior, both for the load carrying capacity after buckling and for assessing the buckling load of imperfect structures, has already been discussed and emphasized in the preceding lectures. The expected physical behavior and the analysis and numerical evaluation of judiciously chosen mathematical models of the structures have been discussed. It is, however, up to the experiments to verify the predicted behavior and validate our calculations. However, as stated by Koiter in the Opening Lecture of the 1974 IUTAM Symposium on Buckling of Structures [1]: "To put it mildly, buckling theory and experiments have not always co-existed in harmony". Furthermore, the experiments may bring out elements of behavior of real structures, which have not been considered in our, by necessity, simplified models. This second role of experimentation is often overlooked.

The pattern of research in structural stability for many years has been one of deep theoretical studies combined at the most with corroborating experiments. As pointed out by Chilver [2], this has been very useful in the study of essentially neutral equilibrium problems of elastic stability. But in cases of extreme instability, theory has only been a guide to practical behavior, and much of our present useful design knowledge is based on careful experiments. The important problems of stability, and in particular, postbuckling behavior, are not always amenable to complete analysis and accurate analyses may be rather difficult and the computations very cumbersome. This has motivated the simplified models discussed in the other lectures, and has brought about renewed emphasis on experimental studies.

For example, in the case of shells, whose buckling and postbuckling behavior sometimes present considerable problems, more prominence has recently been given to experimental studies, as is apparent in reviews of the state of the art (for example [3-8]).

The present course on the postbuckling behavior of structures therefore includes discussions of experimental methods. It is of great interest to note that a similar CISM course ten years ago [9] already emphasized the importance of experimental studies in the postbuckling region.

2. MOTIVATION FOR EXPERIMENTS

Before embarking on the discussion of experimental techniques, it may be worth while to reflect on the purpose of experiments in the computer era. The question of "why continue to do experiments " has recently been asked in many fields of applied mechanics. It should also be asked in relation to buckling and postbuckling behavior of structures.

In two recent reviews on shell buckling experiments [7] and [10] the question was examined in detail. Let us briefly recapitulate and reexamine the eight primary motives, in the broader context of buckling and postbuckling behavior of structures:

a. Better Understanding of Buckling and Postbuckling Behavior and the Primary Factors Affecting It.

In addition to the buckling loads, careful experiments in which the parameters are varied one at a time yield the behavior of the structure just before, at and after buckling, and accentuate the main parameters affecting this behavior. Such a philosophy of "research type experimental programs" has been strongly advocated for shells by Sechler [11] for many years, and has been implemented in some test programs, for example in [12]. Based on these observed parameters numerical schemes can be developed, verified, and can also be employed for "experiments on the computer" to extend the range of the parameters tested.

b. To Find New Phenomena

This reason is a direct extension of the first one. In buckling and postbuckling experiments, the new phenomena are likely to be unexpected behavior patterns or mode interactions.

c. To Obtain Better Inputs for Computations

The mathematical models employed in modern large multi-purpose computer programs can simulate real structures fairly closely at least for buckling, but the simulation depends very much on the input of correct boundary conditions, in particular joints or bonds, on imperfections and load applications. This has been emphasized by recent experience and definitely applies also to postbuckling. Better inputs can often be provided by appropriate nondestructive tests: for example, boundary conditions by vibration correlation techniques, imperfection shapes and amplitudes by imperfect scans, load transfer and eccentricities by strain measurements and vibration correlation techniques, etc. Here automated recording in experiments has just begun and much closer interaction between test and computation is developing.

d. To Obtain Correlation Factors Between Analysis and Test and for Material Effects.

Even when large powerful programs are employed, test results still differ considerably from predictions. These differences are partly due to inaccuracies of inputs and partly to variations in buckling and postbuckling behavior of the mathematical model and the structures tested. They can all be lumped for design purposes in a "correlation factor". The advantage of such a correlation factor is the overall correlation it provides for the designer, but its weakness is that it is completely reliable only for the structures tested. One can statistically evaluate a large number of tests to obtain overall lower bound correlation factors, a method employed extensively for shells, where they are called "Knock-down" factors, but this results in very conservative design. Hence "correlation factors" should be more specialized. Since many experiments are on laboratory scale structures, extensive studies comparing the results of laboratory scale and large scale tests are needed to reassure the experimenter and to guide the designer. Correlation type experiments will therefore continue to be a major task of research and industrial laboratories for quite some time to come, as they provide the designer with esential correction factors which include the effects of new materials and manufacturing techniques and, to some extent, bridge the gap between the buckling and post-buckling behavior of the computation model and the realistic structures.

e. To Build Confidence in Multipurpose Computer Programs

Extensive experimental verification is an essential element for confidence in a large computer program. This is therefore a primary motive for buckling and postbuckling experiments, which becomes more important, as the programs become more sophisticated and ambitious. Though some developers of programs have promoted and applied extensive experimental confirmation, more correlations of the results obtained from computer programs with test results are required, as pointed out for example for shells in [13].

f. To Test Novel Ideas of Construction or Very Complicated Elements of a Structure

Exploratory tests of new concepts have been used extensively by aeronautical, civil, mechanical and ocean engineers, and will continue to be an important tool. Furthermore, if the structure is elaborate and has many openings with complicated stiffening and load diffusion elements, model testing may be less expensive and faster than computation with a large multipurpose program, even in the detail design state.

g. For Buckling Under Dynamic Loading and in Fluid-Structures Interaction Problems

These are areas where computation is cumbersome, expensive, and difficult to interpret reliably. Experiments are therefore preferable, though they too present many difficulties.

h. For Certification Tests of Full Scale Structures

This is the typical industrial task, which will continue till model experiments are sufficiently advanced and integrated with computation to eliminate the necessity for them.

Examination of these motives, proposed for shells in 1980 [10], and recent experience reinforces the conclusion that the computer does not replace the experiments. It may change their purpose somewhat, it may modify the techniques, it broadens the capability to acquire results and it can use the experimental results to improve the computations. The experiment remains an essential link in the analysis also in the computer era, and its scope and usefulness are even greater today.

Though it is accepted today that experiments are essential tools for buckling and postbuckling research, many investigators still shy away from them. One reason may be the initial difficulties. To quote, for example, from a primarily theoretical doctoral thesis: "Postbuckling Behavior of Tee Shaped Aluminum Columns", by R. Hariri [14], "The author experienced a great deal of technical difficulties and some experiments in the early stages yielded surprising and unexpected results". He then concludes this paragraph, "However, the experiences gained and the guidance obtained cannot be overlooked".

As one who came to experimental buckling research from theoretical studies, I can definitely confirm that the insight and guidance to analysis and computation obtained from careful experiments is invaluable.

3. ELEMENTS OF A SIMPLE BUCKLING TEST - A COLUMN UNDER AXIAL COMPRESSION

In order to appreciate the nature of the experimental approach let us first consider a simple buckling test - a column under axial compression. The column is not only the earliest and classic example of elastic instability, dating back to Euler's work in 1744 [15], and of postbuckling studies, but is also the element that has been the subject of the most extensive experimental studies for nearly a century. If one studies the experimental work on buckling of columns carried out at the turn of the century one is amazed at its high quality. Maybe one should not be be surprised, since it was performed by some of the giants of mechanics, like von Kármán [16] and Prandtl [17], who in their doctoral dissertations combined outstanding analysis with outstanding experiments.

It is certainly illuminating to review such a classic set of tests, for example von Kármán's experiments [16]. In the introduction, the main purpose of the work is stated to be the experimental proof of the formulae for the buckling strength of shorter columns, which buckle inelastically. It is pointed out, however, that the experiments presented an opportunity to investigate the influence of exact centering of the columns, as well as the postbuckling behavior after the peak load has been exceeded.

The tests were carried out in a 150-ton hydraulic press with a 1000 mm long working section, permitting slender specimens with a convenient cross section of the order of 20 x 30mm^2. Careful compression tests on very short specimens to evaluate the compressive material properties, preceded the buckling tests. Von Kármán designed special end fixtures (Fig. 1) which facilitated accurate placing of the center line of the column in the loading line. This fixture permitted readjusting of the position of the column under load, which lead to more accurate centering. Von Kármán also evaluated the change in effective length of the tested columns due to the rigidity of the end fixture and corrected his results accordingly. He also considered the possible error due to the friction of the knife edges in their bases and argued that it can be neglected, since the tests in the elastic range showed only very small deviations from the predictions by the well established Euler formula.

These minor details are pointed out here, since they signify some elements of the methodology of careful buckling experiments: readjustment of specimen positions in end fixtures under load, compensation for end fixture rigidity and consideration of secondary effects like friction and justification of their neglect by comparison with well established earlier results.

Von Kármán continued many of his tests well into the postbuckling region (see Fig. 2), since he realised that the postbuckling behavior is of importance for the understanding of the buckling behavior. The results (see Fig. 2) show that for the long columns, the theoretically predicted elastic behavior of constant load with increasing deflection is indeed confirmed experimentally (for example in column No. 1). For short columns, however, the behavior differs and the load decreases very significantly with increasing deflection, due to inelastic effects (for example in column No. 6). These inelastic effects were the prime

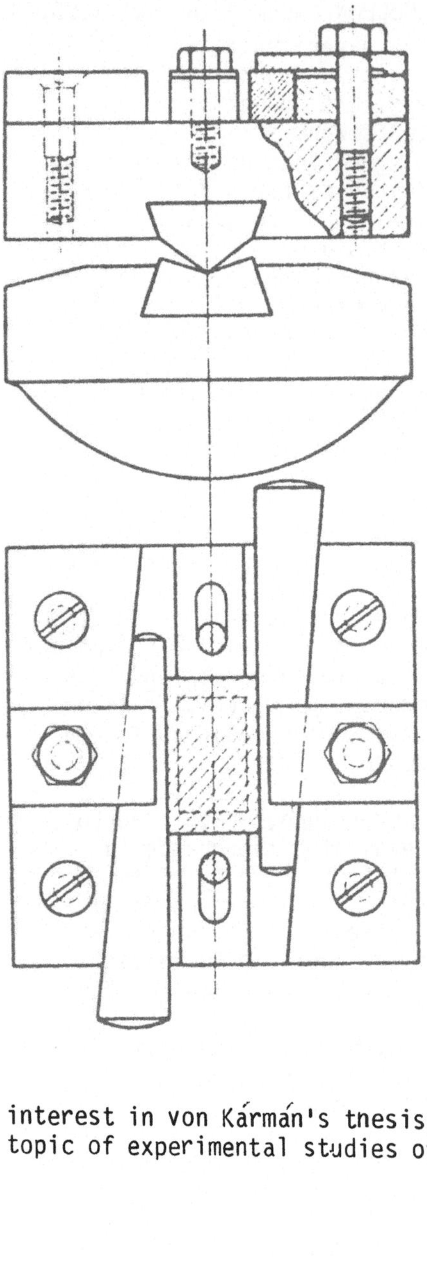

Fig. 1. Von Karman's End Fixtures [16].

interest in von Kármán's thesis, and indeed have since been the main topic of experimental studies of the buckling of columns.

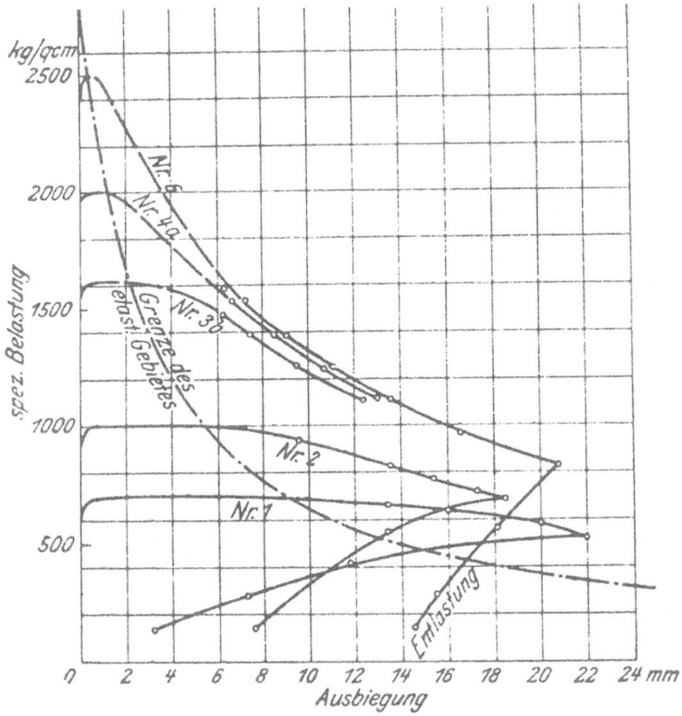

Fig. 2. Von Karman's Load-Deflection Curves beyond the Max. Load [16].

4. BUCKLING AND POSTBUCKLING OF COLUMNS

In the 75 years since von Kármán's thesis the buckling and postbuckling of columns has been studied extensively, with a significant portion of the efforts devoted to experimental investigations. Many of these deal with the interaction between material properties, shape cross section and the postbuckling behavior, and many are design oriented. The Structural Stability Research Council (formerly the Column Research Council) is a USA (but essentially international) body, that has for over 40 years fostered research and developed design and test procedures for column stability. Hence we naturally turn to the SSRC "Guide to Stability Design Criteria for Metal Structures" [18] for a guide to modern column testing, which indeed appears there in the form of Technical Memorandums. It should be pointed out that the SSRC Guide is aimed primarily at civil, mechanical and marine engineers, but is a most authoritative source covering also work carried out in other fields.

As mentioned already, the bulk of the theoretical and experimental studies on buckling of columns in recent decades dealt with inelastic behavior of steel columns and the influence of yield strength, of geometry, of residual stresses (resulting from the different manufacturing processes), and of out-of-straightness (as the geometric imperfections are called in columns). Tall [19] summarizes these experiments and the resulting design curves from the point of view of civil engineers.

Residual stresses occur in a structural member as a result of plastic deformations during manufacture. They may be due to differential cooling after hot-rolling, or due to fabrication processes like flame-cutting or cold-bending or due to localized heat input in welding operations. Welded columns usually have higher residual stresses than rolled columns and their magnitude depends on the geometry of the cross section. They also tend to have a greater out-of-straightness. Hence welded columns have lower strengths than corresponding rolled columns (see Fig. 3), and this strength has to be assessed by a more complicated analysis of the behavior in the inelastic range.

Both size and yield strength influence the strength of a steel column. The process of cooling in heavy shapes (large size cross sections) yields larger residual stresses than in small size shapes, whether rolled or welded, and hence heavy columns have reduced strengths. Since the residual stresses are mainly a function of geometry, they are of the same order of magnitude in high-strength steels as in mild steels. Thus the effect of residual stresses is smaller in columns made of steels with higher yield strength.

This summary of the main characteristics of the buckling strength of steel columns, derived from decades of extensive steel column tests, emphasizes the important interaction of material properties, geometry and fabrication processes on the buckling behavior and strength of columns. It furthermore indicates that one has to be very careful to consider also such "secondary" effects in the design and evaluation of buckling and postbuckling tests of all structural elements.

For the design of steel columns this wealth of empirical information, reinforced by many theoretical studies, has resulted in a concept of multiple column curves that has been adopted both by the USA Structural Stability Research Council (SSRC), [18], and the European Convention for Constructional Steelwork (ECCS), [20].

Having briefly discussed the main "secondary" effects in column experiments, one can proceed to the test procedures. In its Appendix B, the SSRC "Guide to Structural Stability" [18] first presents recommended test procedures for compression testing of metals (pp. 555-559) and then for stub-column tests (pp. 559-569). The object of a compressive stub-column test is to determine the average stress-strain relationship over the complete cross section, which can then be employed as the actual material properties for the column test. The Technical Memorandum "Procedure for Testing Centrally Loaded Columns" ([18], pp. 569-584) is

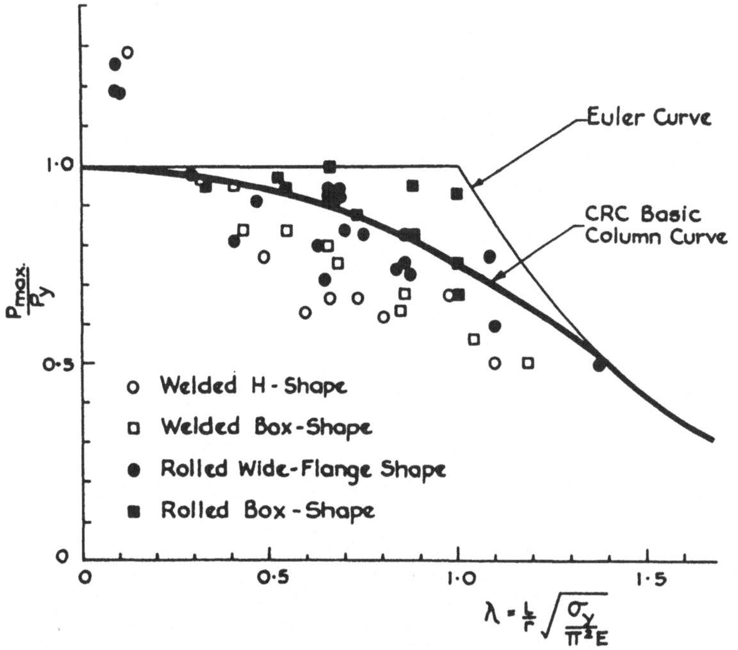

Fig. 3. Test Results for Small to Medium Rolled and Welded Shapes [19].

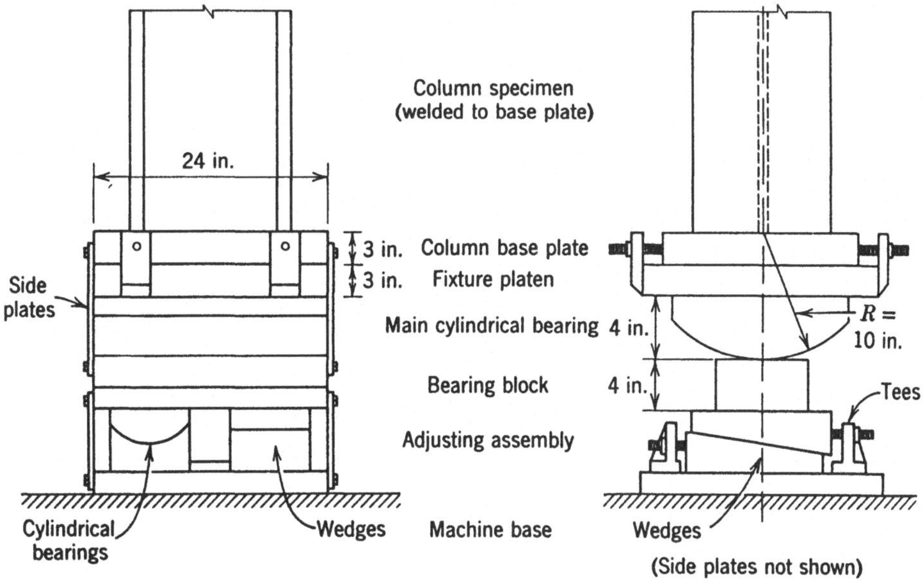

Fig. 5. Standard Large Column End Fixture at Fritz Engineering Laboratory [18].

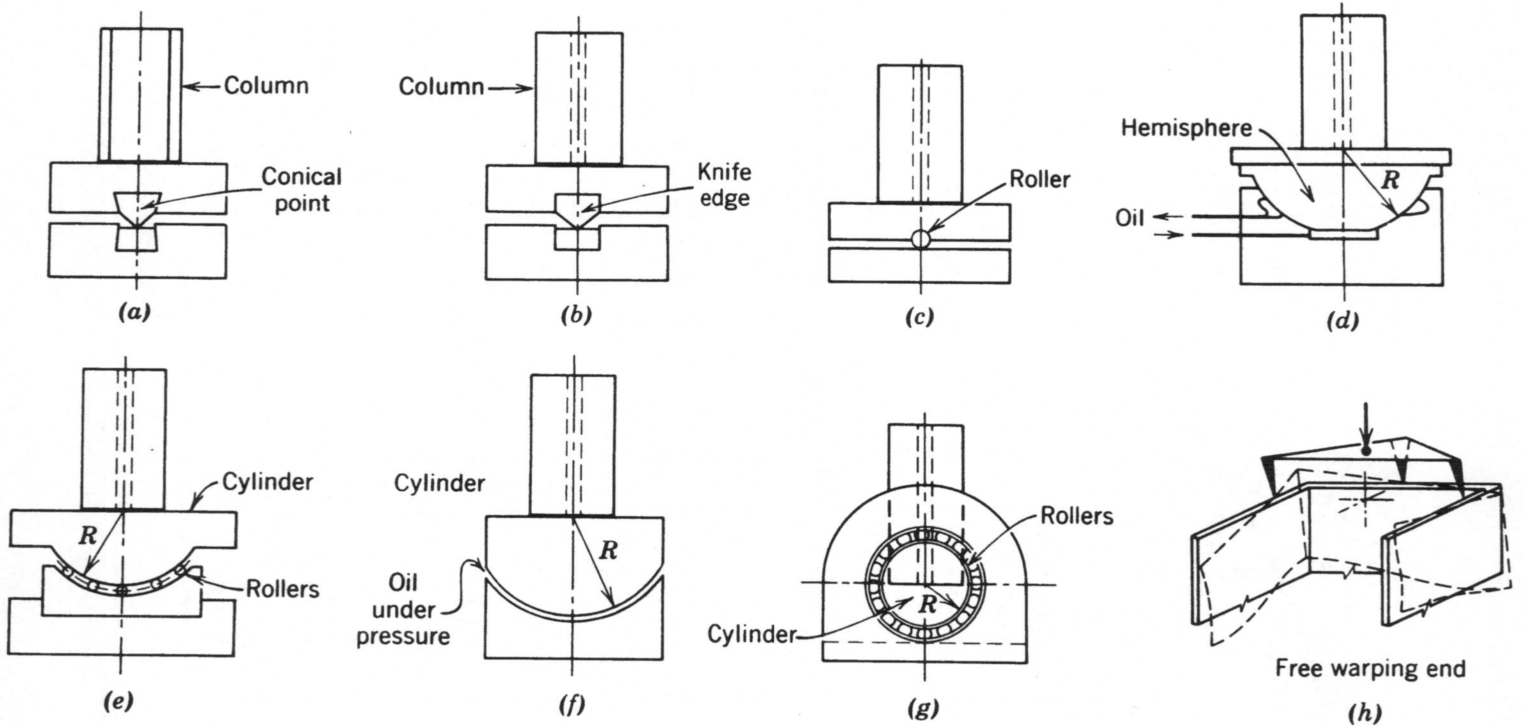

Fig. 4. End Fixtures for Pin-Ended Columns [18].

based on a 1970 Lehigh University Fritz Engineering Laboratory Report and presents a suggested procedure.

The reasons for the wide scatter band of the experimentally determined values of column strength when plotted versus the effective slenderness ratio (KL/r), in which KL denotes the effective column length and r the appropriate radius of gyration of the cross section, are enumerated as:

(1) geometrical imperfections (out-of-straightness)
(2) eccentric application of load
(3) nonhomogeneity of material
(4) residual stresses
(5) variation in the action of the loading machines
(6) imperfections in end fixtures.

These effects have already been discussed, except the two last which are directly related to the tests themselves. The buckling and postbuckling behavior of a column, or of any other structural element, is influenced by the action of the loading devices. These may be categorized as gravity, deformation (screw-type) and pressure (hydraulic) testing machines, each differing in its force-deflection characteristic. The gravity type has the simplest characteristic, which can be represented by straight lines parallel to the deflection axis. The screw-type load deflection characteristic is also well defined, and its shape depends on the elastic response of the loading system. The hydraulic testing machine is the most common today, but its load-deflection characteristic is not easily defined and testing is always conducted under some finite loading rate, which influences the results.

End conditions can vary from full restraint (fixed) to zero restraints (pinned, simple supports) with respect to end rotation and warping. The pinned-end conditions are recommended for column tests, since then the critical cross section is located near the mid-height of the column and is therefore little influenced by end effects. With pinned-end conditions it is however necessary to provide end fixtures with minimum restraint to column end rotation. Under fixed-end conditions, on the other hand, there are often problems of variation of the end restraints, and hence the effective length, with load, which make the tests less reliable.

Figure 4 shows several practical pinned-ends (from [21]), some are "position-fixed" and some are "direction-fixed".

A recommended scheme to reduce end restraint is by means of a relatively large hardened cylindrical surface bearing on a hard flat surface. Such end fixtures used at the Fritz Engineering Laboratory are shown in Fig. 5 (from [22]).

In the actual test procedure some important points should be remembered:

a. Preparation of Specimens

Both ends of the specimen should be milled. Columns may be tested with the ends bearing directly on the loading fixtures, provided the material of which the loading fixtures are made is sufficiently harder than that of the column to avoid damaging the fixtures. Otherwise, base plates should be welded to the specimen ends, matching the geometric center of the specimen to the center of the base plate. The welding procedure should be such that compressive residual stresses at the flange tips caused by the welding are minimized. For columns initially curved, the milled surfaces may not be parallel to each other, but will be perpendicular to the centerline at the ends because milling is usually performed with reference to the end portions of the columns. For relatively small column specimens, it is possible to machine the ends flat and parallel to each other by mounting the specimens on an arbor in a lathe. For small deviations in parallelism, the leveling plates at the sensitive crosshead of the testing machine may be adjusted to improve alignment.

b. Initial Dimensions

The variation in cross-sectional area and shape, and the initial curvature and twist, will affect the column strength. Therefore, initial measurement of these parameters of the specimen is important.

c. Aligning the Column Specimen

Aligning the specimen within the testing machine is the most important step in the column testing procedure, prior to loading. Two approaches have been used to align centrally loaded columns. In the first approach the column is aligned under load such that the axial stresses are essentially uniform over the mid-height and the quarter-point cross sections. The objective in this alignment method is to maximize the column load by minimizing the bending stresses caused by geometrical imperfections of the specimen.

In the second alignment method, the column is carefully aligned geometrically, but no special effort is made to secure a uniform stress distribution over the critical cross section. Geometric alignment is performed with respect to a specific reference point on the cross section. The method of geometric alignment is recommended for columns as it is, generally, simpler and quicker.

In other structural elements, however, the "uniform stress" approach is usually preferable.

d. Instrumentation

It is usually desirable to measure the more important deflections and twists to compare the behavior of the column specimen under load, also well into the postbuckling range, with theoretical predictions of behavior. The instrumentation for column tests has changed markedly in recent years due to progress made in measuring techniques and data acquisition systems, and it is now possible to obtain automatic recordings and plotting of the measurements. Such recordings have been found to be more convenient and more precise than manual readings.

The most important records needed in column testing are the applied load and the corresponding lateral displacements, twist, and overall column shortening. A typical column set-up and instrumentation are shown in Fig. 6 (from [18]).

Lateral deflections normal to both principal cross-sectional axes may be automatically recorded by means of potentiometers attached at quarter points of the column (more points may be used for longer

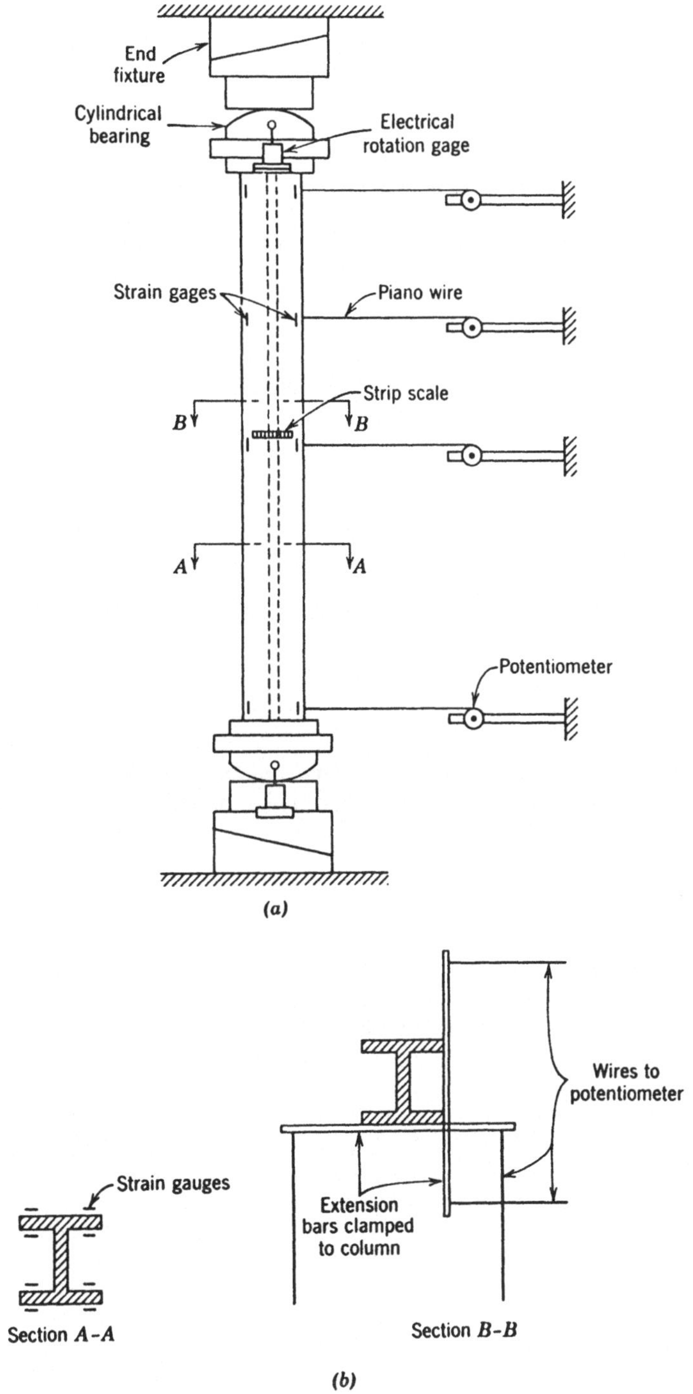

Fig. 6. Typical Test Set-Up [18].

columns). Lateral deflections may also be measured from strip scales attached to the column and read with the aid of a theodolite.

Strains are measured using electric-resistance strain gages. For ordinary pinned-end column tests, it is sufficient to mount strain gages at each end and at the midheight level. As shown in Section A-A of Fig. 6, the gages should be mounted in pairs "back-to-back" to enable the local flange bending effects to be cancelled by averaging the readings of each pair of "back-to-back" gages. In the fixed-end test condition more strain gages are mounted below and above the quarter- and three-quarter levels. This is done to determine the actual effective length of the column by locating the inflection points using the strain gage measurements.

End rotations are measured by mechanical or electrical rotation gages. The angles of twist are determined at mid-height and at the two ends by measuring at each level the differences in lateral deflections of the two flanges. For better accuracy, measurements may be taken at points located at the ends of two rods attached transversely on the adjacent sides of the column, as shown in Section B-B of Fig. 6.

The overall shortening is determined by measuring the movement of the sensitive crosshead relative to the fixed crosshead using a dial gage or potentiometer.

Steel column specimens are usually whitewashed with hydrated lime. During testing, the whitewash cracking pattern indicates the progression of yielding in the column (the cracking reflects the flaking of the mill-scale at yielded zones).

e. Testing

After the specimen is aligned in the testing machine, the test is started with an initial load of 1/20 to 1/15 of the estimated ultimate load capacity of the column. This is done to preserve the alignment established at the beginning of the test. At this load all measuring devices are adjusted for initial readings.

Further load is applied slowly, typically at a rate of 1 ksi/min (6.9MPa/min), and the corresponding deflections are recorded instantly. This rate is established when the column is still elastic. The dynamic curve is plotted until the ultimate load is reached, immediately after which the "maximum static" load is recorded. (The procedure for determining a "static" load is described subsequently.) After the maximum static load is recorded, compression of the specimen is resumed at the "strain rate" which was utilized for the elastic range. In hydraulic testing machines this may be accomplished, approximately, by using the same bypass valve and load valve settings as had been used in the elastic range. The specimen is compressed in the "unloading range" until the desired load-displacement curve is attained. An example of such a curve is shown in Fig. 7 (from [18]).

A static condition, as is needed to obtain the "maximum static" load, is when the column shape is unchanged under a constant load for a period of time. This means that the chord length of the column must remain constant, or practically, the distance between the crossheads must remain constant during the period. For screw-type testing machines the criteria can normally be satisfied by maintaining the

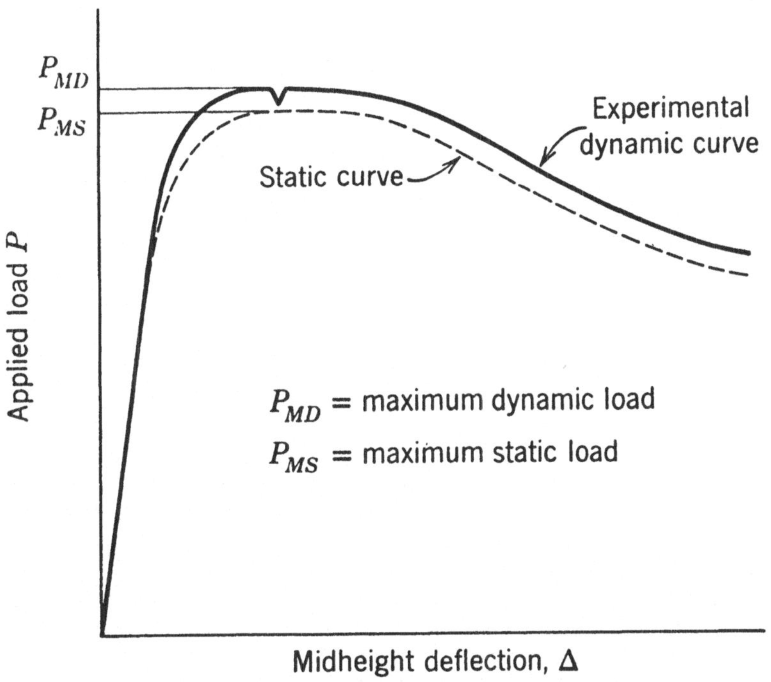

Fig. 7. Typical Load-Deflection Curve of Column [18].

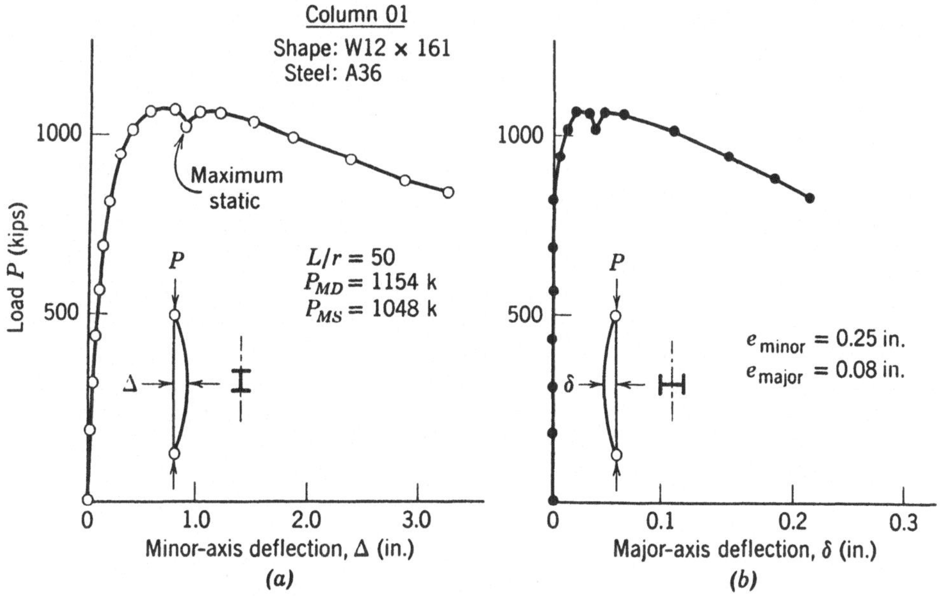

Fig. 8. Load-Deflection Curves [18].

crossheads in a stationary position. However, it is difficult to maintain the distance between crossheads in hydraulic machines because of oil leakage and changes in oil properties due to the temperature changes that accompany pumping and throttling. To attain the static condition in the hydraulic machine, from the dynamic state, the bypass valve is further opened slowly until further lateral deflection of the column at mid-height ceases. The cessation of lateral deflection amounts to the condition of constant chord length.

One should note that the dynamic load is larger than the static one. this means that a column can sustain a considerably higher buckling load if the load is applied rapidly, i.e. under impact (see for example [23]).

f. Presentation of Test Data

The behavior of the test specimen under load well into the post-buckling region is determined with the assistance of measurements of lateral deflections at various levels along the two principal directions, rotations at the ends, strains at selected cross sections, angles of twist, and the column shortening. These measurements are compared to theoretical predictions. The results of the test are best presented in diagrammatic form.

Figure 8a (from [18]) shows the midheight load deflection curve of a typical structural steel H-section column, along the minor axis, where the primary bending occurs. Figure 8b shows the similar curve along the major axis. These curves present the most significant data of the column test.

Similar curves are usually presented for strains, end rotations, angles of twist and overall shortening versus load.

g. Evaluation of Test Results

The test results may be evaluated by comparing the experimental load-deflection behavior and the theoretical prediction. A preliminary theoretical prediction can be made based on simplified assumptions of material properties, residual stresses, and measured initial out-of-straightness. The prediction may be improved if the actual residual stresses and the variations in material properties are used in the analysis. These properties should be determined from preliminary stub-column tests of specimens obtained from the original source stock.

Another recent review of the state-of-the-art of buckling for a particular type of structures - offshore structures [24], considers only tubular columns in the discussion of columns, since both the main and the bracing members of a typical offshore structure are usually circular cylinders. It is of interest to note that the emphasis in this very recent design guide is on information based on experimental investigations. The discussion of column buckling experiments in [24] is essentially similar to that in the SSRC Guide [18], except that the comparisons are with the special codes and design recommendation developed for offshore structures and that interaction with local buckling (shell buckling) is considered in detail.

Since offshore platforms are usually designed as highly redundant space frames, where buckling of an individual member will not

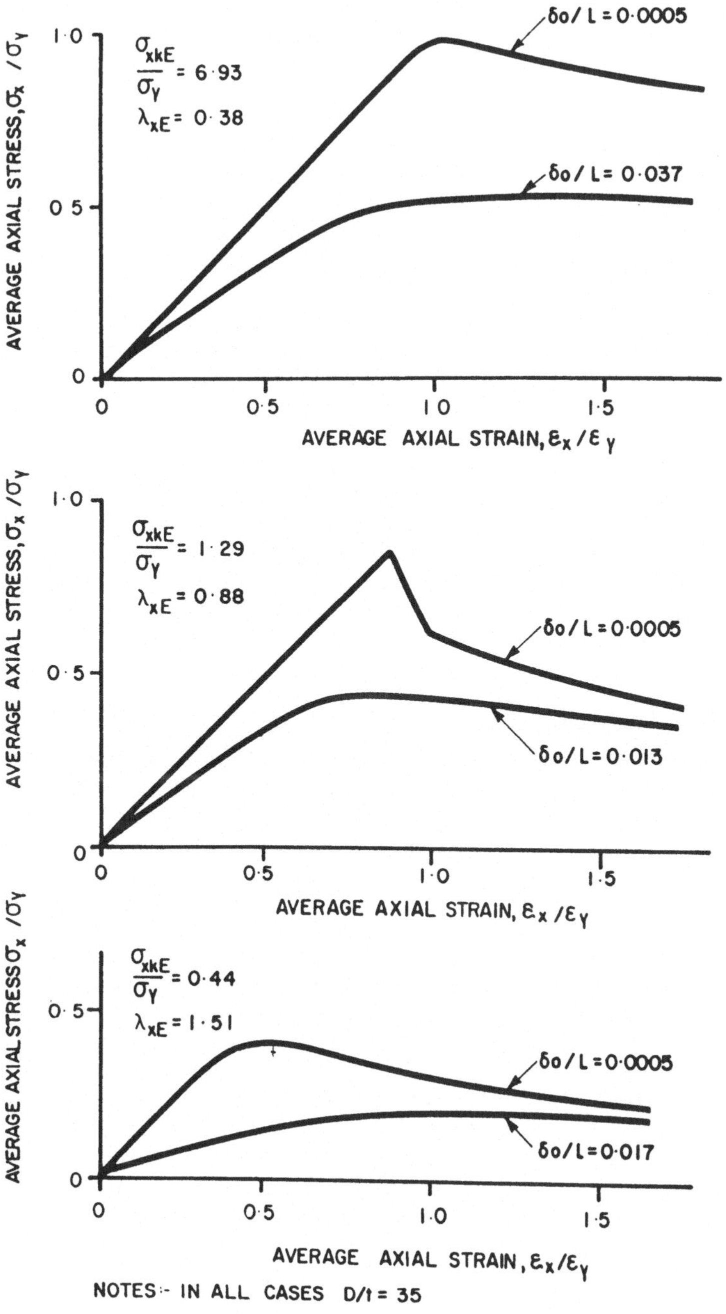

Fig. 9. Typical Average Axial Stress - Strain Curves for Turbular Columns with Small and Large Imperfections [24].

necessarily lead to failure of the structure as a whole, and since they are subjected on rare occasions to extreme loads, post-collapse characteristics are very important for assessment of the survivability of offshore structures.

The post-collapse behavior of tubular columns strongly depends on whether collapse is initiated by local instability. If local stability predominates, the post-buckling behavior is that of a cylindrical shell, which is highly unstable, as has been shown in the theoretical lectures, and will also be stressed later in the discussion on shells. If local buckling is avoided, as in tubes with low (D/t) ratio, the post collapse behavior is controlled by the ratio of the Euler critical stress σ_{xkE} to the yield stress σ_Y, and by the magnitude of initial imperfections (out-of-straightness). Figure 9 (from [24]) shows typical average axial stress-strain curves for tubular columns. (These curves are typical presentations corresponding to experimental results.) For the two extreme cases (a) $\sigma_{xkE} >> \sigma_Y$ and (c) $\sigma_{xkE} << \sigma_Y$, collapse is gradual and a significant load carrying capacity is retained. However, in case (b), when $0.7 < (\sigma_{xkE}/\sigma_Y) < 1.3$, postbuckling can be highly unstable for tubes with small imperfections, with significant reduction in load-carrying capacity in the post-collapse range. The presence of large imperfections considerably reduces the pre-buckling stiffness and the buckling load all cases, but collapse occurs very gradually and with little reduction in load carrying capacity.

5. MODELING - THEORY AND PRACTICE

Most experimental studies on buckling and postbuckling behavior of structural elements are carried out on models of the actual elements. Appropriate modeling is therefore an essential part of experimental investigations. Mathematical modeling, and primarily dimensional analysis, can guide the experimenter in the choice of his models. The principles of mathematical modeling are usually well known to scientists and engineers and are given in many textbooks (a recent example is [25]). Let us apply dimensional analysis to elastic stability problems, following the presentation given by Chilver in [2].

Consider first the general problem of elastic buckling of a structure, under a well-defined loading system. It is assumed that buckling can be defined in some appropriate form, such as the development of gross deformations. Where the loading is due to some external force, such as an external point load, stress or pressure, and interest centers on some critical value of this force at which buckling develops, the dimensional analysis of the problem is relatively simple. It is supposed that the material of the structure is isotropic and homogeneous with Young's modulus, E, and Poisson's ratio, ν. Then, for geometrically similar structures, the critical value of the load, P

$$P_{cr} = F(E,\nu,L)$$

is where L is a typical linear dimension, and F is an arbitrary function. A simple dimensional analysis then gives

$$\frac{P_{cr}}{EL^2} = F(\nu) \ .$$

Buckling is usually a structural problem at the extremes of the geometric forms; in many such cases, as for example slender columns or thin plates, there is not a strong dependence of buckling on the value of Poisson's ratio. In these cases it is probably justifiable to ignore ν, and thus, in its simplest form,

$$\frac{P_{cr}}{EL^2} = \text{constant}$$

for materials of the same Poisson's ratio; the relationship will also hold approximately in situations where dependence on Poisson's ratio is weak.

Now compare the elastic buckling behavior of a given structure and that of a geometrically similar structure, not necessarily made of the same material but of an elastic material of the same Poisson's ratio. If these two structures are denoted by subscripts 1 and 2 respectively, we have

$$\left[\frac{P_{cr}}{EL^2}\right]_1 = \left[\frac{P_{cr}}{EL^2}\right]_2 \ .$$

Thus

$$\frac{[P_{cr}]_1}{[P_{cr}]_2} = \left[\frac{E_1}{E_2}\right]\left[\frac{L_1}{L_2}\right]^2$$

which show that, in general, critical loads are directly proportional to Young's modulus and to the square of the geometrical scale. This is a very useful and simple scaling law for experimental studies; a small-scale model, in a relatively flexible material, can be used effectively to enable useful experiments to be carried out at relatively low loads.

If the structures are considered in terms of stresses, σ_{cr}, rather than loads, P_{cr},

$$\frac{[\sigma_{cr}]_1}{[\sigma_{cr}]_2} = \left[\frac{E_1}{E_2}\right]$$

may be written. Thus, the critical stresses for elastic buckling of a full-size structure and a model in the same material, are equal.

Hence, when complete geometrical similarity is preserved, and there is no dependence on Poisson's ratio,

$$\frac{\sigma_{cr}}{E} = \text{constant} .$$

Now,

$$\frac{\sigma_{cr}}{E} = \left(\frac{\sigma_{cr}}{\sigma_y}\right)\left(\frac{\sigma_y}{E}\right)$$

may be written where σ_y is the yield stress of the material. Then, for two similar structures 1 and 2,

$$\left(\frac{\sigma_{cr}}{\sigma_y}\right)_1\left(\frac{\sigma_y}{E}\right)_1 = \left(\frac{\sigma_{cr}}{\sigma_y}\right)_2\left(\frac{\sigma_y}{E}\right)_2 ,$$

or

$$\left(\frac{\sigma_{cr}}{\sigma_y}\right)_2 = \left(\frac{\sigma_{cr}}{\sigma_y}\right)_1 \frac{(\sigma_y/E)_1}{(\sigma_y/E)_2} .$$

If structure 2 is regarded as a model of structure 1, then a suitable choice of material for the model can usually eliminate plastic buckling effects in the model. The yield stress, or rather the ratio of (σ_y/E), is therefore of prime importance in designing a model. For example, a high-strength model will give

$$\left(\frac{\sigma_{cr}}{\sigma_y}\right)_2 = \text{order } \; 5 \times \left(\frac{\sigma_{cr}}{\sigma_y}\right)_1$$

compared with a mild-steel full-scale structure.

The need to increase (σ_y/E) to study elastic buckling phenomena, and especially postbuckling behavior, is reflected in the materials used for recent postbuckling studies. For example, high-strength steel strip and sheet have been used successfully in the study of the stability of framed structures and flat plates. Polyester films have been used extensively to study the stability of shells, and many shell buckling problems have been studied by using elecrolytically deposited nickel or copper shells.

Some typical examples for (σ_y/E) are:

Structural steel $(\sigma_y/E)=0.0013$.
Medium strength steel (AISI 4130 drawn tubes, used for test specimens of stringer-stringer stiffened cylindrical shells - [26]) $(\sigma_y/E)=0.0022$.

High strength steel (17-7 PH heat treated after hydrospinning, used for test specimens of ring-stiffened conical shells - [27]) $(\sigma_y/E)=0.0057$.
Aluminium alloy (7075-T6 drawn tubes used, for example, for test specimens of stringer-stiffened cylindrical shells - [26]) $(\sigma_y/E)=0.0072$.
Mylar polyester films (used, for example, for test specimens of cylindrical shells - [28]) $(\sigma_y/E)=0.0210$.
Electroformed nickel (used for test specimens of complete spherical shells - [29]) $(\sigma_y/E)=0.0018$.

The values indicate the relative suitability of the materials for models in postbuckling studies.

The modeling of structures to simulate elastic buckling is relatively simple. When problems of collapse involving inelastic effects are considered, the situation may become extremely complex.

When complete geometric similarity is preserved,

$$\sigma_{max} = F(E, \sigma_y, L)$$

can be written, where σ_{max} is a maximum external stress of the system, such as the average compressive stress for collapse of a column or plate, and σ_y is the yield stress; it is assumed that the material is sharp yielding and that strain hardening after yielding may be ignored. Dimensional analysis then yields $= (\sigma_{max}/\sigma_y)\ (E/\sigma_y)$, if one uses Buckingham's Pi theorem, or in a more convenient form

$$\frac{\sigma_{max}}{\sigma_y} = F\left(\frac{\sigma_y}{E}\right) .$$

Hence if full-size structure and model are of the same material, σ_{max} is the same. Furthermore, where (σ_y/E) is the same for the full-size structure and the model, the values of (σ_{max}/σ_y) are the same. Where modeling in the same material is difficult, a suitable choice of (σ_y/E) can lead to model results which are relevant to the full-size structure.

In cases where plastic collapse follows the initial development of elastic buckling, some progress can be made by assuming that the elastic buckling stress, σ_{cr}, for a perfectly-elastic material plays a role in determining the maximum stress, σ_{max}. Suppose

$$\sigma_{max} = f(\sigma_y, \sigma_{cr}),$$

then dimensional analysis yields, as before,

$$\frac{\sigma_{max}}{\sigma_y} = f\left(\frac{\sigma_y}{\sigma_{cr}}\right) .$$

For columns, this suggests a simple interaction curve between (σ_{max}/σ_y) and $(\sigma_{max}/\sigma_{cr})$, since

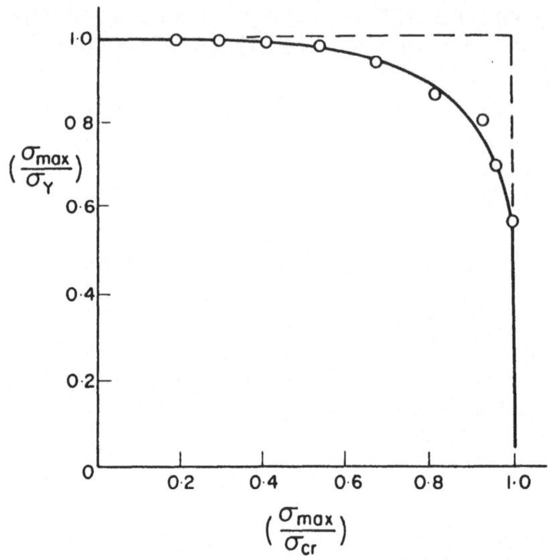

Fig. 10. Pin-Ended Column Tests on a Light-Alloy Material (σ_y = 0.2 Proof Stress) [2].

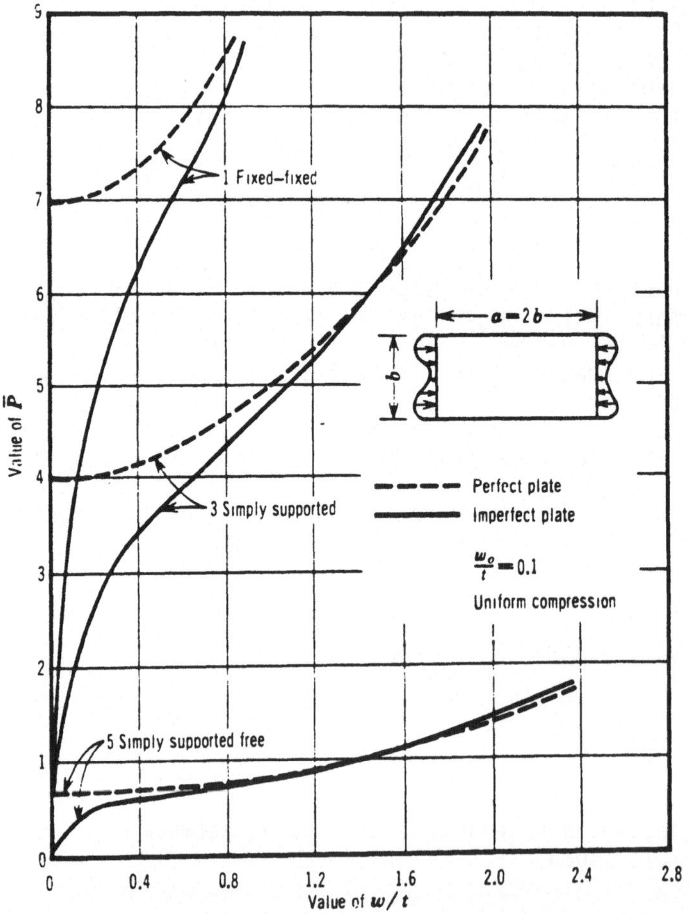

Fig. 11. Load-Deflection Curves for Imperfect Plates [31].

$$\frac{\sigma_{max}}{\sigma_y} = f \left[\frac{\sigma_{max}}{\sigma_{cr}}\right] \times \left[\frac{\sigma_y}{\sigma_{max}}\right]$$

may be written.

Figure 10 shows the results of some pin-ended column tests by Chilver [2] on a light-alloy material; here the yield stress is taken as reasonably well-defined by the 0.2% proof stress of the material. A well-defined interaction curve emerges between (σ_{max}/σ_y) and (σ_{max}/σ_{cr}), where the right side represents the region where elastic buckling predominates and the left side the shorter columns that fail primarily by yielding.

One notes that simple dimensional analysis can be a helpful guide in the design of meaningful experiments and that it can be extended also to deal with yielding and collapse conditions. One of the weaknesses of the dimensional analysis approach is that geometric imperfections, which have a significant effect on buckling behavior, are not included in it. If a structure is strongly imperfection sensitive, even careful experiments will demonstrate strong scatter. The experiments themselves therefore may present an indication or warning of imperfection sensitivity even if theoretical consideration have not brought it out.

6. BUCKLING AND POSTBUCKLING OF PLATES

Whereas in the case of buckling and collapse of columns much of the emphasis in recent decades has been on experimental investigations, for plates the great majority of papers in this century have been concerned with theoretical developments. Walker notes this with regret, in a recent review on plate buckling research [30]. I not only share his view that the most important developments in understanding the phenomenon of plate buckling and generating information for designers have been made when theory and experiment proceeded hand-in-hand, but believe that this applies more generally to the buckling and postbuckling of all structural elements.

As has been shown in the theoretical discussions, plates have a stable postbuckling behavior (see Fig. 11, from [31]). This characteristic, which was first pointed out by Schuman and Back in 1930 [32], who observed that the failing load of a flat panel was materially higher than its buckling load, makes plates very suitable structural-elements in lightweight structures. The interest of aeronautical engineers in the thirties and forties in lightweight plate structures motivated extensive theoretical studies as well as experimental investigations, (for example, [32] and [33]). The experimental results were brilliantly interpreted by von Karman [34] with a simple concept - the effective width, which became a universally used design formula.

The stable postbuckling behavior, as presented in Fig. 11, shows the inherent difficulty in defining the buckling load for plates with imperfections. The strains, measured with strain gages, in a typical

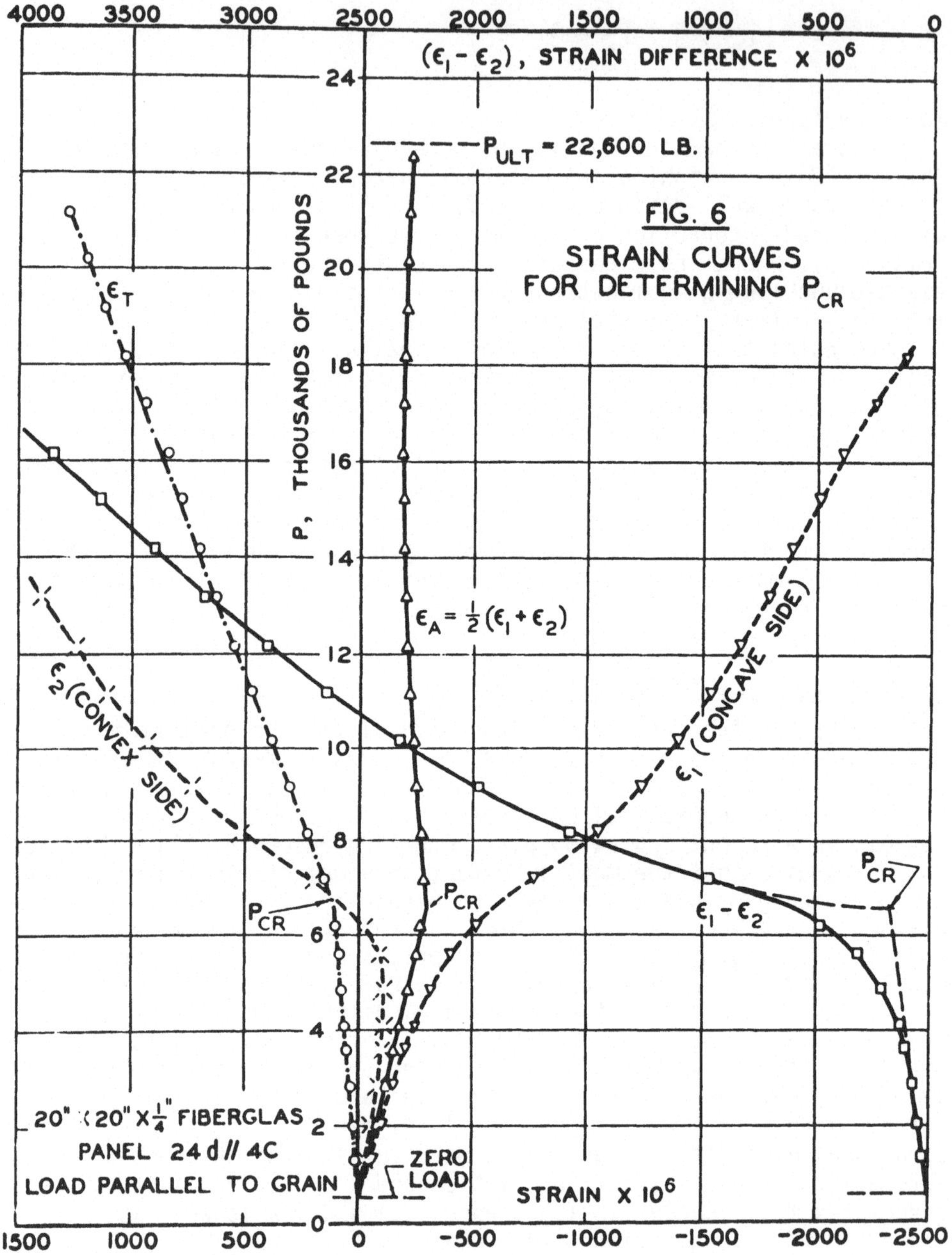

Fig. 12. Strain Curves for Determining P_{cr} [35].

carefully performed plate buckling test on a flat rectangular fiberglass panels [35] are presented in Fig. 12. Three possible methods of defining P_{cr} from the strain records are shown in the figure: (a) a sufficiently sharp break in the algebraic mean compressive strain ε_A, (b) a sufficiently sharp break in transverse strain curve ε_T and (c) the extrapolation of the parts of the strain difference $\varepsilon_1 - \varepsilon_2$ curve below and above the buckling load. Another method is shown in Fig. 13 (another plot of test data from [35]) - the inflection point method, which consists in locating the least slope of the load deflection curve. The application of the first two methods is also shown in this figure. The many methods are an indication that none of them is too precise in defining P_{cr}. Hence much effort has been devoted to apply the Southwell Plot [36], which is strictly applicable only to structures with a neutral postbuckling path-like columns, also to plates.

Let us briefly rederive the Southwell plot for a simply supported column (see, for example, [37] or [38]). The equilibrium equation of an imperfect (initially crooked) column is

$$w^{iv} + \alpha^2 w'' = -\alpha^2 w_0''$$

where $w(x)$ = the additional lateral deflection measured in tests

$w_0(x)$ = the initial deflection (imperfection)

$\alpha^2 = P/EI$

and the boundary conditions for simple supports are

$$w(0) = w''(0) = w(L) = w''(L) = 0$$

Representing the initial deflection by a Fourier series

$$w_0(x) = \sum_{n=1}^{\infty} W_{on} \sin\left(\frac{n\pi x}{L}\right)$$

leads to

$$w_{(x)} = \sum_{n=1}^{\infty} W_{on}\left[\left(\frac{n^2 P_E}{P}\right) - 1\right]^{-1} \sin\left(\frac{n\pi x}{L}\right)$$

where $P_E = \left(\frac{\pi^2 EI}{L}\right)$, the Euler Load.

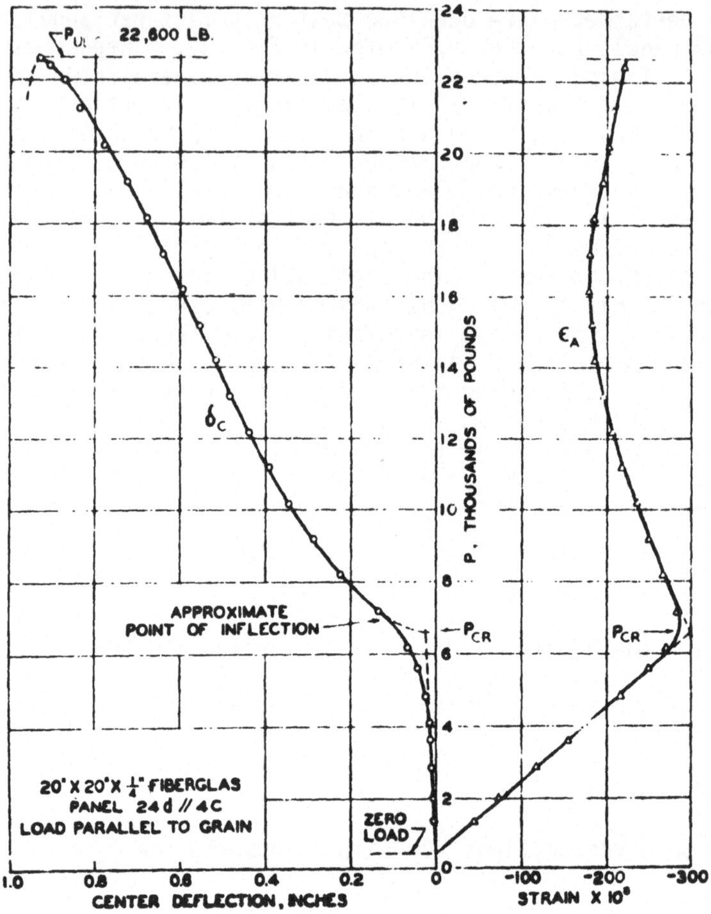

Fig. 13. Typical Test Plot [35].

The maximum deflection $W = w(L/2)$ is therefore

$$W = W_1 - W_3 + W_5 - \ldots.$$

where

$$W_n = W_{on} \left[\left(\frac{n^2 P_E}{P}\right) - 1\right]^{-1}$$

When the buckling load is approached, or as Southwell noted "if P is a fairly considerable function of P_E"

$$W \simeq W_1 = W_{o1} \left[\left(\frac{P_E}{P}\right) - 1\right]^{-1}$$

and the fundamental mode predominates. Hence as $P \to P_E$, the imperfection component that represents the buckling mode is the one that is primarily magnified. Hence we can write

$$W_o \simeq W_{o1}$$

and the expression for W can be rearranged as

$$W \simeq P_E \frac{W}{P} - W_o$$

The slope of the plot of W versus $\frac{W}{P}$, the Southwell plot, yields the buckling load of the corresponding perfect column. Figure 14 (from [36]) shows the Southwell plots for columns tested by von Karman [16]. Note that only at the higher values of W (or δ in the figure) the relation is linear and the procedure justified. As a matter of fact Southwell rejected all points for $P<0.8P_C$. The results of Fig. 14 were excellent - in no case did the critical load derived from the Southwell plot differ by more that 2.5% from the classic Euler load.

Considerable effort has been devoted to extend this simple and effective method to the buckling of other structures (see, for example, [39]) and its applicability to plates has been shown for many cases. However, Spencer and Walker [38] pointed out that in plates there may be significant nonlinearities at higher loads which cast some doubts on the applicability of the Southwell Plot. For example, Fig. 15a shows the Southwell Plot for an aluminum plate test [40], that is strongly curved and hence not amenable to useful interpretation. By a reinterpretation of plate theory, Spencer and Walker devised another type of plot (see Fig. 15b) which yields P_C. If the alternative plot proposed in [38] is nonlinear, more complicated techniques have to be employed.

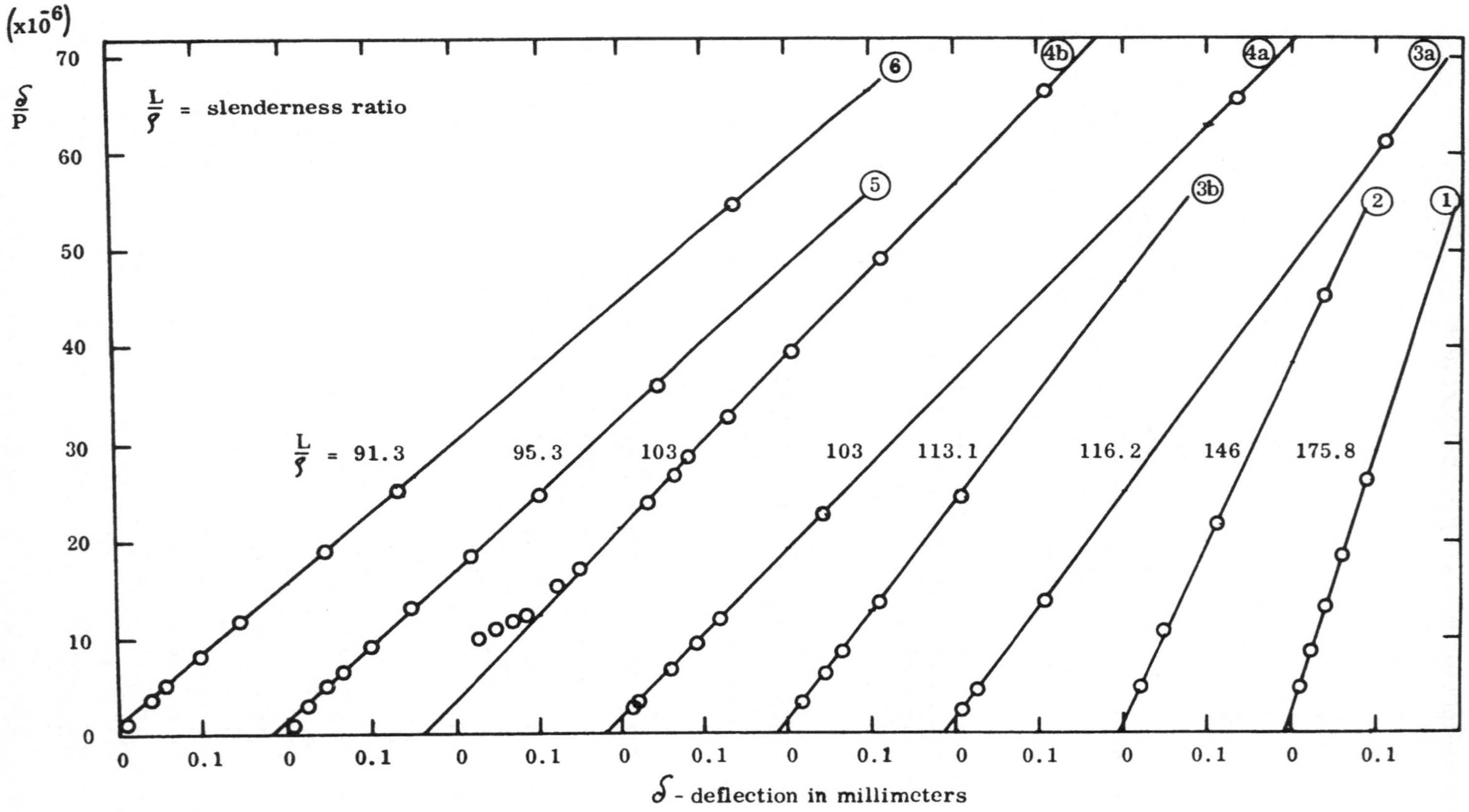

Fig. 14. Von Karman Data on Compressed Columns Plotted in the Linear Form by Southwell [36].

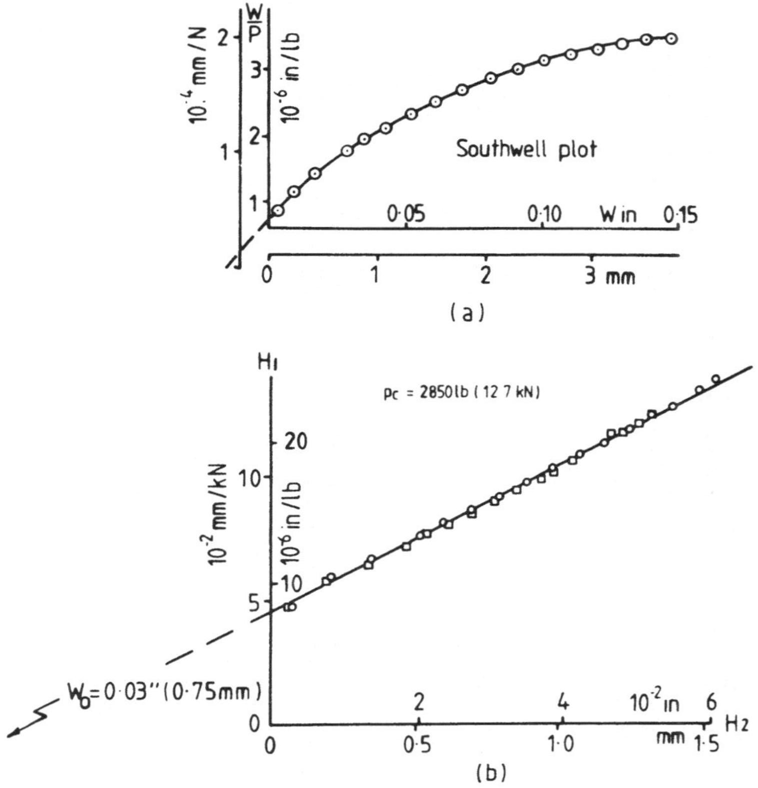

Fig. 15. Southwell Plot and Modified Southwell Plot for Plate Test in Ref. 40 [38].

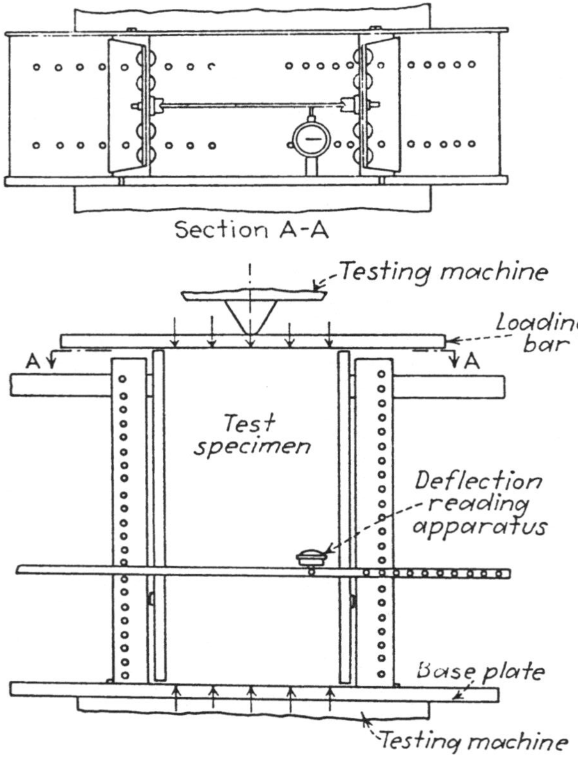

Fig. 16. U.S. Bureau of Standards Plate Testing Apparatus [41].

Buckling and postbuckling behavior of plates depends very strongly on the boundary conditions, both for loaded and unloaded edges (see, for example, Fig. 11). Hence much of the experimental efforts have been focussed at the definition and precision of the test boundary conditions, in particular in the postbuckling range, where variation of the boundary conditions with load may occur. A typical test set-up of the thirties is that used for rectangular plates at the U.S. Bureau of Standards, shown in Fig. 16 (from [41]). The test were aimed to realize conditions of simple supports along the unloaded edges, by using V notches to allow freedom of rotation. It was realized, however, that simple supports were not precisely achieved. Sometimes the loaded edges were made semi-circular to ensure introduction of the load in the mid-plane. This type of test set-up is still used today in many tests, but for study of the postbuckling behavior the boundary conditions vary too much.

A significant improvement in test boundary conditions was made by Hoff, Boley and Coan [35] in a series of very careful tests on rectangular fiberglass plates in which the simple support on the loaded edges was provided by an arrangement of split needle bearings. The bearing assemblies and the small bearings, which had a quadrant cut from their shells to permit the insertion of the slotted rods which supported the test specimen, were a fairly complicated set-up but ensured a close approximation to a continuous simple support. This approach has since been employed by many investigators who aimed at precise simple supports. A typical example were Walker's experimental studies on the strength of rectangular plates under eccentric in-plane loads [42]. Figure 17a shows the loaded edge in Walker's tests and Fig. 17b the split bearings. The unloaded longitudinal edges in both [35] and [42] were knife edges, (through slotted tubes were also tried in [35]) whose friction no doubt influenced the plate behavior.

Another more recent example is a special plate testing rig developed at Cambridge University for steel and aluminium plates in compression ([43] and [44]). As shown in Fig. 18 there is a comb-like array of "fingers" along the unloaded edges of the specimen. The role of these fingers is to provide a determinate edge condition, namely simple support with free pull-in, without taking any appreciable part of the longitudinal load. It may be pointed out that, as shown by theoretical studies, not only the out-of-plane and torsional boundary conditions but also the in-plane boundary conditions (the pull-in mentioned above) strongly influence the postbuckling behavior of the plates, and therefore their definition will significantly affect the precision of the experiments.

For the postbuckling behavior of plates in general, if the experiments are carefully performed and the boundary conditions well defined, and advantage is taken of modern theoretical developments, the correlation between experimental and theoretical results is fairly good. For example in Fig. 19 (from [31]), which shows the comparison between the theoretical load-deflection results of [31] and those of the tests of Walker [42], the correlation is good. In Fig. 20 (from [44]), which compares the predicted elastic-plastic stress-strain

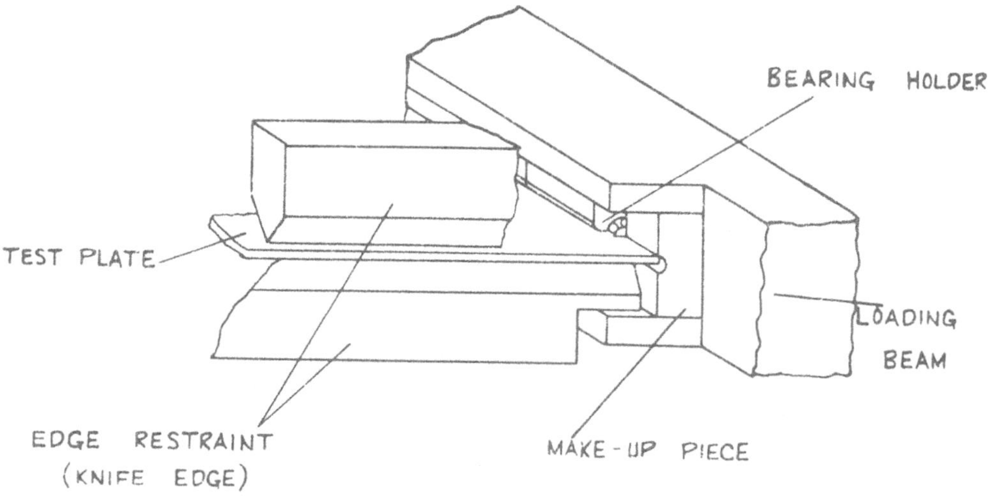

Fig. 17a. Loaded-Edge at the Extremity of the Plate [42].

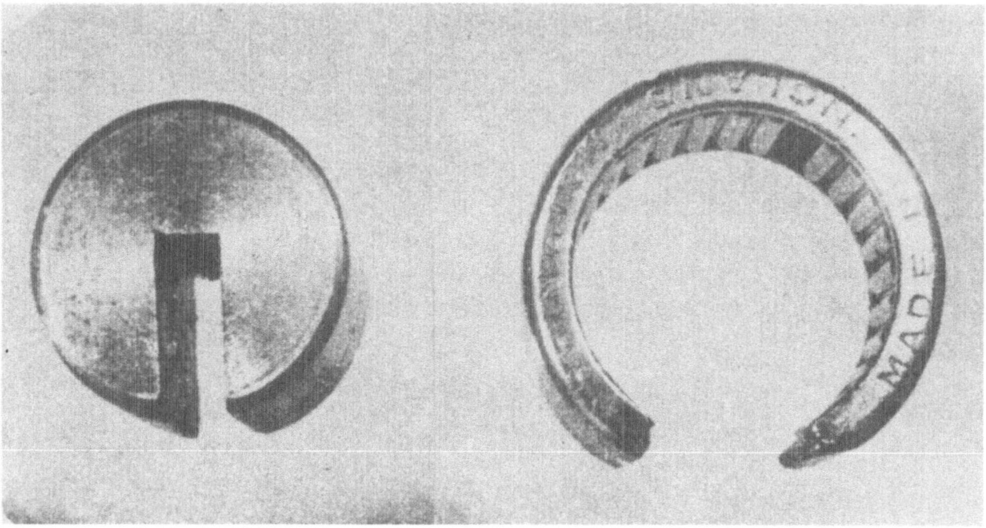

Fig. 17b. A Needle Bearing and Slotted Roller for the Loaded Edge Condition of Simple-Support [42].

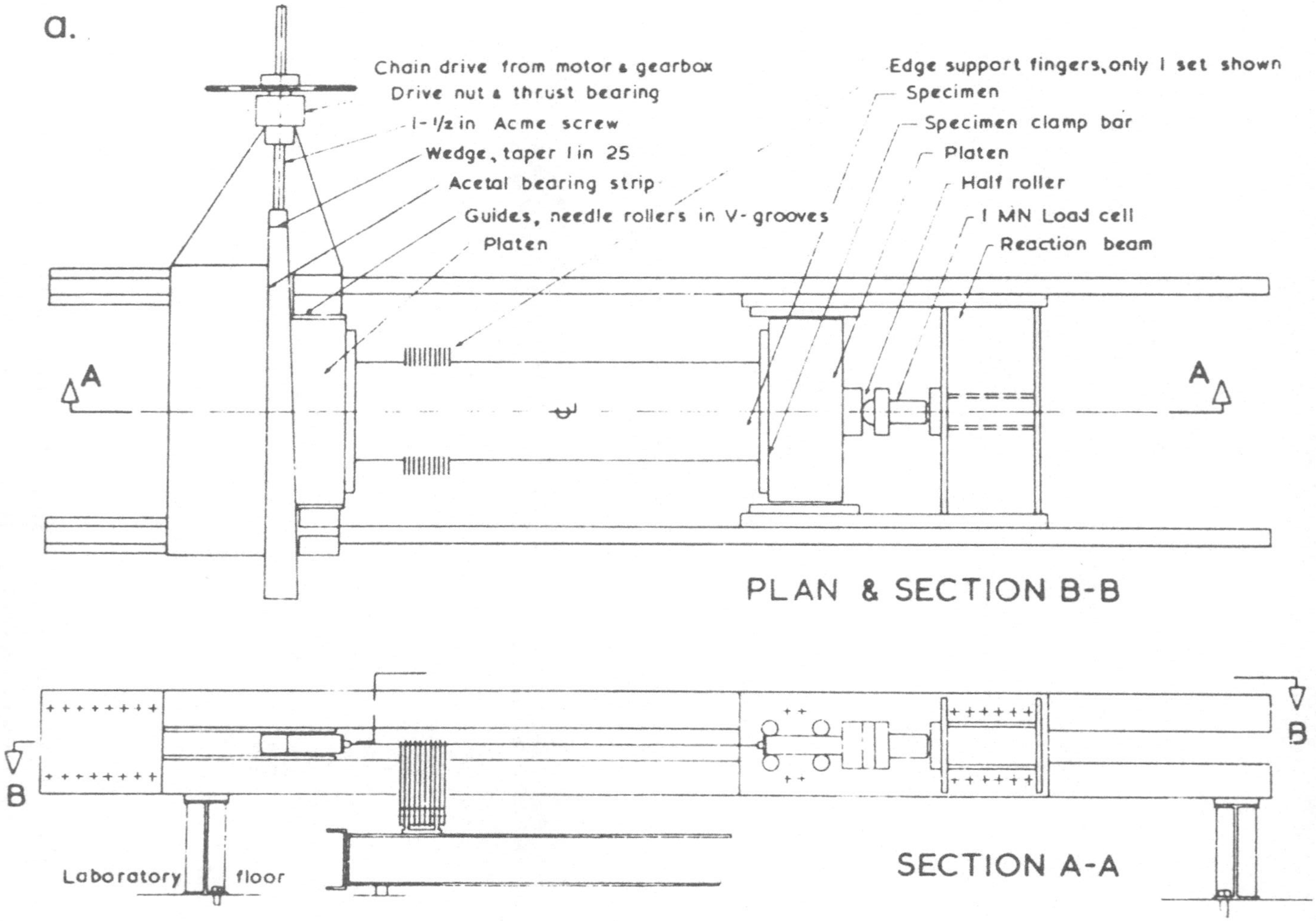

Fig. 18. Test Arrangement for Plate Buckling at Cambridge [30].

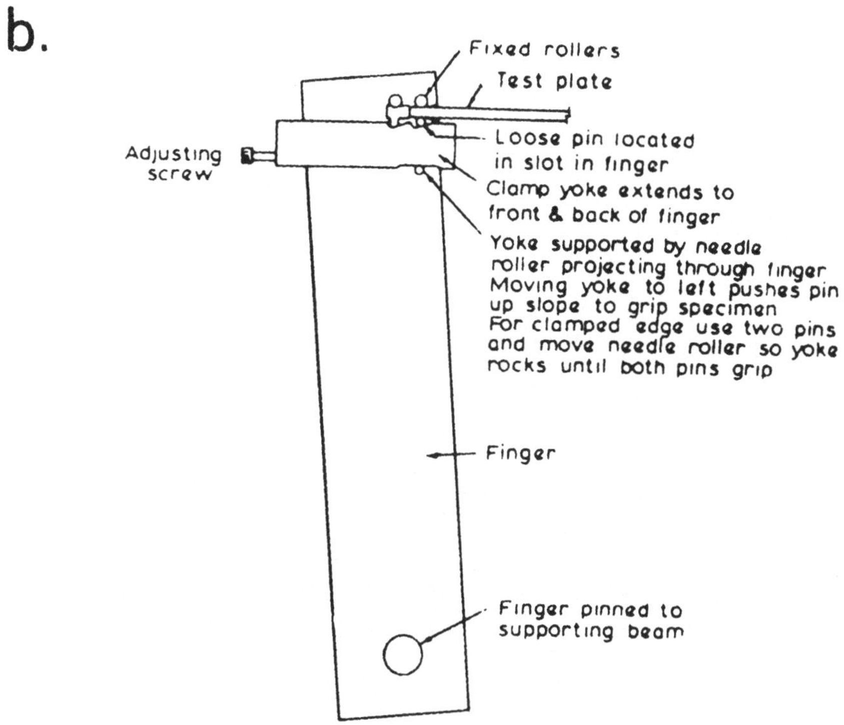

Fig. 18 (continued)

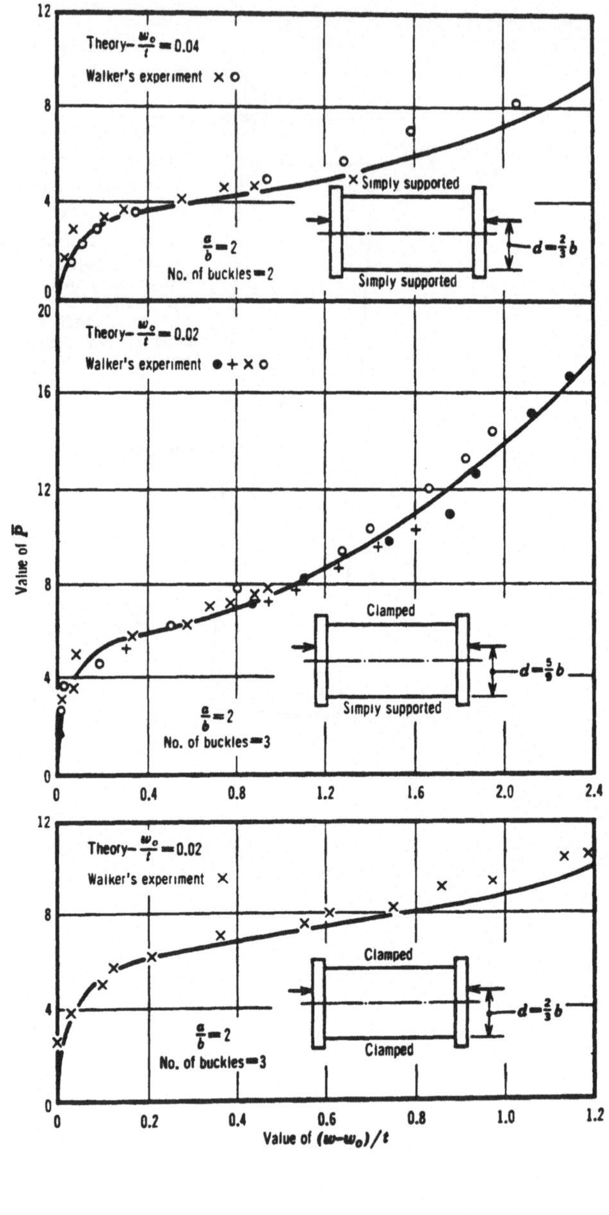

Fig. 19. Comparison of Theoretical and Experimental Load-Deflection Results [31].

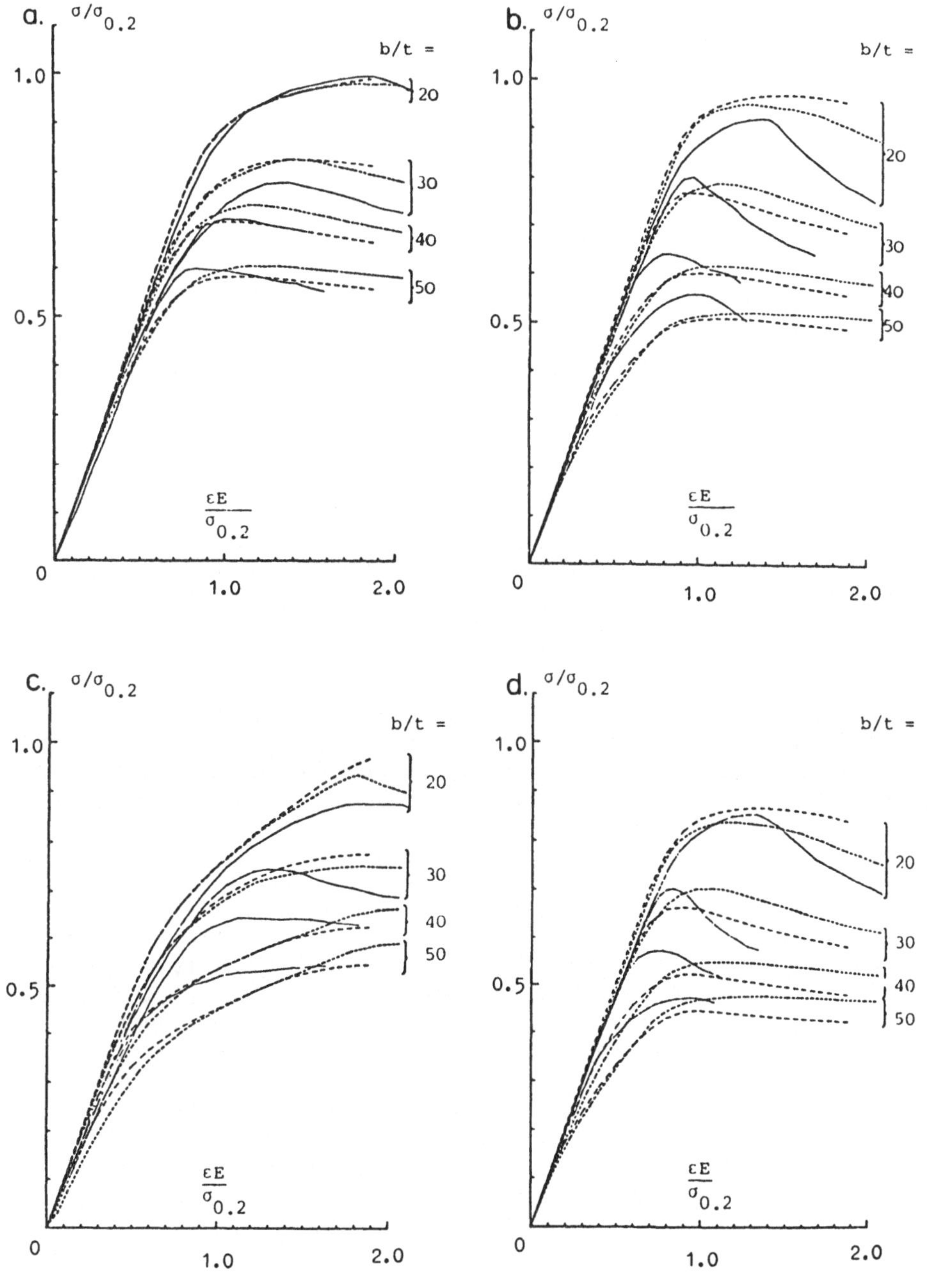

Fig. 20. Comparison between Theoretical and Experimental Stress-Strain Curves. a. 5083-M, $w_0/b = 0.005$, non-welded; b. 6082-TF, $w_0/b = 0.00$, non-welded; c. 5083-M, $w_0/b = 0.005$, heavily welded; d. 6082-TD, $w_0/b = 0.005$; heavily welded
___, Experimental; ---, Finite Strip; ..., Simple Theory [44].

curves for aluminium plates with the experimental results, the correlation again is fairly good.

One important aspect of plate postbuckling behavior should be mentioned, and that is the abrupt changes in the buckle pattern (number of waves) as one advances into the deep postbuckling region. This was emphasized by Stein [45] when he compared his perturbation analysis with experimental results carried out at NASA Langley. Figure 21 (from [45]) shows that the longitudinal wave number increases with load in the postbuckling region. The change in buckled form in the postbuckling range has also been discussed by other investigators, for example, in [31].

Before leaving plates, it is important to stress that the discussion has dealt with single plates only, has not included shear panels and has not touched stiffened plates and the important aspects of stiffener-plate interaction in buckling and postbuckling behavior.

7. BUCKLING AND POSTBUCKLING OF SHELLS

In contrast to columns which have a neutral postbuckling path and plates which exhibit a stable postbuckling behavior, shells usually have a very unstable postbuckling behavior that strongly influences their buckling characteristics. Thin shells, however, are very efficient structures that can support very high buckling loads and hence their buckling and postbuckling have presented scientific and engineering challenges for decades. The extensive theoretical studies, discussed already in the earlier lectures, have clarified the phenomena, connected initial postbuckling behavior with imperfection sensitivity, have developed analysis procedures and established the imperfection sensitivity of some shell/load combinations. Unfortunately their impact on engineering practice has been very small, and empirical "knock-down" factors are still primarily relied on in the design of buckling-critical shells (see discussion in [3] or [8]).

One of the reasons for this incomplete "technology transfer" from researcher to designer is probably the relative complexity of the analysis as well as the difficulties encountered in correlating theory with experimental results. Another reason may have been the lack of experimental investigations that were closely coordinated with theoretical studies. As a matter of fact, up to the sixties most shell buckling tests were for design data only and usually performed with insufficient care that resulted in wide scatter and nonrepeatability (see also [46]). In the last two decades, however, more careful experiments have been carried out that not only helped the understanding of buckling and postbuckling behavior, but also have began to influence the designers.

The center of attention for half a century has been the perplexing behavior of the isotropic circular cylindrical shell under axial compression (see, for example, [47]). Hence also considerable experimental effort has been devoted to this problem. Let us therefore first consider a typical experimental study of the postbuckling behavior of circular cylindrical shells under axial compression.

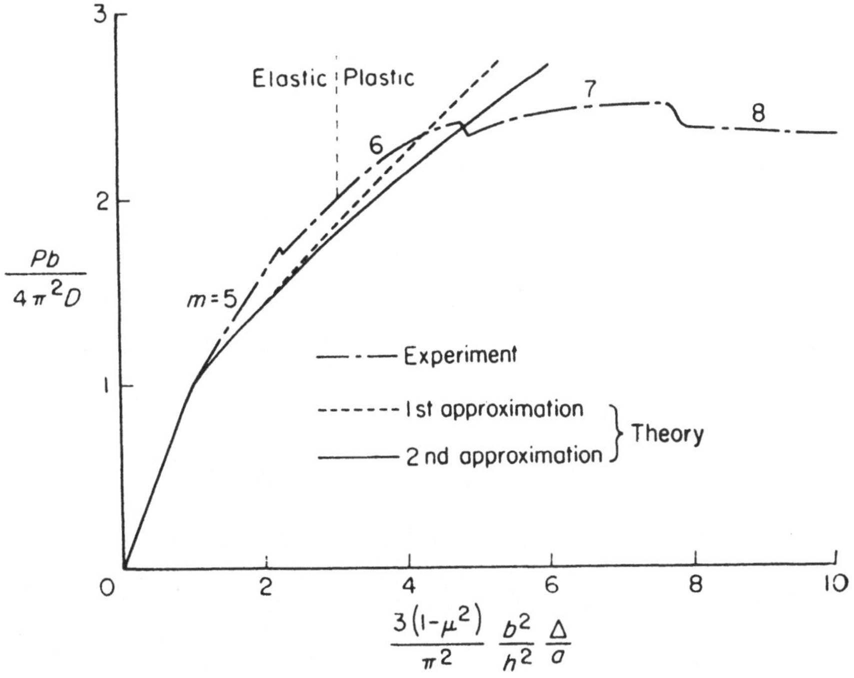

Fig. 21. Comparisons of Non-Dimensional Load-Shortening Curves as given by (Elastic) Theory and Experiment [45].

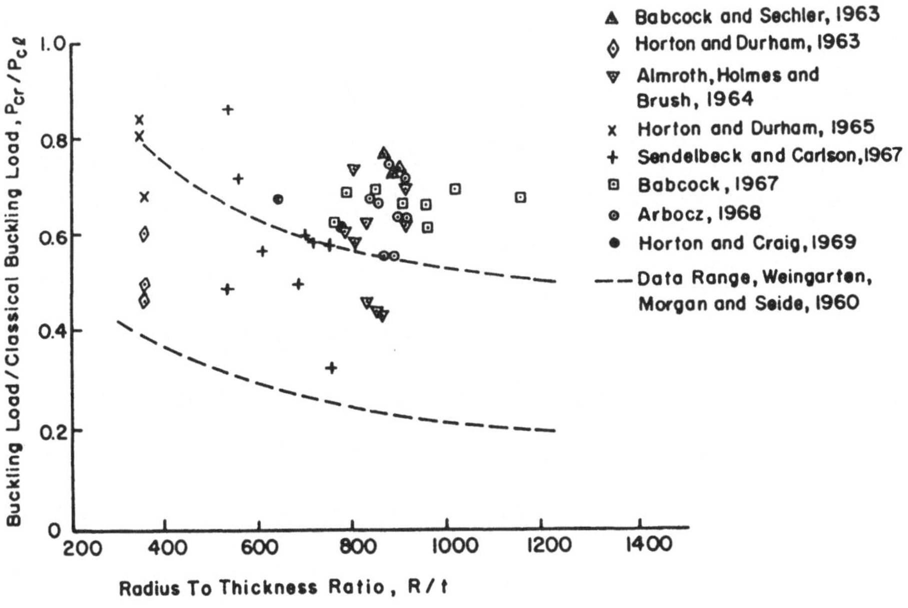

Fig. 22. Axial Buckling Load for Electroformed Cylindrical Shells [46].

Before describing the test set-up and procedure one has to consider model fabrication, which for shells is of major importance because their buckling behavior is often very sensitive to details, in particular initial imperfections. As pointed out by Babcock [46], the main problem with shell model making is that models must be made such that the difference in buckling loads and behavior from one nominally identical specimen to another must be less than the parameter in the experiment under investigation. This led to the quest for much better shell models.

Electroforming was one of the methods for making better shells favored by many investigators in the sixties (see [46]). The materials have usually been copper (for example, [48] or [49]) or nickel (for example [29] or [50]) and the shells are plated on either reusable or fusible mandrels. The quality of the shell is that of the mandrel, but plating is usually accompanied by initial stresses in the plated material which tend to relieve themselves when the shell is removed from the mandrel and cause imperfections. The magnitude of these initial stresses and the proportional limit depends on the plating bath conditions and can be reduced by the addition of plating additives, which makes the process somewhat of an art or black magic. The electroformed specimens were indeed better shells (with smaller geometrical imperfections), which when tested yielded significantly better "knock-down" factors, as can be seen for cylindrical shells in Fig. 22 (from [46]). However, the electroformed specimens did often not come up to the high expectations regarding their quality and have practically disappeared. For postbuckling studies, the suitability of the electroformed shells is doubtful, since their (σ_y/E) values are not much above that of mild steel, as was pointed out in Section 5 above.

Another method initiated in the sixties was fabrication of shells from Mylar and other similar polyester films, primarily on account of the high (σ_y/E) values, by an order of magnitude larger than that of the usual structural material, which facilitates postbuckling experiments. The method is most suitable to developable surfaces, like cylindrical or conical shells. Mylar specimens are inexpensive and can be buckled, deep into the post buckling region, many times without noticeable degradation of the shell quality, due to its high (σ_y/E) value. In the sixties, one disadvantage found in Mylar was its anisotropic material properties. For example, as much as ± 15% variation in tensile modulus, depending on the orientation of the specimen with the full sheet axis, was observed [51] but usually ignored.

Much effort was invested in the seventies in making the Mylar specimens more accurate and the bonding at the seams and the mounting in the end plate more precise. For example, Yamaki and Otomo [52], who used a similar polyester film (Mitsubishi Diafoil), checked the degree of anisotropy of Young's modulus by mirror extensometers, parallel and perpendicular to the rolling direction, and found about 6% difference. Similar measurements of Poisson's ratio with resistance strain gages also revealed very small differences. They therefore decided to regard the material as isotropic and used it extensively in these and later postbuckling tests [53] and [54]. At the DFVLR in Germany, where Mylar foil has been used extensively for cylindrical and conical specimens (for example, [55] and [56]), anisotropy and thickness variation of Mylar foil were also carefully measured. The anisotropy was found not to exceed 12% and the thickness variations were within 5%. With the improved quality of Mylar and similar polyester films (like Melinex, Diafoil, etc.), these foils have become the favorite of many investigators for postbuckling studies of thin shells with (R/t)=400-1000 (see [28], [52]-[57] and others). One should, however, point out that many other avenues of making near perfect, or intentionally imperfect shell models have been pursued (see [7] and [10]).

The experimental studies by Yamaki and his coworkers [52]-[54] are typical of carefully performed postbuckling tests of cylindrical shells. The specimens in these experiments are made of Diafoil polyester film for the reasons pointed out above. We will discuss the tests under axial compression [53]. The test apparatus is shown schematically in Fig. 23 (from [53]). In the figure, 1 is the table of a tension test machine, 2 is a base plate, 3 is a test cylinder and 4 is a stiffening steel plate fixed to the cover plate. To the crosshead 5 of the test machine, a reduction gear box 6 is fixed and, by turning the lever 7, the cylinder 8 can be moved in the axial direction. A strain-gage load cell 9 with the capacity 3000 N (= 300 kg) is connected to 8, which is directly calibrated with the test machine. The loading head 10 has three equidistant legs, on both faces of which strain gages 11 are bonded. The compressive load on each leg can be finely adjusted by turning the bolt 12, observing the strain output from 11. The loading head is constrained to move in the axial direction by a frictionless ball bushing 13, supported by a three-leg stand 14. With this loading system, tests are performed under controlling the edge shortening of the shell. However, it is to be noted that the shell is also subjected to a dead load of 50.2 N (= 5.12 kg), due to the weights of the loading head, stiffening as well as cover plates.

The deflection-measuring apparatus 16, consists of the contact point 17 and strain gages 18 bonded on both sides of a 0.2 mm thick phosphor bronze plate, attached to the end of the aluminum lever. The calibration of the strain output with deflection is made by the micrometer 19. With this system displacements up to 8 mm can be measured exerting negligible forces on the surface of the shell. The contact point 17 can be moved both axially and circumferentially by the rack mechanism 20 and the rotating disc 15, respectively, whose positions are detected electrically by potentiometer circuits. With this apparatus, both axial and circumferential distributions of the deflection can be recorded precisely without distorting the buckled surface of the shell. The edge shortening of the shell is measured utilizing the strain gages 21 bonded to a thin cantilever fixed at the end of a dial gage 22, which, in turn, is used for calibration. Two sets of these are used in series to obtain the average value of edge shortenings at the diametrically opposite points on the stiffening

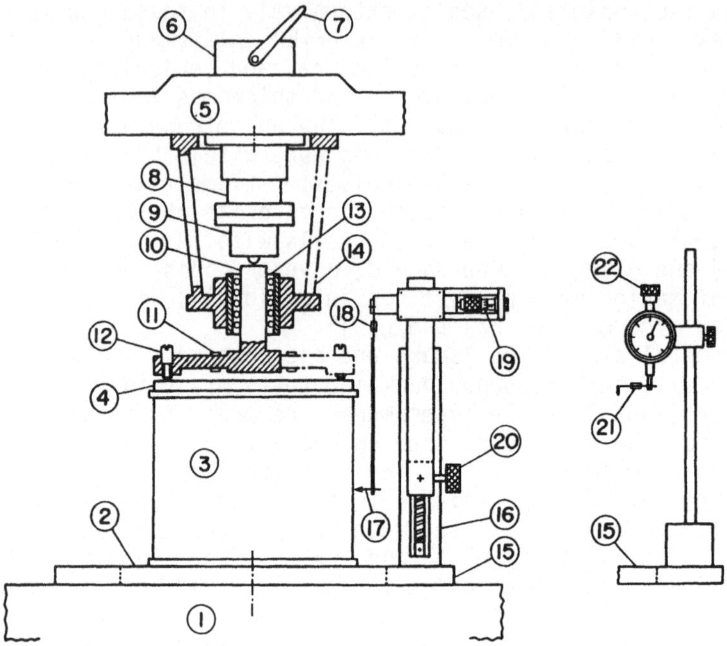

Fig. 23. Schematic Diagram of the Test Setup [53].

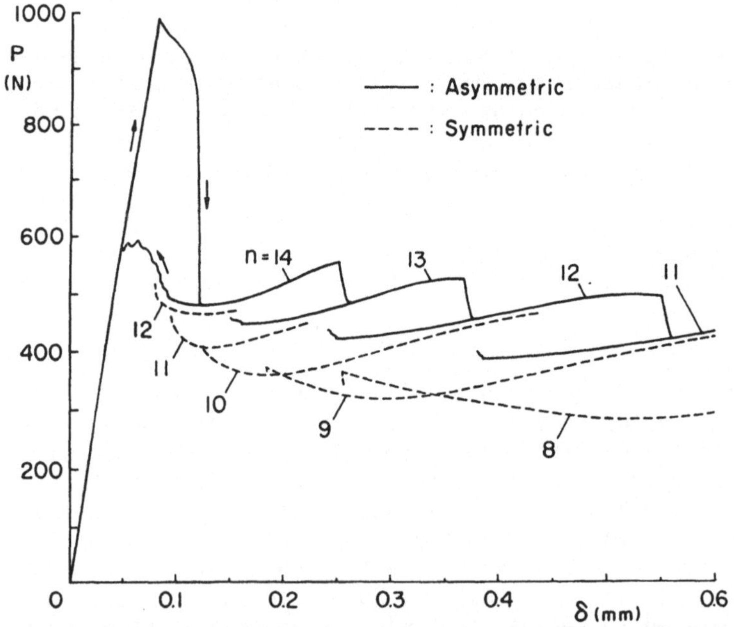

Fig. 24. Variations of the Load and Wave Number with Edge Shortening [53].

plate. A six-element dynamic strain meter is used for strain measurements while a X-Y recorder is used for recording the results.

In the tests, the relation between the compressive load P and edge shortening δ is obtained by lowering or returning the loading head deliberately. Typical results for the specimen No. 200-3 are shown in Fig. 24. As shown by solid lines, P is increased linearly with δ until the primary critical load (= 980 N) is reached, at which a snap-through buckling occurs with a sudden load reduction. The postbuckling configuration is of asymmetric type with two tiers of staggered buckles, having the circumferential wave number n = 14. With further increases in δ, secondary snap-through buckling occurs successively towards the new equilibrium states, with n reduced by one each time. On the other hand, when the shortening is reduced, similar snappings take place at the lower critical shortenings towards immediate previous paths, until the shell resumes the prebuckling state. It is observed that, under excessive edge compression with δ greater than 0.5 mm, the foregoing asymmetric configuration becomes locally unstable, with a tendency to degenerate into symmetric central buckles. In this case, a fully symmetric postbuckling configuration with uniformly distributed one-tier buckles can be realized easily with a slight adjustment of the shell wall. The additional postbuckling relations thus obtained are shown by dotted lines in the figure.

To see the overall distribution of buckled waves, circumferential distributions of w (positive inward) are recorded along various circular sections, with a typical result as shown in Fig. 25. In the figure, x and θ stand for axial and angular coordinates, respectively. From these measurements contour lines can be obtained for typical postbuckling configurations. Some of these are shown in Fig. 26. In Fig. 27 (from [54]) similar typical postbuckling contour lines are shown for the case of torsion.

The details of the test setup and procedure as well as the relevant measurements have been presented to indicate the care and effort necessary to perform a meaningful postbuckling experiment for a thin shell, and to point out the precise information that can be obtained. This information not only deepens our understanding of the postbuckling behavior of the shell, but can be used to evaluate and verify postbuckling analyses.

Regarding the physical behavior, the experiments considered (of [53]) clearly show for the case of axial compression the decrease in the number of circumferential waves as one proceeds into the postbuckling region. This concurs with the results of other postbuckling experiments on Mylar cylindrical shells under axial compression (for example, those summarized in Chapter 3 of [9]). Under hydrostatic pressure [52], however, the number of circumferential waves remains constant until the torsional pattern occurs in which the cylinder finally fails at excessively high pressures. Similarly under torsion [54] the number of buckled waves remains unchanged as one proceeds further into the postbuckling region. Again Yamaki's results for these two loading cases concur with those observed in other postbuckling tests (see, for example, [9]). It may be of interest to

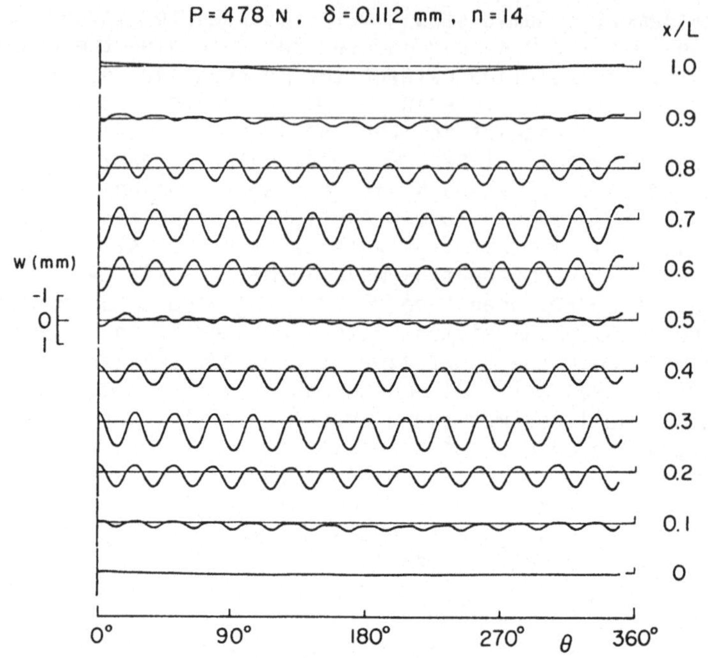

Fig. 25. Distribution of Buckled Waves Around the Shell [53].

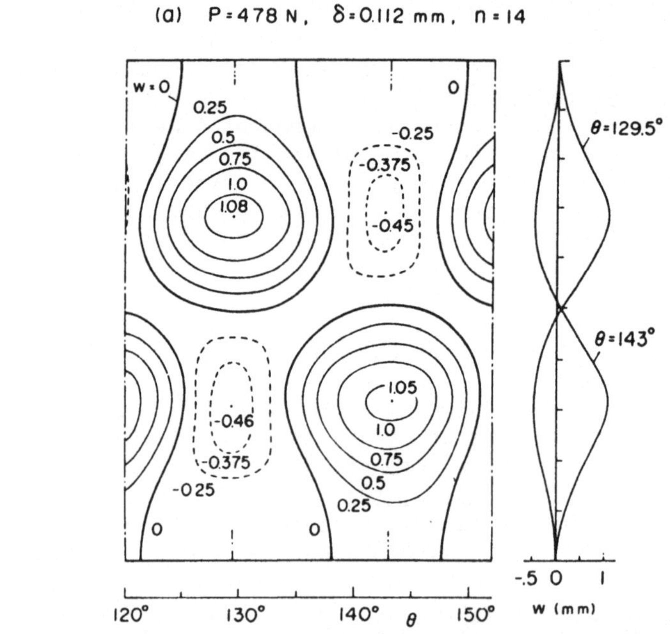

Fig. 26. Contour Lines for Typical Postbuckling Configurations under Axial Compression (in mm) [53].

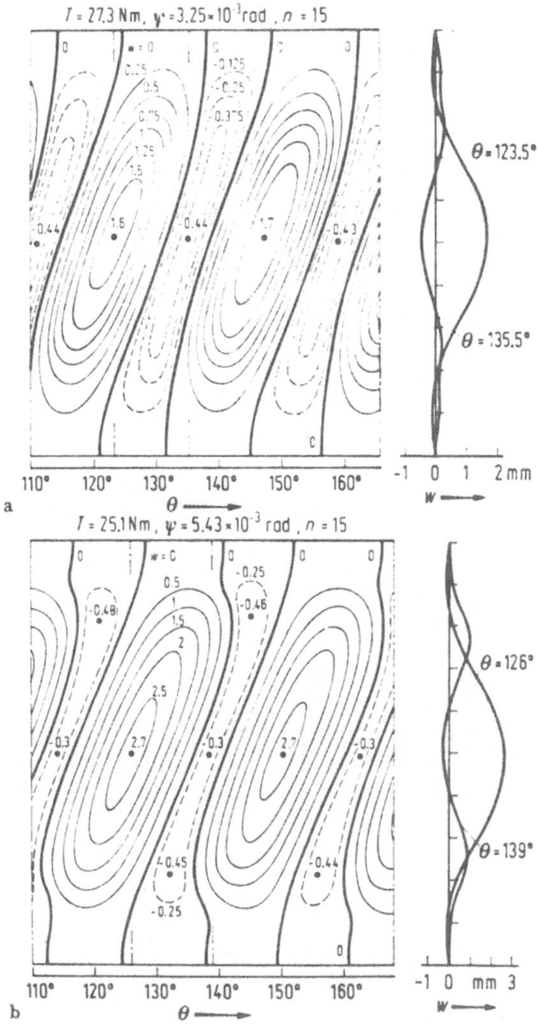

Fig. 27. Contour Lines for Typical Postbuckling Configurations under Torsion (in mm) [54].

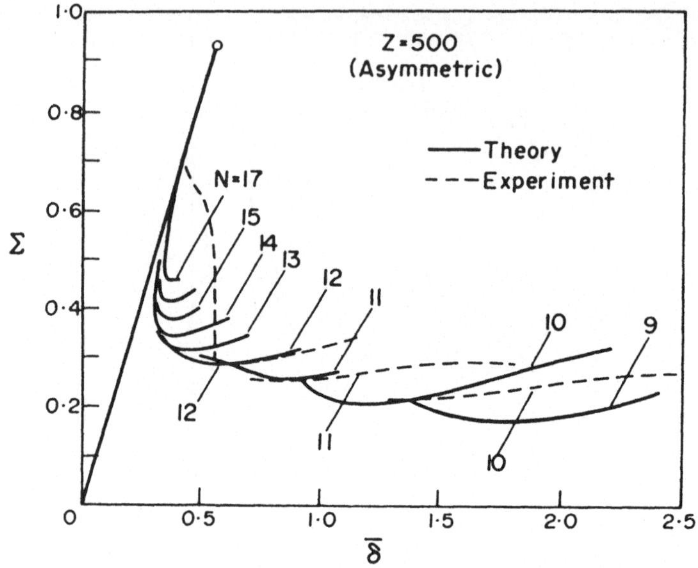

Fig. 28a. Load-Shortening Relations for the Asymmetric Postbuckling Mode: Z = 500, R/h = 405, ν = 0.3 [58].

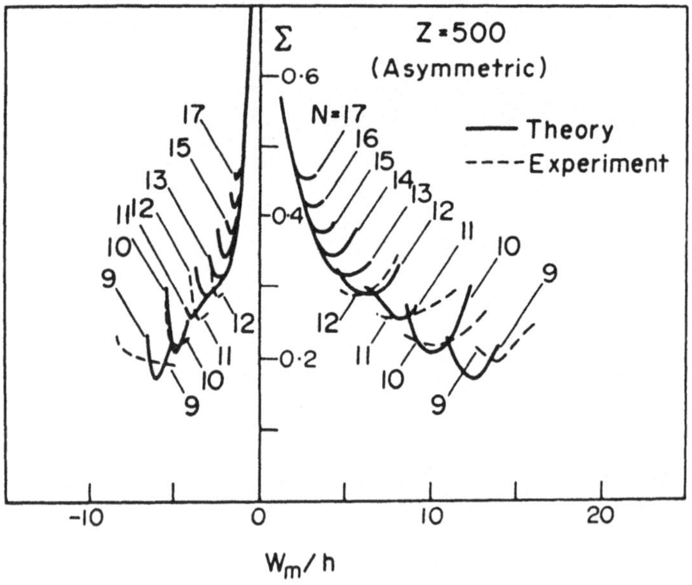

Fig. 28b. Load-Deflection Relations for the Asymmetric Postbuckling Mode: Z = 500, R/h = 405, ν = 0.3 [58].

recall that for rectangular plates under axial compression the number of longitudinal waves increases as one advances into the postbuckling region.

Regarding the comparison between theory and experiment, Yamaki and his coworkers obtained fairly good correlation between their carefully measured experimental results and their calculations (for example, [58] or [59]) performed later. Such a comparison is shown for one case in Fig. 28 (from [58]). Two typical presentations are shown in the figure: (a) load-shortening relations and (b) load-deflection relations. Both sets of curves are for the asymmetric postbuckling mode. There exists also a symmetric postbuckling mode for the same case, which has been measured and calculated (see [58]). Esslinger and her coworkers [55] also showed fair correlation between experiment and theory in most of the postbuckling region. The fairly good correlation between experiment and theory in the deep postbuckling region, which is better than that obtained at buckling and in its immediate neighbourhood, is an indication that the geometric imperfections, which are the prime cause of scatter in the buckling loads of isotropic shells lose their dominant influence in the far postbuckling region.

Spherical shells under uniform external pressure are the second type of shells whose buckling and postbuckling behavior has challenged and sometimes perplexed the investigators (see [60]). Extensive theoretical studies and extensive experimental investigations were carried out, unfortunately usually quite separately. The large scatter in experimental results motivated also here a quest for good models, which for spherical shells or caps are even more difficult to achieve. As a matter of fact, the first electroformed shells were spherical shells [61]. Very careful buckling studies on electroformed nickel complete spheres were carried out at Stanford University [62] and followed by a detailed study, including high speed photography, of the postbuckling behavior [29].

One should note that, whereas for axially compressed cylindrical shells, or longitudinally compressed plates, the end shortening is used as the deformation parameter here for spherical shells under pressure it is usually the volume increment. Hence, whereas for cylindrical shells and plates one plots load-shortening curves, for spheres one presents load-volume-change curves.

The experiments of [62] include an interesting technique developed in the early sixties at Stanford University: that of buckling of a shell on a mandrel, that limits the deflection, see Fig. 29 (from [62]). By not removing the wax mandrel (used in electroplating process) from the nickel shell, a small gap is formed during cooling after electroplating (which occurs at an elevated temperature). The deflection limitation permits the study of the development of the buckle wave pattern under forced "ideal" conditions (see Fig. 30).

For realistic postbuckling studies the wax mandrel was removed. The pressure ratio versus volume change ratio obtained during unloading is shown in Fig. 31, where also the observed change in wave pattern as one advances into the deep postbuckling range is indicated.

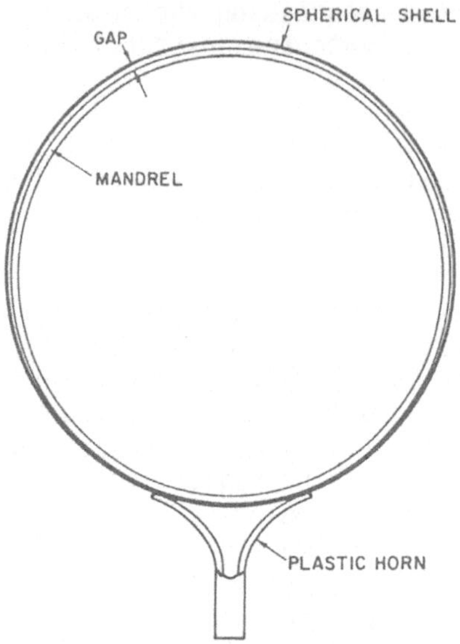

Fig. 29. View of Section Through Specimen with Mandrel [62].

Fig. 30. Air System Test with Wax Mandrel Inside Specimen [62].

A recent careful experimental and theoretical study of the large-deflection behavior of clamped shallow spherical shells under external pressure [63] shows very good agreement between experiment and theory, see Fig. 32. The specimens in these experiments were formed by thermovacuum molding with rigid polyvinylchloride sheets. Here two types of curves are presented: the customary pressure (Q) versus volume-change (V_c) curves and pressure versus deflection at the apex (Δ_0) curves.

8. INITIAL IMPERFECTIONS IN SHELLS

One of the primary problems of experimental studies of shell buckling is the influence of initial imperfections on the buckling loads obtained experimentally. For many shell-load combinations, like cylindrical and conical shells under axial compression or spherical shells under external pressure, the initial imperfections are the most dominant factor affecting the actual buckling load and the scatter in results. For other cases the imperfections are of lesser importance and in the deep postbuckling their effect is small, as has already been pointed out.

Probably the most important change in buckling experiments on shells in the last decade is therefore the extent of geometric imperfection meaurements carried out by the experimenters. Whereas previously imperfection measurements were usually considered useless and bothersome, except by a few investigators who believed in their importance (see [3], [11], [48], [49] or [64]), some type of geometric imperfection measurement is nowadays considered to be an integral part of a properly carried out shell buckling test, be it on a laboratory scale or on a large scale. A typical statement that illustrates this radical change is, for example, the laconic sentence appearing in the short summary of two recent Det Norske Veritas test reports (see, for example, [65]): "The imperfections are measured and stored on tape".

It is now widely accepted that significant advances towards more accurate predictions of the buckling load of thin shells depend on the availability of extensive data of realistic initial imperfections and their correlation with manufacturing techniques. For example, at the IUTAM Symposium on Buckling of Structures held at Harvard University in 1974, this was one of the main conclusions of the closing round-table discussion.

Once the initial imperfection shapes are known, and the boundary conditions are well defined - which still may be a problem, fairly adequate analytical tools and computer codes are available for calculation of the buckling load and of the shell structure (for example, [3] or [66]), as has been discussed in the other lectures, though adaptions and improvements are still necessary.

With the realization of the importance of the shape and amplitudes of initial imperfections, measurement tecnniques have been developed both for small and large shells, as discussed in detail in [10], [12], [49], [64] and [67-70]. In a typical imperfection scanning system, the measurement can be carried out on an unloaded shell, yielding the

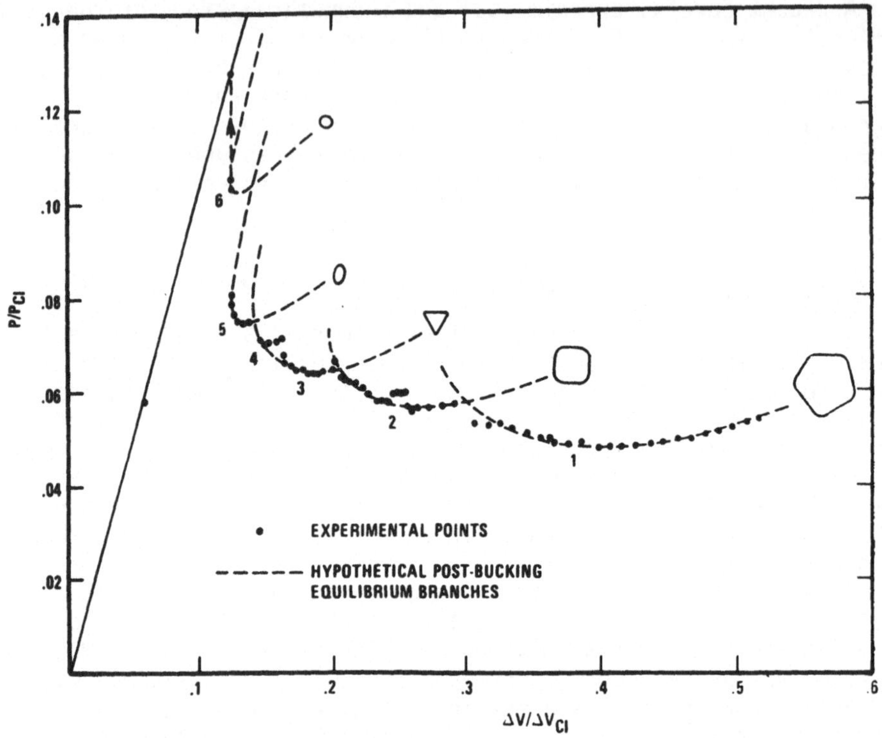

Fig. 31. Pressure Ratio vs. Volume Change Ratio Obtained During Unloading [29].

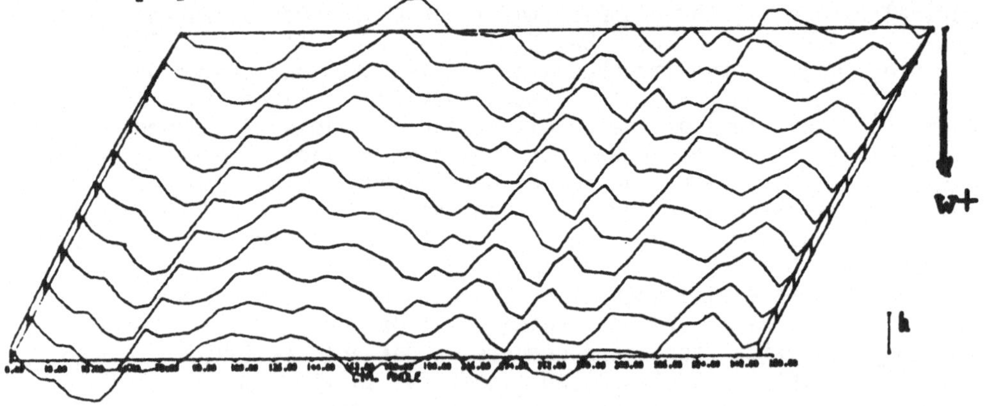

Fig. 33. Measured Initial Imperfection Shapes of Integrally Stringer-Stiffened Cylindrical Shells (Adjusted Scans Related to "Perfect" Cylinder) [68].
a. Laboratory Scale 250 mm (10 in) - Diameter Aluminum Alloy Technion Shell AB6 [64].

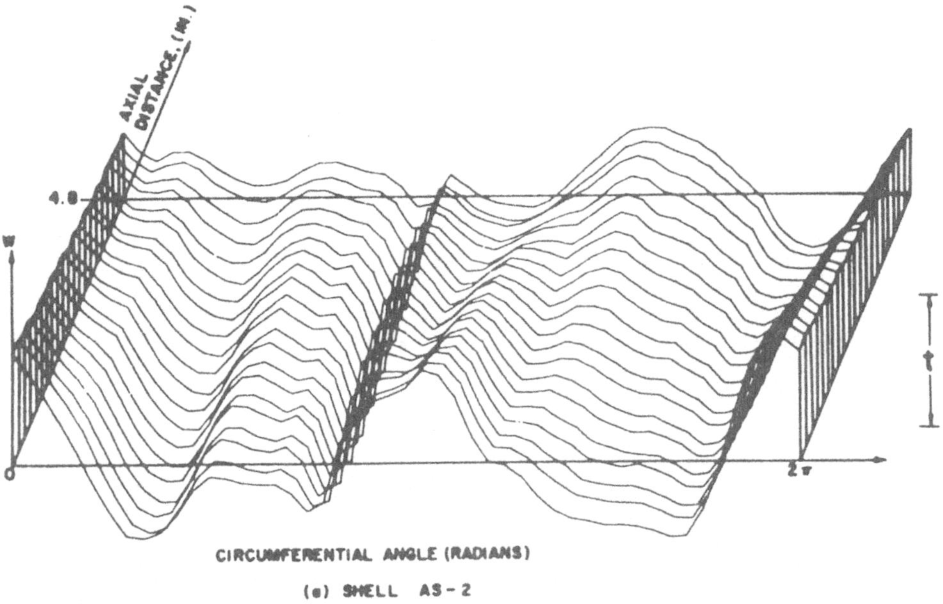

b. Laboratory Scale 250 mm (8 in) - Diameter Aluminum Alloy Caltech Shell AS2 [64].

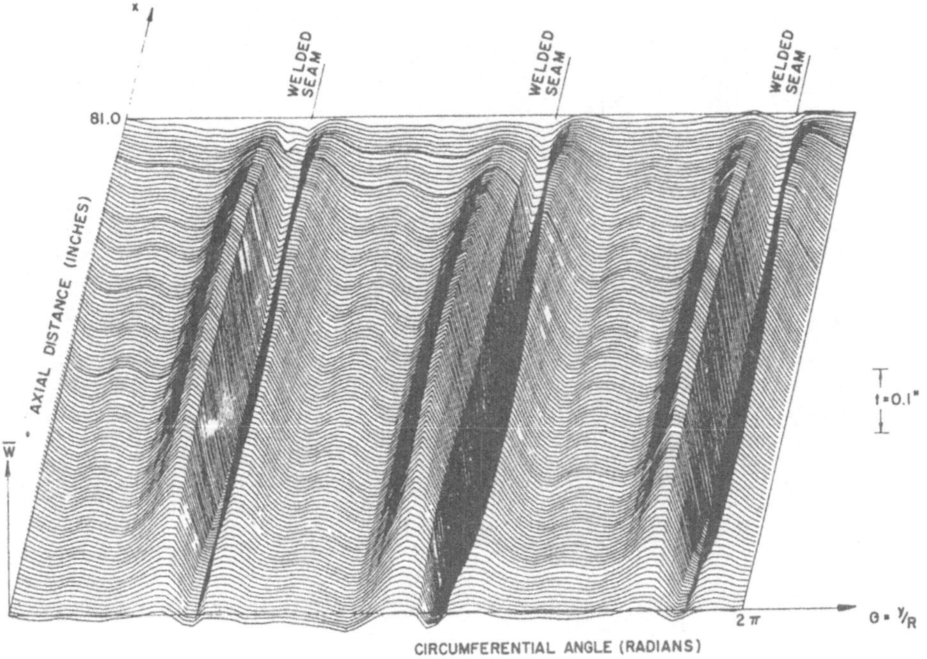

c. Large 3.05 m (10 ft) - Diameter Aluminum Alloy NASA Cylinder [68].

Fig. 33 (continued)

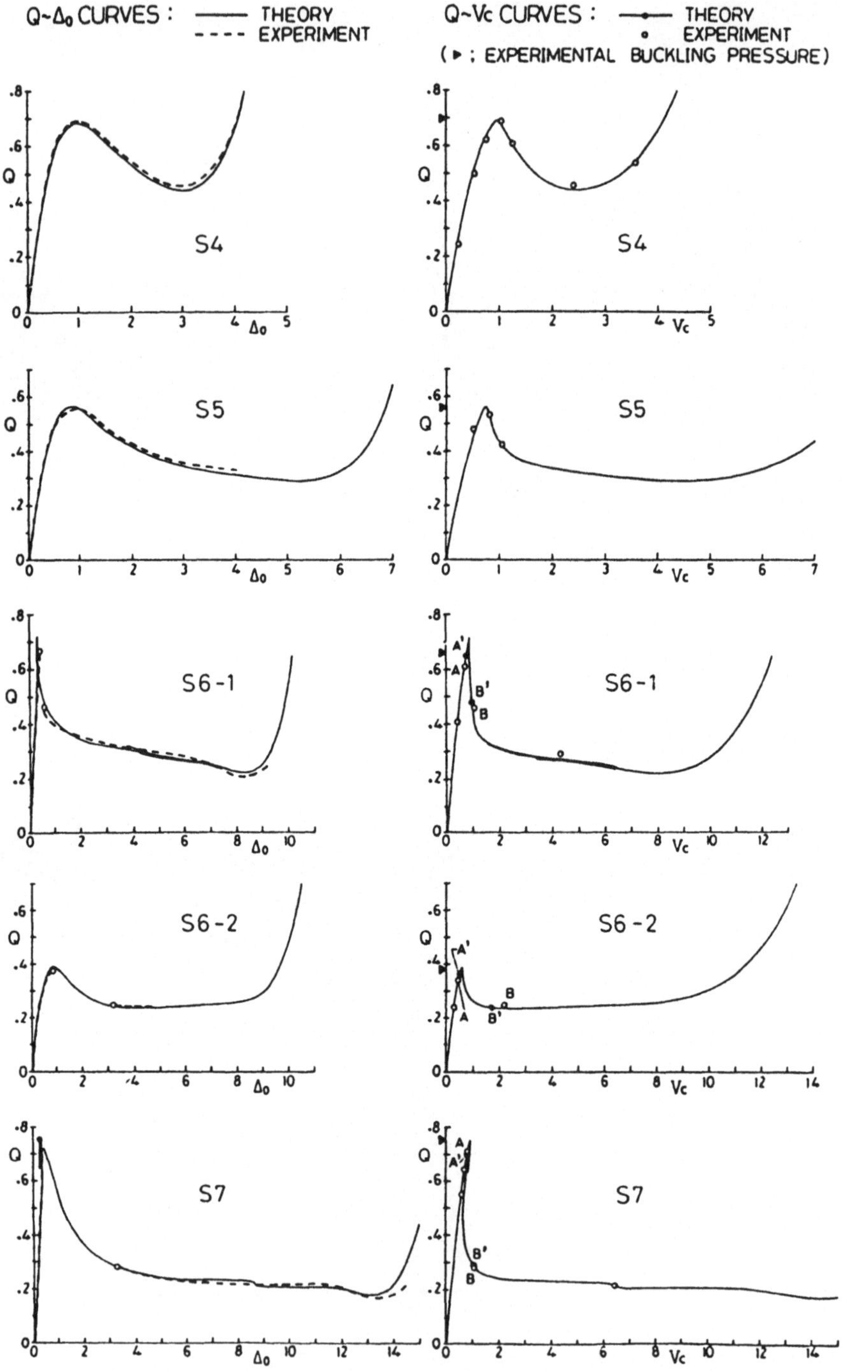

Fig. 32. Pressure Q vs. the Deflection at the Apex Δ_0 and Volume Change V_c [63].

initial geometrical imperfections, or on the shell under a specific axial load measured by a special load cell, yielding the growth of imperfection under load. In measurements on unloaded shells, the shell to be measured is placed in the scanning rig with boundary conditions as close as possible to those in the ensuing test.

The records of the measured radial displacements at the chosen points (along the circumference and at different heights) are the deviations of the inner surface of the shell from the cylindrical surface generated by the probe during its scanning path. However, the actual initial geometric imperfections are taken to be the deviations from an imaginary cylindrical reference surface, defined as the best-fit cylinder to the measured data, computed by a least square method. The recorded measured displacements are therefore usually recalculated with respect to this best-fit reference surface (using, for example, the DATA REDUCTION program developed at Caltech). In this manner any rigid body displacements of the shell with respect to the scanning system are removed. A typical "adjusted scan" (the initial imperfections calculated in this manner) is shown in Fig. 33 (from [10]) for two laboratory scale stiffened circular cylindrical shells and one large stiffned shell.

One recent addition to these measurement systems, however, deserves further discussion. It is a unified multipurpose scanning and measurement system for imperfections and vibrations developed at Technion jointly with Instrument and Control Ltd., Haifa [71]. As pointed out in [10] the deficiency of some of the imperfection measurement systems in use today is that imperfections are measured before the shell is fixed in its final boundary conditions. This was particularly emphasized in relation to vibration correlation studies, which will be discussed later. The new system [71] has been designed to overcome this drawback.

A closed loop noncontact probe is used to measure the vibrations and imperfections of cylindrical shells. Measurements include natural frequencies, modes of vibration, and mapping of imperfections. All measurements and mapping are carried out by the same probe inside the closed shell, which is fixed in its final boundary conditions, and an electronic control permits automatic execution of the different modes of operation. Figure 34 shows the complete system in operation on a spot welded shell AAC-1, of the joint RWTH Aachen-Technion project: (a) without the shell and (b) with the shell in position. This system can accommodate specimens of diameter 240-500 mm and length up to 470 mm. In the imperfection mode the digital data is recorded on a Nova 2 minicomputer connected to the experimental system, and analog plots of the imperfection shapes of the shell are also obtained for real time control. The digital results are stored on a computer disk which enables further numerical analysis. The scan can be taken at different load levels. In the vibration mode, the test procedure is essentially similar to that employed on the smaller integrally stiffened shells (see, for example, [64]) though operation of the new system differs considerably. The vibration plots obtained are very clear.

a. The Scanning and Measurement System.

b. Test System in Operation [7].

Fig. 34.

Test Setup for Vibration Correlation Test on Spot-Welded Shell AAC-1, with Multipurpose Scanning and Measurement System for Vibrations and Imperfections.

As mentioned in [10] an International Imperfection Data Bank, with branches in Delft and Haifa, has been established in order to facilitate evaluation of imperfection measurements and correlation studies.

It should be mentioned that, as for columns and plates, imperfections other than geometric, such as material defects or residual stresses, are also of importance, but for the buckling of shells the geometric imperfections occupy the dominant position except in the plastic range. In the case of welded shells, however, the residual stresses may also be of similar importance in the elastic range. However, though considerable efforts have been devoted to their measurement, the integrated effect of residual stresses on the buckling of shells can still not be adequately predicted.

9. BOUNDARY CONDITIONS AND LOADING FOR SHELLS

As pointed out in [46] and [10] the influence of boundary conditions has been intensively studied in the seventies, in particular for stiffened shells. For closely stiffened shells (as prescribed by optimization), governed largely by general instability of the entire shell, the effects of imperfections on buckling behavior have been found to be less pronounced, but the influence of boundary conditions becomes predominant. This fact has motivated closer studies of the boundary effects and also of the correlation between vibration and buckling, discussed in detail in [10], [72] and [73], and briefly summarized in the next section.

In recent decades most experimentalists have employed shell mountings which approach fully clamped conditions. In fact, these boundary conditions (C-4) are sometimes referred to as "test" boundary conditions. In the experiments, continuous clamping is usually obtained by casting the end of the shell in some material, such as a low-melting temperature alloy or epoxy. Complete clamping (C-4) is then assumed, but recent results obtained by vibration correlation (for example, [74]), cast some doubts on this assumption. The additional imperfections introduced by clamping (see [64]) are still ignored by most investigators, though some recent tests reconfirm their occurrence (see, for example, [75]) and underline the need for further study. For large shells the 3-bay specimens, with the middle bay representing the test shell, continue to be used in some experiments (see [10]), but this method is very expensive and does not assure complete simulation of actual boundary conditions. Some significant advances in shell mounting procedures are therefore still needed. The vibration correlation technique, which is also being applied now to larger shells, is one promising direction.

The problems of load nonuniformity are still present in recent shell buckling experiments, though more attention is given to load distribution measurements. Very recent tests at the Technion also reaffirm the importance of load distribution measurement even in closely stiffened shells.

Eccentricity of loading, usually defined as the radial distance between the line of axial load application and the shell mid-skin, has been shown to have considerable influence on the buckling load of stringer-stiffened shells. Weller et al. [76] amplified the theoretical investigation by tests on integrally stringer-stiffened cylindrical shells loaded eccentrically and having different boundary conditions. These studies have been extended (see [10], [73] and [76]) to consider the influence of load eccentricity on the vibrations of axially loaded stringer-stiffned shells both theoretically and experimentally, and correlate the results with those for buckling.

It should be pointed out that the load eccentricity is usually not well defined, as it depends on the detailed behavior of the joint under load, and its nondestructive determination would therefore significantly improve buckling load predictions. Joint efficiency and behavior under load represent more general problems of definition to the experimenter and designer, which may become even more severe with the bonded joints that will be introduced in composite structures from cost effectiveness considerations - another motivation for nondestructive testing.

10. NONDESTRUCTIVE TESTING FOR SHELLS

The seventies marked a turning point in the feasibility of nondestructive methods in shell buckling experiments. After having been practically discarded earlier by researchers, though applied effectively to columns, they were successfully applied in the past decade and look promising today. The nondestructive test methods can be divided here into two groups: (a) those for determination of boundary conditions, and (b) those for direct determination of buckling loads.

a. Determination of Boundary Conditions

Though for columns other methods have also been successful, for shells only the vibration correlation technique has so far yielded reliable results, and this method is at present limited to closely stiffened shells. The technique (see, for example, [72-75] or [77]) consists of an experimental determination of the lower natural frequencies for a loaded shell, and evaluation of equivalent elastic restraints that represent the actual boundary conditions. It is based on the similarity of the strong influence of axial and rotational restraints on free vibrations of stiffened shells, in particular for the lower natural frequencies - whose mode shapes resemble the buckling modes, to that observed for buckling loads.

The technique is described in detail in [72-75], [77] and [10]. It was applied to shells of different Technion tests series for axial compression loading. Figure 35 (reproduced from [7]) presents the results for 31 shells, and shows the significant reduction in scatter as a result of the experimental determination of the boundary conditions by nondestructive vibration tests. The scatter of the knockdown factor is reduced from 0.6-1.3 to 0.6-0.9, the low values of ρ_{sp} relating to clamped shells (probably on account of the additional imperfections

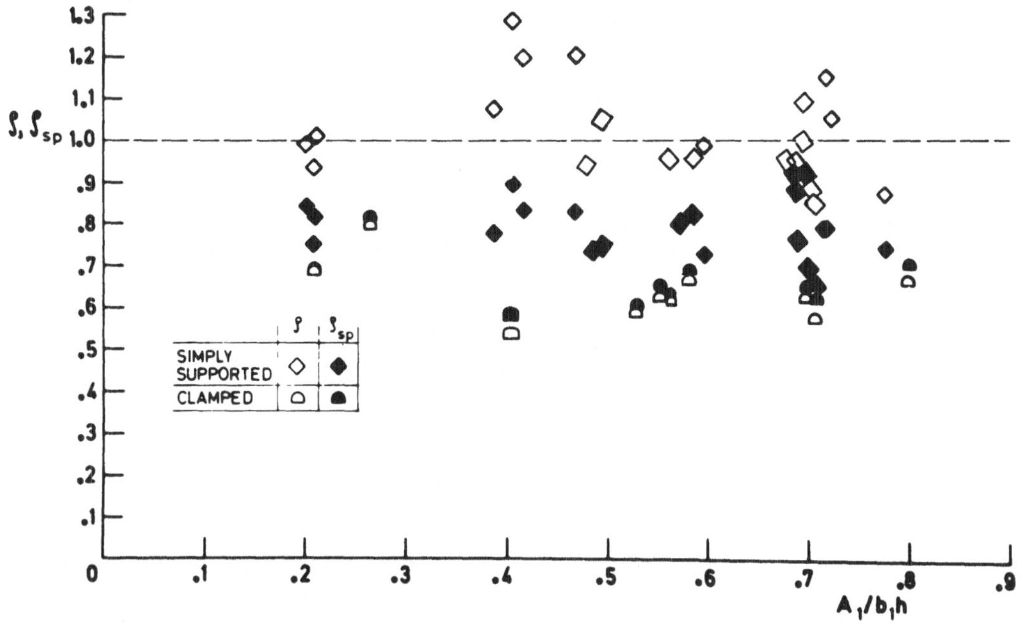

Fig. 35. "Knock-Down Factor" of 31 Simply Supported and Clamped Stringer-Stiffened Cylindrical Shells Corrected for Experimentally (nondestructively) Determined Boundary Conditions [7].

introduced by clamping, which have already been mentioned).

Since in the vibration tests an actual imperfect shell is measured, initial imperfections are indirectly included or "lumped" in the correlation. Theoretical studies of the influence of imperfections on vibrations of unstiffened and stiffened cylindrical shells have shown that imperfections have a strong influence on the frequency of vibration, similar to that on buckling (see, for example, [78]). The influence of boundary conditions and that of imperfections are therefore similar for vibrations and buckling and can be "lumped". However, when realistic end supports are employed they may also have load eccentricity, which may not be well defined a priori, and depends on the tolerances and behavior of the joints under load. The effects of load eccentricity are different in vibrations and buckling and could therefore obscure the correlation. This difficulty was overcome when careful evaluation of test data revealed a salient property which distinguishes vibrations in the presence of significant load eccentricities [77]. This important property is the large increase in the frequency ratio squared $(f/f_{SS4L})^2$ with the number of circumferential waves n of the vibration pattern, which does not occur in the absence of load eccentricity. If one now plots the corresponding theoretical frequency ratio squared for some likely values of load eccentricity (for SS4 boundary conditions in the presence of load eccentricity $\bar{e}$), the same property becomes evident. One can easily find a load eccentricity having the same slope as the experimental one.

This frequency slope property therefore presents a tool for a nondestructive identification of significant unknown load eccentricity in stiffened cylindrical shells, and once the load eccentricity has been identified, a modified vibration correlation method is available (see [77]). Since this frequency slope property is a crucial element of the technique, it was verified on shells of prescribed load eccentricity (see, for example, [79]).

The vibration correlation method has also been extended to other loading cases for stiffened shells, external pressure and combinations of external pressure and axial compression, and the applicability of the method has been verified in a substantial test program (see [72-75], [77] and [7]).

b. Direct Determination of Buckling Loads

Recent progress in the development of nondestructive test methods for direct prediction of buckling loads of shells has been mainly in two directions: the static-stiffness method, and the vibration approach.

The static-stiffness approach developed at Georgia Institute of Technology (see, for example, [80]) relates the variation in the shell wall lateral (or normal) stiffness as a function of applied compressive force to the buckling load. This method extends the lateral-stiffness method used successfully on columns. Lateral stiffness is determined by growing levels of axial load by a stiffness probe applying small normal force, and the circumferential distribution of this stiffness is plotted at a number of axial stations. "Weak" regions are identified and finally lateral stiffness-axial compression plots, at regions of least stiffness are extrapolated to yield the predicted buckling load.

Other investigators have developed similar methods. But since the load has to approach buckling failure load in these methods, they cannot be considered true nondestructive techniques.

The vibration approach is probably the most promising. At Technion the same experimental technique used to define boundary conditions was extended to direct prediction of buckling loads. The method is essentially curve fitting to the experimental points of the frequency squared versus axial load, but using only those points below 50-60% of the buckling load, to make the procedure truly nondestructive. The curve fitting is actually carried out with respect to a straight line, representing the experimental points to a certain exponent q, obtained from previous vibration and buckling tests on similar shells. This exponent was related to the dominant geometric properties of stringer-stiffened cylindrical shells, the Batdorf shell parameter Z and the stringer area ratio (A_1/bt), and a functional relation with two empirical constants was found (see [81-73]). Being empirical, this relation is obviously limited to the range of geometries tested, and requires many more tests for wider applicability. In [81] and [73] the buckling load ratios for direct prediction ρ_{extrap} of all the shells studied are summarized and compared with the theoretical predictions for effective boundary conditions found with the vibrations correlation technique ρ_{sp}. The scatter of ρ_{extrap} is found to be about 2/3 that of ρ_{sp}. Similar comparisons for additional shells tested since confirm this further reduction in scatter. Additional studies are needed to make the method applicable to a wider range of shells.

REFERENCES

1. Koiter, W.T., Current trends in the theory of buckling, in Buckling of Structures, Proceedings of IUTAM Symposium on Buckling of Structures, Harvard University, Cambridge, USA, June 17-21, 1974, (Ed. B. Budiansky), Springer-Verlag, Berlin, 1976, 1-16.
2. Chilver, A.H., The role of experimentation in the study of elastic stability of structures, in Stability, Solid Mechanics Division, SM Study No. 6, University of Waterloo, Ontario, Canada, 1972, 63-84.
3. Arbocz, J., Past, present and future of shell stability analysis, Zeitschrift fur Flugwissenschaften und Weltraumforschung, 5, 6, 1981, 335.
4. Tennyson, R.C., Interaction of cylindrical shell buckling experiments with theory, in Theory of Shells, (Eds. W.T. Koiter and G.K. Mikhailov), North-Holland Publishing Co., 1980, 65-116.
5. Valsgard, S. and Foss, G., Buckling research in Det norske Veritas, in Buckling of Shells in Offshore Structures, (Eds. J.E. Harding, P.J. Dowling and N. Agelidis), Granada, London, 1982, 491-548.
6. Schulz, U., Der Stabilitatsnachweis bei Schalen, Berichte der Versuchsanstalt für Stahl, Holz und Steine der Universität Fridericiana in Karlsruhe, 4. Folge, Heft 2, 1981.

7. Singer, J., The status of experimental buckling investigation of shells, in Buckling of Shells, (Ed. E. Ramm), Proceedings of a State-of-the-Art Colloquium, Stuttgart, Springer-Verlag, Berlin, Heidelberg, New York, 1982, 501-533.
8. Babcock, C.D., Shell stability, Journal of Applied Mechanics, 50, (1983), 935-940.
9. Esslinger, M. and Geier, B., Postbuckling behavior of structures, CSIM Courses and Lectures, No. 236, Springer-Verlag, Wien-New York, 1975.
10. Singer, J., Buckling experiments on shells - a review of recent developments, Solid Mechanics Archives, 7, 1982, 213-313.
11. Sechler, E.E., The role of experimentation in shell research, in Mechanics Today, 5, (Ed. S. Nemat-Nasser), Pergamon Press, Oxford, 1980, 439-449.
12. Singer, J., Arbocz, J. and Babcock, C.D., Buckling of imperfect stiffened cylindrical shells under axial compression, AIAA Journal, 9, 1, 1971, 68-75.
13. Buchert, K.P., Practical application of shell research, in Buckling of Shells in Offshore Structures, (Eds. J.E. Harding, P.J. Dowling, and N. Agelidis), Granada, London, 1982, 257-283.
14. Hariri, R., Post buckling behavior of tee shaped aluminum columns, Doctoral Thesis, University of Michigan, 1967, University Microfilms International, Ann Arbor, Michigan.
15. Euler, L., De curvis elastics, 1744, Leonhard Euler's "Elastic Curves", translated and annotated by Oldfather, W.A., Ellis, C.A. and Brown, D.M., reprinted from Isis, 20, 58, 1933, The St. Catherine Press, Bruges, Belgium.
16. Von Kármán, Th., Untersuchungen uber Knickfestikeit, Mitteilungen uber Forschungsarbeiten auf dem Gebiet des Ingenieurwesens, Verein Deutscher Ingenieure, Heft 81, Berlin, 1910.
17. Prandtl, L., Kipperscheinungen, Dissertation, München, 1899.
18. Johnston, Bruce, A., Ed., Structural Stability Research Council, Guide to Stability Design Criteria for Metal Structures, (3rd edition), John Wiley and Sons, New York-London, 1976, 569-584.
19. Tall, L., Centrally compressed members, in Axially Compressed Structures, Stability and Strength, (Ed. R. Narayanan), Applied Science Publishers, London, 1982, 1-40.
20. European Convention for Constructional Steelwork, European Recommendations for Steel Constructions, Milan, ECCS, March, 1978.
21. Estuar, F.R. and Tall, L., Testing pinned-end steel columns, in Test Methods for Compression Members, ASTM STP 419, American Society for Testing Materials, 1967.
22. Huber, A.W., Fixtures for testing pin-ended columns, ASTM Bulletin, No. 234, December 1958.
23. Ari-Gur, J., Weller, T. and Singer, J., Experimental and theoretical studies of columns under axial impact, International Journal of Solids and Structures, 18, 1982, 619-639.
24. Ellinas, C.P. Supple, W.J. and Walker, A.C., Buckling of Offshore Structures, Granada, London, 1984.

25. Dym, C.L. and Ivey, E.S., Principles of Mathematical Modeling, Academic Press, New York, 1980.
26. Weller, T. and Singer, J., Experimental studies on the buckling under axial compression of integrally stringer-stiffened circular cylindrical shells, Journal of Applied Mechanics, 44, 4, 1977, 721-730.
27. Weller, T. and Singer, J., Experimental studies on buckling of ring-stiffened conical shells under axial compression, Experimental Mechanics, 10, 11, 1970, 449-457.
28. Kodama, S., Otomo, K. and Yamaki, N., Postbuckling behavior of pressurized circular cylindrical shells under torsion - 1. Experiment, International Journal of Non-Linear Mechanics, 16, 3.4, 1981, 337-353.
29. Berke, L. and Carlson, R.L., Experimental studies of postbuckling behavior of complete spherical shells, Experimental Mechanics, 8, 12, 1968, 548-553.
30. Walker, A.C., A brief review of plate buckling research, Behaviour of Thin-Walled Structures, (Eds. J. Rhodes and J. Spence), Elsevier Applied Science Publishers, London, 1984, 375-398.
31. Rhodes, J. and Harvey, J.M., Examination of plate post-buckling behavior, Journal of the Engineering Mechanics Division, ASCE, 103, EM3, 1977, 461-477.
32. Schuman, L. and Back, G., Strength of rectangular plates under edge compression, NASA Technical Report, No. 356, 1930.
33. Sechler, E.E., The ultimate strength of thin flat sheet in compression, GALCIT Publication 27, Guggenheim Aeronautics Lab., California Institute of Technology, 1933.
34. von Karman, T., Sechler, E.E. and Donnell, L.H., The strength of thin plates in compression, ASME Transactions, 54, 1932, 53-57.
35. Hoff, N.J., Boley, B.A. and Coan, J.M., The development of a technique for testing stiff panels in edgewise compression, Proceedings of the Society for Experimental Stress Analysis, 5, 1948, 14-24.
36. Southwell, R.V., On the analysis of experimental observations in problems of elastic stability, Proc. Royal Society London, Ser. A, 135, 1932, 601-616.
37. Simitses, G.J., An Introduction to the Elastic Stability of Structures, Prentice-Hall, Englewood Cliffs, N.J., 1976, 68.
38. Spencer, H.H. and Walker, A.C., Critique of Southwell plots with proposals for alternative methods, Experimental Mechanics, 15, 8, 1975, 303-310.
39. Horton, W.H., Cundari, F.L. and Johnson, R.W., The analysis of experimental data obtained from stability studies on elastic column and plate structures, Israel Journal of Technology, 5, 1/2, 1967, 104-113.
40. Schlack, A.L., Experimental critical loads for perforated square plates, Proceedings of the Society for Experimental Stress Analysis, 25, 1968, 69-74.
41. Timoshenko, S.P. and Gere, J.M., Theory of Elastic Stability, McGraw-Hill, New York, 1961, 424.

42. Walker, A.C., Flat rectangular plates subjected to a linearly-varying edge compressive loading, Thin-Walled Structures, (Ed. A.H. Chilver), Chatto and Windus, London, 1967, 208-247.
43. Bradfield, C.D., Tests of plates loaded in in-plane compression, Journal of Constructional Steel Research, 1, 1, 1980, 27- .
44. Mofflin, D.S. and Dwight, J.B, Buckling of aluminium plates in compression, Behaviour of Thin-Walled Structures, (Eds. J. Rhodes and J. Spence), Elsevier Applied Science Publishers, London, 1984, 399-427.
45. Stein, M., Loads and deformations of buckled rectangular plates, NASA Technical Report, TR R-40, 1959.
46. Babcock, C.D., Experiments in shell buckling, Thin-Shell Structures, Theory, Experiments and Design, (Eds. Y.C. Fung and E.E. Sechler), Prentice-Hall, Englewood Cliffs, N.J. 1974, 345-369.
47. Hoff, N.J., The perplexing behavior of thin circular cylindrical shells in axial compression, Israel Journal of Technology, 4, 1, 1966, 1-28.
48. Babcock, C.D. and Sechler, E.E., The effect of initial imperfections on the buckling stress of cylindrical shells, NASA Technical Note, TN D-2005, 1963.
49. Arbocz, J. and Babcock, C.D., Experimental investigation on the effect of general imperfection on the buckling of cylindrical shells, NASA Current Report, CR-1163, 1968.
50. Singer, J. and Bendavid, D., Buckling of electroformed conical shells under hydrostatic pressure, AIAA Journal, 6, 12, 1968, 2332-2337.
51. Ishay, O., Weller, T. and Singer, J., Anisotropy of Mylar A sheets, ASTM Journal of Materials, 3, 2, 1968, 337-351.
52. Yamaki, N. and Otomo, K., Experiments on the postbuckling behavior of circular cylindrical shells under hydrostatic pressure, Experimental Mechanics, 13, 9, 1973, 299-304.
53. Yamaki, N., Otomo, K. and Matsuda, K., Experiments on the postbuckling behavior of circular cylindrical shells under compression, Experimental Mechanics, 15, 1, 1975, 23-28.
54. Yamaki, N., Experiments on the postbuckling behavior of circular cylindrical shells under torsion, Buckling of Structures, Proc. IUTAM Symp. Harvard Univ., Cambridge, Mass., U.S.A., June 17-21, 1974, (Ed. B. Budiansky), Springer-Verlag, Berlin, 1976, 312-330.
55. Esslinger, M. and Geier, B., Calculated postbuckling loads as lower limits for the buckling loads of thin walled circular cylinders, Buckling of Structures, Proc. IUTAM Symp. Harvard Univ., Cambridge, Mass. U.S.A., June 17-21, 1974, (Ed. B. Budiansky), Springer-Verlag, Berlin, 1976, 274-290.
56. Geier, B. and Heidemann, U., Experimental investigations on the buckling of thin walled isotropic cylinders subjected to external hydrostatic pressure, DFVLR, DLR-FB, 77-46, 1977.
57. Foster, C.G., Interaction of buckling modes in thin-walled cylinders, Experimental Mechanics, 21, 3, 1981, 124-128.

58. Yamaki, N. and Kodama, S., Postbuckling behavior of circular cylindrical shells under compression, International Journal of Non-Linear Mechanics, 11, 1976, 99-111.
59. Kodama, S. and Yamaki, N., Postbuckling behavior of pressurised circular cylindrical shells under torsion - II. theory, International Journal of Non-Linear Mechanics, 16, 3/4, 1981, 355-370.
60. Kaplan, A., Buckling of spherical shells, Thin-Shell Structures, Theory, Experiments and Design, (Eds. Y.C. Fung and E.E. Sechler), Prentice-Hall, Englewood Cliffs, N.JU., 1974, 247-288.
61. Thomson, J.M.T., Making of metal shells for model stress analysis, Journal of Mechanical Engineering Science, 2, 1960, 105-108.
62. Carlson, R.L., Sendelbeck, R.L. and Hoff, N.J., Experimental studies of the buckling of complete spherical shells, Experimental Mechanics, 7, 7, 1967, 281-288.
63. Yamada, M. and Yamada, S., Agreement between theory and experiment on large-deflection behavior of clamped shallow spherical shells under external pressure, Collapse, the Buckling of Structures in Theory and Practice, (Eds. J.M.T. Thompson and G.W. Hunt), Cambridge University Press, Cambridge, 1983, 431-441.
64. Singer, J., Abramovich, H. and Yaffe, R., Initial imperfection measurements of stiffened shells and buckling predictions, Israel Journal of Technology, 17, 1979, 324-338.
65. Grove, T., and Didriksen, T., Buckling experiments on 4 large ring-stiffened cylindrical shells subjected to axial compression and lateral pressure, Det norske Veritas, Report No. 77-431, 1977.
66. Arbocz, J. and Babcock, C.D., Prediction of buckling loads based on experimentally measured initial imperfections, Buckling of Structures, Proc. of IUTAM Symp. Harvard University, Cambridge, Mass., U.S.A., June 17-21, 1974, (Ed. B. Budiansky), Springer-Verlag, Berlin, 1976, 291-311.
67. Walker, A.C., Andronicou, A. and Shridharan, S., Experimental investigations of the buckling of stiffened shells using small scale models, Buckling of Shells in Offshore Structures, (Eds. J.E. Harding, P.J. Dowling and N. Agelidis), Granada, London, 1982, 45-72.
68. Arbocz, J. and Williams, J.G., Imperfection surveys on a 10-ft. diameter shell structure, AIAA Journal, 15, 7, 1977, 949-956.
69. Dowling, P.J. and Harding, J.E., Experimental behaviour of ring and stringer stiffened shells, Buckling of Shells in Offshore Structures, (Eds. J.E. Harding, P.J. Dowling and N. Agelidis), Granada, London, 1982, 73-108.
70. Oldland, J., An experimental investigation of the buckling strength of ring-stiffened cylindrical shells under axial compression, Norwegian Maritime Research, 9, 1981, 22-39.
71. Rosen, A., Singer, J., Grunwald, A., Nachmani, S. and Singer, F., Unified noncontact measurement of vibrations and imperfections of cylindrical shells, Proceedings of the 7th International Conference on Experimental Stress Analysis, Haifa, Israel, August 23-27, 1982, Technion - Israel Institute of Technology, 1982, 524-538.

72. Singer, J., Recent studies on the correlation between vibration and buckling of stiffened cylindrical shells, Zeitschrift fur Flugwissenschaften und Weltraumforschung, 3, 6, 1979, 333-343.
73. Singer, J., Vibrations and buckling of imperfect stiffened shells - recent developments, Collapse, the Buckling of Structures in Theory and Practice, (Eds. J.M.T. Thompson and G.W. Hunt), Cambridge University Press, Cambridge, 1983, 443-481.
74. Singer, J. and Rosen, A., Influence of boundary conditions on the buckling of stiffened cylindrical shells, Buckling of Structures, Proc. of IUTAM Symp. Harvard Univ., Cambridge, U.S.A., June 17-21, 1974, (Ed. B. Budiansky), Springer-Verlag, Berlin, 1976, 227-250.
75. Abramovich, H., Singer, J. and Grunwald, A., Nondestructive determination of interaction curves for buckling of stiffened shells, TAE Report 341, Technion - Israel Inst. of Technology, Dept. of Aeronautical Eng., Haifa, Israel, Dec. 1981.
76. Weller, T., Singer, J. and Batterman, S.C., Influence of eccentricity of loading on buckling of stringer-stiffened cylindrical shells, Thin-Shells Structure, Theory, Experiment and Design, (Eds. Y.C. Fung, and E.E. Sechler), Prentice-Hall, Englewood-Cliffs, N.J., 1974, 305-324.
77. Singer, J. and Abramovich, H., Vibration techniques for definition of practical boundary conditions in stiffened shells, AIAA Journal, 17, 7, 1979, 762-769.
78. Singer, J. and Prucz, J., Influence of imperfections on the vibrations of stiffened cylindrical shells, Journal of Sound and Vibration, 80, 1, 1982, 117-143.
79. Rosen, A. and Singer, J., Vibrations and buckling of eccentrically stiffened shells, Experimental Mechanics, 16, 3, 1976, 88-94.
80. Horton, W.H. Nassar, E.M. and Singhal, M.K., Determination of critical load of shells by nondestructive methods, Experimental Mechanics, 17, 1977, 154-160.
81. Singer, J., Vibration correlation techniques for improved buckling predictions of imperfect stiffened shells, Buckling of Shells in Offshore Structures, (Eds. J.E. Harding, P.J. Dowling and N. Agelidis), Granada, London, 1982, 285-330.

Lecture Notes in Physics

Vol. 257: Statistical Mechanics and Field Theory: Mathematical Aspects. Proceedings, 1985. Edited by T.C. Dorlas, N.M. Hugenholtz and M. Winnink. VII, 328 pages. 1986.

Vol. 258: Wm. G. Hoover, Molecular Dynamics. VI, 138 pages. 1986.

Vol. 259: R.F. Alvarez-Estrada, F. Fernández, J.L. Sánchez-Gómez, V. Vento, Models of Hadron Structure Based on Quantum Chromodynamics. VI, 294 pages. 1986.

Vol. 260: The Three-Body Force in the Three-Nucleon System. Proceedings, 1986. Edited by B.L. Berman and B.F. Gibson. XI, 530 pages. 1986.

Vol. 261: Conformal Groups and Related Symmetries – Physical Results and Mathematical Background. Proceedings, 1985. Edited by A.O. Barut and H.-D. Doebner. VI, 443 pages. 1986.

Vol. 262: Stochastic Processes in Classical and Quantum Systems. Proceedings, 1985. Edited by S. Albeverio, G. Casati and D. Merlini. XI, 551 pages. 1986.

Vol. 263: Quantum Chaos and Statistical Nuclear Physics. Proceedings, 1986. Edited by T.H. Seligman and H. Nishioka. IX, 382 pages. 1986.

Vol. 264: Tenth International Conference on Numerical Methods in Fluid Dynamics. Proceedings, 1986. Edited by F.G. Zhuang and Y.L. Zhu. XII, 724 pages. 1986.

Vol. 265: N. Straumann, Thermodymamik. VI, 140 Seiten. 1986.

Vol. 266: The Physics of Accretion onto Compact Objects. Proceedings, 1986. Edited by K.O. Mason, M.G. Watson and N.E. White. VIII, 421 pages. 1986.

Vol. 267: The Use of Supercomputers in Stellar Dynamics. Proceedings, 1986. Edited by P. Hut and S. McMillan. VI, 240 pages. 1986.

Vol. 268: Fluctuations and Stochastic Phenomena in Condensed Matter. Proceedings, 1986. Edited by L. Garrido. VIII, 413 pages. 1987.

Vol. 269: PDMS and Clusters. Proceedings, 1986. Edited by E.R. Hilf, F. Kammer and K. Wien. VIII, 261 pages. 1987.

Vol. 270: B.G. Konopelchenko, Nonlinear Integrable Equations. VIII, 361 pages. 1987.

Vol. 271: Nonlinear Hydrodynamic Modeling: A Mathematical Introduction. Edited by Hampton N. Shirer. XVI, 546 pages. 1987.

Vol. 272: Homogenization Techniques for Composite Media. Proceedings, 1985. Edited by E. Sanchez-Palencia and A. Zaoui. IX, 397 pages. 1987.

Vol. 273: Models and Methods in Few-Body Physics. Proceedings, 1986. Edited by L.S. Ferreira, A.C. Fonseca and L. Streit. XIX, 674 pages. 1987.

Vol. 274: Stellar Pulsation. Proceedings, 1986. Edited by A.N. Cox, W.M. Sparks and S.G. Starrfield. XIV, 422 pages. 1987.

Vol. 275: Heidelberg Colloquium on Glassy Dynamics. Proceedings, 1986. Edited by J.L. van Hemmen and I. Morgenstern. VIII, 577 pages. 1987.

Vol. 276: R.Kh. Zeytounian, Les Modèles Asymptotiques de la Mécanique des Fluides II. XII, 315 pages. 1987

Vol. 277: Molecular Dynamics and Relaxation Phenomena in Glasses. Proceedings, 1985. Edited by Th. Dorfmüller and G. Williams. VII, 218 pages. 1987.

Vol. 278: The Physics of Phase Space. Proceedings, 1986. Edited by Y.S. Kim and W.W. Zachary. IX, 449 pages. 1987.

Vol. 279: Symmetries and Semiclassical Features of Nuclear Dynamics. Proceedings, 1986. Edited by A.A. Raduta. VI, 465 pages. 1987.

Vol. 280: Field Theory, Quantum Gravity and Strings II. Proceedings, 1985/86. Edited by H.J. de Vega and N. Sánchez. V, 245 pages. 1987.

Vol. 281: Ph. Blanchard, Ph. Combe, W. Zheng, Mathematical and Physical Aspects of Stochastic Mechanics. VIII, 171 pages. 1987.

Vol. 282: F. Ehlotzky (Ed.), Fundamentals of Quantum Optics II. Proceedings, 1987. X, 289 pages. 1987.

Vol. 283: M. Yussouff (Ed.), Electronic Band Structure and Its Applications. Proceedings, 1986. VIII, 441 pages. 1987.

Vol. 284: D. Baeriswyl, M. Droz, A. Malaspinas, P. Martinoli (Eds.), Physics in Living Matter. Proceedings, 1986. V, 180 pages. 1987.

Vol. 285: T. Paszkiewicz (Ed.), Physics of Phonons. Proceedings, 1987. X, 486 pages. 1987.

Vol. 286: R. Alicki, K. Lendi, Quantum Dynamical Semigroups and Applications. VIII, 196 pages. 1987.

Vol. 287: W. Hillebrandt, R. Kuhfuß, E. Müller, J.W. Truran (Eds.), Nuclear Astrophysics. Proceedings. IX, 347 pages. 1987.

Vol. 288: J. Arbocz, M. Potier-Ferry, J. Singer, V.Tvergaard, Buckling and Post-Buckling. VII, 246 pages. 1987.